Ex Libris

Randy Manning

B292 © APCo

# MANUAL OF
# *Grasses*

THE NEW **ROYAL HORTICULTURAL SOCIETY** DICTIONARY

# MANUAL OF
# *Grasses*

*Consultant Editor* RICK DARKE
*Series Editor* MARK GRIFFITHS

**TIMBER PRESS**
*Portland, Oregon*

Derived from
**The New Royal Horticultural Society Dictionary of Gardening**
Editor in Chief Anthony Huxley
Editor Mark Griffiths, Managing Editor Margot Levy
in four volumes, 1992

First edition published 1994 by
THE MACMILLAN PRESS LTD,
London and Basingstoke

Associated companies in Auckland, Delhi, Dublin, Gaborone,
Hamburg, Harare, Hong Kong, Johannesburg, Kuala Lumpur,
Lagos, Manzini, Melbourne, Mexico City, Nairobi, New York,
Singapore, Tokyo.

First published in North America in 1994 by
Timber Press, Inc.
The Haseltine Building
133 S.W. Second Avenue, Suite 450
Portland, Oregon 97204-3527, U.S.A.

ISBN 0-88192-300-1

Printed and bound in Great Britain by
Mackays of Chatham PLC

# Contents

# Preface

Once in a while, horticulture is beset by mania. The first and most famous attack was Tulipmania, where bulbs were bankable and fortunes made and lost. Pteridomania was less flamboyant but no less frenzied, as scores of fern-lovers combed dell and dale for tasteful freaks of nature. Orchidmania offered travel, exoticism, expense and rarity – an elite sport. A history of gardening for the closing decades of the twentieth century will probably take note of our deepening involvement with the grasses and plants that resemble them. It would be an overstatement to talk of Graminomania (or Poamania, as some might have it). Besides, the uniqueness of ornamental grasses seems to reside not in any capacity to inspire a monocultural obsession, but in the endlessly varying ways in which they can enhance almost any garden. What group of plants is better qualified to meet our prevailing tastes – quiet harmonies, fineness of line, careful if seemingly artless contrasts of colour and form? That said, ornamental grasses are at present enjoying an unparalleled vogue. This autumn, the Royal Horticultural Society's Halls have been transformed by the pearly banners of *Miscanthus*, spangled *Chasmanthium*, coppery *Carices* and the burgundies and clarets of *Pennisetum* and *Imperata*. This sudden prominence derives not only from a general revival of interest in herbaceous perennials, but also from a new philosophy of garden design. Gardening is going back to nature. The romance of a lakeside in autumn, the underwood in spring, or a summer meadow would be hard to evoke without grasses.

The Grass Revolution has been seeded by plant explorers, plant breeders, ecologists and designers – a lively interaction of different horticultural disciplines and styles, supported by strong lines of communication and exchange between the Far East, Europe and North America. One of the major centres for this activity is Longwood Gardens, Pennsylvania, where U.S. native grasses are developed and assessed, as are other introductions, new and old, from all over the world. A prime mover in this activity, no one could be better placed to oversee the production of this manual and to write its introduction than Rick Darke, Longwood's Curator of Plants.

This manual describes the principal ornamental members of the true grasses (Gramineae, including Bamboos), sedges (Cyperaceae), rushes (Juncaceae), cat tails (Typhaceae) and sweet flags (Acoraceae). It does not include lawn grasses, cereal crops, or 'grassy' plants such as *Liriope*, *Ophiopogon*, *Isoetes*, *Bobartia* or *Butomus*. It takes as its starting point the accounts of ornamental grasses and grass-like plants that appear in *The New Royal Horticultural Society Dictionary of Gardening*. For these we were very fortunate to receive the expert advice of Thomas Cope and David Simpson at the Royal Botanic Gardens, Kew. The essay on bamboos and their descriptions are the work of David McClintock and are reproduced here with all gratitude. Guidance on general principles of coverage and cultivars was kindly provided by Mervyn Feesey and that pioneer of ornamental grasses, Roger Grounds, and, in the United States, by Rick Darke and his colleagues.

In the accounts which follow, certain qualities peculiar to ornamental grasses are often described. Airiness, grace, translucency, intricacy, boldness – these are all difficult to capture in line drawing, but the artists whose work appears in these pages, Clare Roberts, Vana Haggerty, Shirley Wheeler and Camilla Speight, have succeeded brilliantly. Warmest thanks are also due to Dr. Brent Elliott and the staff of the Lindley Library at the Royal Horticultural Society and to the Keeper and Librarian of the Royal Botanic Gardens, Kew. Preparation of this volume would have been impossible without the help of colleagues at Macmillan Publishers, in particular Gabriel Weston, Mick Card and, as ever, Margot Levy.

London 1994                                                                                          Mark Griffiths

(a) *Typha latifolia* (b) *Cortaderia selloana* (c) *Miscanthus sinensis* 'Zebrinus' (d) *Carex pendula* (e) *Panicum virgatum*
(f) *Molinia caerulea* (g) *Eragrostis curvula* (h) *Stipa calamagrostis* (i) *Coix lacryma-jobi* (j) *Hystrix patula*
(k) *Leymus arenarius* (l) *Briza maxima* (m) *Cyperus longus* (n) *Zea mays* (o) *Juncus effusus* f. *spiralis*
(p) *Pennisetum villosum* (q) *Lagarus ovatus* (r) *Luzula maxima* (s) *Cynodon dactylon* (t) *Eriophorum latifolium*
(u) *Bouteloua gracilis* (v) *Melica uniflora* (w) *Schoenoplectus lacustris* ssp. *tabernaemontani*

# Introduction

'Of late years public taste has been turned to the advantageous effect of
grasses in landscape gardening. Ferns had the credit of first winning attention from colour to form, and grasses next stepped in to confirm the prefer-
ence for grace and elegance over gaudy colouring.'

— from *British Grasses* by Margaret Plues, London, 1867.

Although more than a century old, these lines capture much of the spirit, enthusiasm and
thrust behind the current fervour for ornamental grasses. Grasses are indeed enjoying a renais-
sance, as gardeners learn to look beyond flower colour to embrace the subtler satisfactions of
line, form, texture and translucency that are characteristic of the lingering beauty of ornamental grasses.

Also responsible for the renewed interest in grasses is a dramatic increase in the number of
species and cultivars available to today's gardener. In the Victorian heyday of ornamental
grasses, a limited few such as *Arundo*, *Cortaderia*, *Miscanthus* and *Pennisetum* were repeatedly employed, most often as specimen curiosities set into broad lawns. In recent years, plant
exploration, introduction, breeding and selection have enriched the modern palette of ornamental grasses so that it now includes myriad variations in size, form, texture and colour to
suit a multitude of purposes in the garden.

The innovative landscape nurseryman Karl Foerster (1874–1970) was an early and constant
promoter of ornamental grasses, and his influence has been wide-ranging. Foerster assembled
plants from around the world and grew them for evaluation in this nursery in Potsdam-
Bornim, Germany. By the 1940s Foerster's catalogue offered more than one hundred types of
ornamental grasses. Foerster also developed a more naturalistic style of garden design based
on his nursery trials and his observation of grasses growing in association with other plants in
their native habitats. His 1957 book *Einzug der Gräser und Farne in die Gärten* (*Using
Grasses and Ferns in the Garden*) provided a record of his experiences and is still one of the
most compelling works on the subject. Foerster's teachings have inspired two generations of
German horticulturists. A superb feather-reed grass, *Calamagrostis* × *acutiflora* 'Karl
Foerster' is named for him, and is now common in gardens around the world.

Richard Simon of Maryland brought some of Foerster's influence and plant palette to
North America in the late 1950s, when few ornamental grasses were available in the United
States. With the help and encouragement of the landscape architect Wolfgang Oehme and the
nurseryman Kurt Bluemel, both German-born advocates of Karl Foerster's philosophies,
Simon began offering ornamental grasses through his Bluemont Nursery catalogue.
Bluemel's own Maryland nursery, founded in 1964, has since become the premier commer-
cial introducer of grasses to the United States. Bluemel has worked to propagate, promote and
develop the market for new grasses introduced by institutions such as the U.S. National
Arboretum and Longwood Gardens. One such example is the feather-reed grass
*Calamagrostis brachytricha*, discovered by Dr. Richard Lighty while on a Longwood-spon-
sored plant collecting expedition to Korea in 1966. U.S. National Arboretum staff Dr. John
Creech and Sylvester March introduced a number of ornamental grasses from Japan in the
mid 1970s, including the variegated *Miscanthus* cultivars 'Cabaret', 'Cosmopolitan' and
'Morning Light', as well as the diminutive green-leaved *Miscanthus* 'Yaku Jima'. These
mainstays of modern horticulture were first offered commercially by Kurt Bluemel Inc., as

was *Miscanthus transmorrisonensis*, introduced from Taiwan by Paul Meyer of the University of Pennsylvania's Morris Arboretum in 1979. In the 1980s a number of stellar introductions such as *Panicum virgatum* 'Heavy Metal' and *Miscanthus sinensis* 'Sarabande' originated with Bluemel. In recent years Bluemel has been an important conduit for European selections such as the British *Phalaris arundinacea* 'Feesey', and the spectacular early-blooming *Miscanthus* cultivars developed by Ernst Pagels of Leer, Germany, including 'Graziella' and 'Malepartus'.

In the 1990s, many of the most important additions to the gardening world's palette of grasses have been native North American species and cultivars thereof. A fresh look at American grasslands by horticulturists from coast to coast is generating an abundance of widely adapted ornamentals. Native plant specialist Roger Raiche at the University of California's Berkeley Botanic Garden has woven many beautiful, drought-tolerant western natives such as *Festuca californica, Muhlenbergia rigens, Calamagrostis foliosa* and *Carex spissa* into the gardens' displays, and has worked with Greenlee Nursery and others to make them available. At the University of California's Santa Barbara Botanic Garden, Carol Bornstein's initiative to explore native grasses has resulted in fine introductions such as *Leymus condensatus* 'Canyon Prince'. Prairie Nursery in the midwest has extolled the virtues of previously obscure but highly ornamental prairie species such as *Sporobolus heterolepis*. In the eastern states, Longwood Gardens' nursery trials of native American grasses have produced *Sorghastrum nutans* 'Sioux Blue', Kurt Bluemel has developed *Panicum virgatum* 'Squaw' and 'Warrior', and Bluemount Nursery has selected a giant, blue-leaved form of *Panicum virgatum* named 'Cloud Nine'.

This manual treats grasses in the broad sense, including the true grasses, Gramineae or Poaceae (including the bamboos), the sedges, Cyperaceae, and the rushes, Juncaceae. Although these families share a fineness of line, they may be distinguished for horticultural purposes by a range of ornamental characteristics and cultural requirements.

The true grasses, Gramineae, are among the most highly evolved plants on the earth. This truly cosmopolitan group is made up of over 9000 species belonging to more than 600 genera. It should be no wonder that grasses are proving such a treasure trove of ornamentals; members of the grass family are found on all the continents in nearly all habitats. Grasses are part of almost all ecological formations and are the dominant vegetation in many, such as prairies, steppes and savannas. Karl Foerster's characterization of grasses as 'Mother Earth's hair' is not just fanciful; grasses are the principal component in more than one-fifth of the planet's cover of vegetation. Immensely important economically, grasses include all the cereal crops as well as sugar cane, bamboos, canes and reeds. The grass family is comprised of herbaceous annuals and perennials, and bamboos, which are semi-woody. Herbaceous perennial grasses are unquestionably the most varied and versatile group for purposes of landscape design.

Perennial grasses are among the easiest to grow of all garden plants. Used properly, they can contribute to richly rewarding landscapes requiring little maintenance. Grasses are adaptable to a wide range of soil, temperature and moisture conditions, and are relatively free of pests and diseases. In native habitats the greatest number of grasses prefer sunny sites, and this is true for the majority of ornamental species in the garden. Sun-loving species often need two-thirds to full-day sun for best performance. Shading these grasses usually results in elongated, lax growth and poor flowering. Light requirements may vary considerably between related cultivars; most *Miscanthus* cultivars demand considerable sun, yet 'Purpurascens' stands upright and flowers well in half shade. Some grasses need both full sun and long growing seasons. For example, many *Miscanthus* fail to develop flowers in the short seasons of the northeastern United States and northern Europe. Again, a careful choice of cultivars will avoid this problem. Ernst Pagels developed his recent *Miscanthus* introductions ('Graziella', 'Malepartus', 'Kleine Fontäne', etc.) with the goal of producing plants that would flower in the relatively cool, short season of northwestern Germany. These plants are extremely useful in much of Great Britain and in cooler parts of the U.S. Grasses imported from warm southern climates sometimes succumb during winters in northern countries. The cause may not be the climates sometimes succumb during winters in northern countries. The cause may not be the ultimate low temperatures in the new environment, but lack of hardiness resulting from a weak sun in the growing season. *Saccharum ravennae* , for example, flowers well and easily tolerates −18°C/0°F winter lows in parts of the U.S. that enjoy hot summers. In England it often does not bloom at all and may fail in winter.

Although fewer than sun-loving species, several grasses are native to moist, shady woodlands and woodland edges. Some of these, such as *Calamagrostis brachytricha*, *Chasmanthium latifolium*, and *Spodiopogon sibiricus*, are highly ornamental choices for the shade garden. Other ornamental species such as *Deschampsia flexuosa* and *Hystrix patula* grow happily in very dry shade, always a difficult niche to fill in the garden.

Grasses are tolerant of many different soil types. An inquiry into the particulars of a species' native habitat often provides insights useful for siting plants in the garden. For example, grasses native to wet areas, such as *Miscanthus* or *Spartina* are demonstrating tolerance of low soil aeration. These species often are ideal choices for poorly aerated garden soils such as heavy clays. Species found on infertile sands in the wild will obviously tolerate similar conditions in the garden; many, however, appreciate a rich garden loam. Some, such as the fescues, demand well drained soils. These will succumb to root rots in soils that stay moist, especially in winter. Although woodland natives respond particularly well to feeding, it is generally unnecessary for most grasses except when they are planted in infertile sands. On rich loams and clays, feeding can produce a superabundance of soft growth and may cause grasses to flop over.

The fibrous root systems of grasses are very efficient, making most extremely drought-tolerant. Grasses roll the edges of their leaves inward in response to moisture stress, and this can be used as an indicator of the need for watering. Once established, most ornamental grasses rarely need watering even in the driest summers. There is sometimes significant variation in the drought-tolerance between closely related cultivars. For example, the narrow-leaved *Miscanthus* 'Sarabande' will go through extended droughts with only minor tip burn. The broader-leaved *Miscanthus* 'Purpurascens' will scorch badly under the same conditions. Shade species such as *Chasmanthium* or *Spodiopogon* will perform well in full sun if given additional water in dry periods.

Grasses are generally pest- and disease-free. Occasionally, in unusually wet seasons, the foliage of some species will suffer from rust diseases. This can be minimized by spacing plantings to provide free air movement and by avoiding overhead irrigation. A mealybug, *Pilococcus miscanthi*, introduced to the United States in the late 1980s, now poses a serious threat to *Miscanthus*. It attacks all parts of the plant including the roots, so above-ground mechanical or chemical methods are not sufficient for control. Drenching with systemic insecticides has proved effective. Grasses are not usually attacked by grazers such as deer, rabbits and squirrels, although these may pose a threat to emerging bamboo culms.

Herbaceous perennial grasses may be grouped loosely into two categories, warm season growers and cool season growers. Warm season grasses like it hot. They tend to sulk in cool spring weather, but once temperatures reach approximately 28°C/82°F they begin vigorous growth which continues unabated through summer. Most bloom toward summer's end and then die back to ground level with the onset of cold temperatures. In colder climates it is risky to divide or transplant warm season grasses in autumn. The plant's food reserves are lowest after flowering and seed-set, the leaves are no longer photosynthesizing, and the roots are relatively inactive in winter. Warm season grasses are best divided or transplanted in spring after strong growth has resumed. Examples of warm season growers are *Miscanthus, Cortaderia, Pennisetum, Panicum* and *Andropogon*. Cool season grasses behave in the opposite way, growing best at temperatures below 28°C/82°F. New foliage growth begins in late winter or early spring, followed by spring or early summer flowers. Cool season growers sulk in summer; some simply interrupt growth, others go into a full summer dormancy, dying to the ground. Growth resumes in autumn, and often continues until winter temperatures drop below

5°C/40°F. Cool season grasses may be divided or transplanted almost any time of year except during their hot summer lull. Examples are *Arrhenatherum*, most *Festuca* species, *Koeleria*, and *Calamagrostis × acutiflora*.

Although all grasses spread to some extent by rhizomes or stolons, they can be segregated into clumping and running types for garden purposes. Each type has its strengths for different design applications. Many perennial species produce only a modest annual increase in girth, effectively remaining in a clump or tuft. Since they stay in place, these types are relatively easy to control. On the other hand, they can result in high landscape maintenance if planted for ground cover, since they will not spread to occupy spaces where individual clumps have weakened or died. The rhizomes of running types such as *Spartina pectinata* or *Glyceria maxima* may travel nearly one metre in a single growing season. This habit can be invasive and maintenance-requiring if plants are comingled with less aggressive companions but is ideal for ground cover use.

Only a minority of perennial grasses have the potential to be seriously invasive; this aspect, however, should be given careful consideration when choosing plants for the garden, especially if it is located near a sensitive native plant community. The popular *Miscanthus sinensis* is rapidly naturalizing areas such as bottomlands in the eastern United States. Some of the early-blooming cultivars are particularly fertile, and may accelerate this trend. The potential for invasiveness of any particular grass varies from climate to climate. *Cortaderia jubata* is a serious problem in coastal California, but it poses no threat in colder climates. Grasses that self-sow prolifically can add substantially to maintenance chores in the garden; fortunately they are few. Some cultivars such as *Calamagrostis × acutiflora* 'Karl Foerster' do not set viable seed, and may be used without concern.

Grasses that fall dormant in winter benefit from being cut back once a year. Although not required for the health of the plant, this practice makes for better garden appearance. Ideal heights and times for cutting back vary with the different species and cultivars. As a general rule, plants that remain attractive during winter dormancy may be cut back to about 10cm very early in spring. Grasses which are unattractive at this time are best cut back in autumn. Pruning shears are adequate for cutting back small grasses or a few large grasses. For cutting back large plantings, such as a screen of *Miscanthus*, electric hedge shears or even a chain saw will be more efficient. If cut back annually, some of the larger grasses may require extra support. This is usually the result of excess moisture or fertilizer, too much shade, or use of inferior cultivars. If the prior year's culms are allowed to stand they provide a natural support – a practice not suited to all tastes.

While it is possible to propagate many ornamental grasses by seed, most cultivars do not come true from seed and must be divided. It is best to lift smaller grasses from the ground with a trowel before dividing them with a sharp knife. Use a sharp spade to lift mid-sized and larger grasses before dividing them with the spade or a hatchet. Old clumps of the largest grasses, such as *Cortaderia*, can be quite a challenge to lift, and may require separation with an axe. As with other garden perennials, some grasses eventually die out in the centre and become floppy and unsightly. Divisions from the outside of each clump can be used for renewal. Some grasses, such as *Pennisetum* and *Panicum*, may be propagated by stem cuttings.

Most grasses are best purchased in spring. Container-grown plants remaining unsold in nursery yards until autumn are frequently pot-bound. Although retail garden centres offer an increasing selection of ornamental grasses, newer or rarer cultivars are often available only through mail-order nurseries. Container-grown grasses are often potted in a light, soilless mix. It is advisable to knock off some of the mix and loosen up the roots to encourage establishment in what is usually the heavier and less hospitable soil of the garden. Field-grown plants, when available, may re-establish more readily.

As with other plant groups that are widely adapted and of easy cultivation, there may be genuine concern that ornamental grasses will eventually suffer from over-use. Some would say that Pampas Grass has long since passed from favorite to cliché. More recently, *Miscanthus* and *Pennisetum* have become staples in the obligatory landscaping that is done to mitigate the monotony of commercial sites, such as shopping malls, automated banking machines, and entrance ways to tract housing developments. This trend may be forestalled through education, as the public and the commercial landscape trade become aware of the greater potential and diversity of grasses.

In aesthetic terms, ornamental grasses are quite unlike other perennials. Lacking typical broad-petalled, brightly coloured flowers, grasses derive much of their beauty from a unique set of attributes centred on line, light and movement. Grasses contribute a strong linear presence to the landscape that results from the close parallel arrangement of so many narrow leaf blades. Though the blades may be upright, arching or even pendulous, the repetition of their lines is uniquely stunning. Individual grass flowers are minute, with perianth parts so reduced that they are not recognizable as sepals or petals. The flowers are typically grouped in one- to many-flowered spikelets which in turn make up the spicate or paniculate inflorescences commonly referred to as flower heads or plumes. In many the inflorescence is delicately translucent, particularly when dry, and will glow brilliantly when back-lit or side-lit by the sun. Coaxed by the wind, the plumes move in and out of sun streams, creating magical flickering effects while the glossy foliage below alternates between translucency and shimmering reflection. Stirring gently in a summer breeze, dancing before an autumn storm, or flying in a spring gale, grasses mirror nature's moods and bring a special dynamism to the garden.

Grasses offer a surprising array of forms and textures, characteristics which are of prime importance when the design emphasis is not on flower colour. Shapes run the full gamut from the strict vertical of *Calamagrostis* × *acutiflora* 'Karl Foerster', through the rounded fountains of *Pennisetum alopecuroides*, to the explosive irregularity of *Calamagrostis brachytricha*. Although most grasses are fine-textured, some, like *Arundo donax*, are extremely coarse. In scale they run from the 4-metre tall *Arundo* to 15cm miniatures such as *Pennisetum* 'Little Bunny'.

Despite the size and delicacy of their flowers, grasses are hardly without colour. Modern cultivars offer summer foliage in countless shades of green as well as white (*Arrhenatherum* 'Variegatum'), yellow (*Millium effusum* 'Aureum') , blue (*Sorghastrum* 'Sioux Blue', and red (*Imperata* 'Red Baron'). These are followed by an autumn array of golds (*Molinia* 'Skyracer'), burnt-umbers (*Sporobolus heterolepis*), and burgundies (*Panicum* 'Hänse Herms'). The spectrum of flower colour is also broad, with snow white (*Melica ciliata*), pearly pink (*Pennisetum orientale*) and purple-bronze (*Pennisetum* 'Moudry') blooms lasting from late spring through autumn. The splendid autumn tones of foliage and flowers weather gracefully to winter hues of chestnut, fawn and russet.

The rigours of winter fail to diminish the beauty of ornamental grasses; in the opinion of many gardeners, this is their peak season. Frost often traces the graceful outlines of grasses on winter mornings. Even though dormant, many species retain their shape and stature through sleet, snow and freezing rain. Little Bluestem, *Schizachyrium scoparium*, and Broomsedge, *Andropogon virginicus*, paint broad golden-orange brush strokes on winter's white canvas. Encrusted in ice, the spikelets of *Chasmanthium* become jewel-like. Summer's downy plumes of *Miscanthus* close winter as stunning filigrees.

The following promotional copy from the 1909 nursery catalogue of the Storrs & Harrison Co. of Ohio. provides an interesting reference point from which to consider the traditional place of ornamental grasses in landscape design: 'In the laying out of lawns and artistic gardens a few of the many beautiful hardy grasses should not be overlooked. Their stateliness, tropic luxuriance, and soft colours harmoniously punctuate the prevailing green, while their graceful, sinuous yielding to every wind gives animation to gardened landscapes too apt to look "fixed".' These lines acknowledge the subtle beauty of grasses and celebrate the movement they bring to the garden. However, they also stereotype grasses as curious afterthoughts useful mostly for providing contrast with the ubiquitous lawn. Other contemporary writings and many of surprisingly more recent vintage suggest that grasses are best grouped by themselves in the garden. It is unfortunate that these two approaches have been so widely adopted, since they rarely realize the potential for grasses' contribution to the landscape.

The stirrings of a more varied, naturalistic approach to using grasses in artistic landscapes can be recognized in the work of landscape architect Jens Jensen in 1890's America and that of Karl Foerster and others in 1930's Germany, yet a more worldwide recognition of its appeal has only come about in the last thirty years. Leading modern proponents of grasses come from two somewhat overlapping schools. The Washington, D.C. landscape architectural firm Oehme, van Sweden Associates puts a priority on the fine-arts aspects of garden-making. Although grasses are frequent signatures in the firm's designs, they are included for their innate beauty, not out of a desire to recreate native assemblages. Andropogon Associates of Philadelphia, Pennsylvania is strongly biased toward a practice of landscape architecture that seeks to emulate and protect regionally unique aesthetics and ecologies, of which grasses are an inherent part. The California landscape architect Ron Lutsko's firm is somewhere in between, often creating landscapes with a finely crafted balance of artistry and ecological realism. These aesthetic objectives, exemplified by American practitioners, are to an increasing degree shared by garden designers in Western Europe and have long been in evidence in the gardens of the Orient.

An unusually versatile group, ornamental grasses can serve infinite capacities in the garden, limited only by the imagination of the designer. Native landscapes can be a rich source of inspiration. It takes little observation to know that grasses often occur naturally in huge sweeps and masses. In savannas and prairies, they are the form and foundation, the matrix of the landscape. Space permitting, many ornamental grasses are most effective when used in these ways in the garden. Meadow gardens, by their nature, should have a consistent framework of grasses through which flowering forms make seasonal appearances. Prairie natives such as *Andropogon* and *Schizachyrium* are obvious choices for massed plantings, especially in naturalistic gardens. Coastal lowlands in Japan are a splendid sight when millions of native *Miscanthus* bloom shoulder to shoulder in autumn. Something of this drama can be recreated in large garden settings by planting *Miscanthus* in mass. In modest gardens, a sweep of a refined grass such as *Calamagrostis* × *acutiflora* 'Karl Foerster' can create a mass effect without actually occupying so much area. This same grass and taller species such as *Miscanthus* 'Giganteus' or *Saccharum ravennae* can also be massed for screening purposes or for hedging to enclose or form garden spaces. Such screens and hedges will disappear temporarily after grasses are cut back in spring, they reappear quickly and are fully functional through summer, autumn, and most of winter.

Many grasses make reasonably good ground covers. Even though they are clumping types, the sturdy *Seslerias*, including *Sesleria caerulea*, *Sesleria autumnalis*, *Sesleria nitida* are low-growing, long-lived, and evergreen in milder climates. Prairie dropseed, *Sporobolus heterolepis*, is another clump-former suited to groundcover massing. Deep-rooted and extraordinarily drought-tolerant once established, *Sporobolus* remains attractive for decades without the need for division or re-setting, a claim that can be made for few perennial flowers.

Spreading or running species such as *Leymus arenarius*, *Glyceria maxima* 'Variegata', and *Hakonechloa macra* often make good ground covers. Flowering bulbs, such as narcissus and tulips, are happy to coexist with groundcover grasses. The bulbs usually flower earlier than the grasses and then their foliage is effectively masked by that of the grasses.

Allowing for their seasonal ebb and flow, grasses can be stunning specimen focal points or accents. For example, the classic symmetry of the variegated giant reed, *Arundo donax* 'Variegata' might serve as a living sculpture. Many have multiple seasons of interest, and can carry a design through much of the year. In these instances it is especially important to take advantage of natural back-lighting or side-lighting to express the plant's luminous qualities. Thoughtful placing of grasses so that they may be viewed from inside the house is also rewarding. Through a window, the movement of the grasses will catch the eye, providing a subtle connection and beckoning the gardener even in winter.

Ornamental grasses often work best when intermingled with other types of annuals, biennials, perennials, shrubs and trees. A border made solely of perennial flowers is a high-maintenance bore, and a border composed entirely of different grasses can be equally laborious and one-dimensional. The characteristic fine texture and linearity of grasses is most effective when visually balanced by other garden elements that contribute strong, solid forms to the landscape picture. These might be companion plants with bold foliage such as *Silphium terebinthinaceum, Rudbeckia maxima, Petasites* or *Gunnera.* A number of coarse biennials such as *Verbascum bombyciferum, Angelica gigas,* and *Cynara cardunculus,* as well as annuals like *Ricinus* and *Helianthus* are also ideal. Large-flowered companions such as *Hemerocallis* and *Hibiscus* provide exciting contrast, as do the dark, massive trunks of trees.

In wild landscapes and in the garden, grasses are especially beautiful near water. Their fine foliage is stunning when mirrored in the broad surface of a dark pool or pond. Many grasses are native to wet habitats. Ornamental variants of these species, such as *Glyceria maxima* 'Variegata' or *Spartina pectinata* 'Variegata', will thrive along pond edges and stream banks. Another winning combination borrowed from native landscapes is that of feathery grasses tumbling over massive boulders. Species such as *Panicum virgatum* and *Deschampsia cespitosa* produce clouds of the finest textured inflorescences. These grasses are dramatic when set among rocks, stones or spilling over paths. They make a superb backdrop for garden sculpture and will soften the outlines of other 'hard' garden features.

Japan has a long tradition of growing grasses in containers. *Imperata* and *Hakonechloa* are rarely planted in the landscape, but are frequently grown in decorative containers as companions to specimen bonsai. These and a host of other grasses deserve more frequent experimentation as container subjects in western gardens. Grasses with coloured foliage such as *Helictotrichon sempervirens* provide steady interest when planted in containers with annual flowers or foliage plants such as *Tradescantia pallida* 'Purpurea' (*Setcreasea purpurea*). Also, many perennial grasses are sufficiently cold-hardy to remain outdoors in unprotected containers through winter. *Calamagrostis* × *acutiflora* 'Karl Foerster', for example, has easily survived −18°C/0°F in a modest-sized concrete urn at Longwood Gardens.

A single garden can certainly serve a multitude of purposes, and it is delightful when a single plant in the garden serves more than one purpose. A few grasses answer to more than just visual duty in the landscape. The flowers of *Sporobolus heterolepis* are pleasantly scented. Not only handsome, Lemongrass, *Cymbopogon citratus,* is a versatile culinary herb. A great majority of ornamental grasses make wonderful cut flowers. When dried, many remain attractive for years. Most grasses bloom so profusely they may be cut repeatedly without fear of diminishing their beauty in the garden. Pampas Grass, *Cortaderia,* has long been a cut-flower favourite; others such as *Miscanthus, Saccharum, Chasmanthium,* and *Spartina* can be equally intriguing.

Unlike perennials, the true annual grasses are rarely grown for their foliage, which is generally undistinguished and often weedy. Their beauty lies in their flowers and seed-heads. Only a few, such as hare's-tail grass, *Lagurus ovatus,* are truly suited to landscape gardening. The majority are grown just for cutting. Shattering of the dried flowers is reduced if inflorescences are cut before they open fully.

Many of the cultivated annual grasses are from warm climates. They require hot sunny places in the garden but are not particular about soils. Seed may be direct-sown in spring, and then seedlings should be thinned to a minimum of 15cm between plants. In climates with short growing seasons, seedlings are best started indoors and transplanted to the garden after the danger of frost is past. Most annual grasses come true from seed. Seed saved germinates well the next year, with the percentage rapidly decreasing with each successive year.

Some species such as *Pennisetum setaceum, Pennisetum villosum,* or *Rhynchelytrum repens* behave as annuals in cold northern gardens; however, they are actually perennials in their native tropical or subtropical habitats. Annual in a horticultural sense only, these grasses are more like cold-hardy perennials in their appearance and use in the garden.

Botanically, bamboos are members of the grass family, Gramineae. For garden purposes,

Botanically, bamboos are members of the grass family, Gramineae. For garden purposes, they are more different than they are like the annual and herbaceous perennial grasses. Most bamboos are shrubby or tree-like with well developed woody stems. Although their individual flowers are similar in detail to other grasses, the inflorescences of bamboos are never as showy. In gardens, bamboos are valued for their beautiful foliage and stems (culms), both of which vary considerably in their colour and size. In terms of design use, there are bamboos for screening, waterside planting, specimen display, groundcovers, and container planting. Most of the larger types benefit from rich, moist soils. The smaller running genera such as *Pleioblastus* tolerate a wide range of soil and moisture conditions. All bamboos suffer from exposure to strong winter winds.

As a group, bamboos are notorious for their invasiveness. Though there are both running and clumping bamboos, the majority of the clumping types are not cold-hardy beyond their native tropical or subtropical regions. In milder climates, the cold-hardy running genera such as *Phyllostachys* often require containment measures beyond the resources of the average gardener. However, with appropriate siting and maintenance, such plants contribute to the gardened landscape like no others. A path cut through a mature, managed grove of one of the larger bamboos can be one of the most compelling parts of a garden. The culms of these types range in colour from deep green to glaucous blue to ebony black. In addition, many cultivars have beautifully variegated culms.

Many bamboos in cultivation have strikingly variegated foliage. Variegation ranges from nearly pure whites to strong yellows. The foliage of *Pleioblastus viridistriatus* begins the season a bright yellow-green with dark green stripes, darkening as the season progresses. The margins of semi-evergreen *Sasa veitchii* desiccate evenly in winter, turning cream-white and taking on a stunning appearance that is even more dramatic than variegated foliage.

One of the exciting recent developments in bamboos has been the rapid increase in commercial availability of the cold-hardy clump formers *Fargesia murielae* and *Fargesia nitida*. Native to upper elevations in the Chinese Himalayas, they will easily withstand low temperatures of −10°F. These Chinese mountain bamboos are graceful and fountain-like in form, reaching three metres in height with a four-metre spread at maturity. The stems are up to 1cm in diameter. The cultivar *Fargesia nitida* 'Ems River' has particularly attractive, purple-black stems.

Sedges are sufficiently distinct botanically from the grasses to be classified as a separate family, Cyperaceae. The stems of sedges are usually solid and triangular in cross section, thus the old adage 'sedges have edges'. In comparison, grass stems are largely hollow and round. The leaf sheaths of sedges are closed, completely surrounding the stems. Grass sheaths are split and open, and are easily pulled away from the stems. Like grasses, the tiny individual flowers of sedges have no recognizable petals or perianth, and they are also arranged in spikelets. Sedge inflorescences are composed of spikelets grouped in heads or spikes. Though many are attractive or interesting, they never have the showy paniculate arrangement of true grasses.

The sedge family is much smaller than the grass family, including approximately 115 genera and 3600 species, nearly all of which are perennial. Sedges are widely distributed throughout the world, especially in temperate and arctic regions. They usually occur on infertile soils that are sometimes dry but most often are damp or even waterlogged. All have a rhizomatous root system; some are clump-forming others running. Horticulturally, sedges are not nearly as important as true grasses. They rarely offer the translucency or movement of the grasses, and they are less often grown for their flowers; however, many are prized for their attractive form and distinctive foliage. Many sedges are evergreen, even in relatively cold climates. In addition, many of the evergreen types are subtly or even boldly variegated.

In general, sedges require more moisture than grasses. Like grasses, most prefer sunny sites, but a number of sedges are woodland species that grow well in shaded garden settings, especially by the edges of ponds and streams. Deciduous types and evergreens grown in cold climates should be cut back once yearly in spring. Division and transplanting of sedges should be done in late spring or in summer when plants are in growth. Many of the species are readily propagated by seed. The special characteristics of most cultivars are retained only through division.

Sedges grown primarily for their flowers are relatively few; however, some are particularly worthy of mention. The cottongrasses, *Eriophorum* are found in bogs in north temperate and arctic regions. The showy flower heads resemble tufts of cotton, due to the presence of numerous whitish bristles. The cream-white, spicate inflorescences of Fraser's sedge, *Carex fraseri*, are held well above the glossy, strap-like foliage on long peduncles, and are unusually showy for a sedge. Woolgrasses, including *Scirpus cyperinus*, grow over 1 metre tall. Their upright stems are topped with attractive flower heads. The spikelets, which are green at first, then reddish brown, are clustered at the ends of delicate, nodding peduncles. Weeping sedge, *Carex pendula*, has gracefully arching flowering stems and pendent catkin-like spikes. Members of the genus *Rhynchospora* have distinctive white bracts subtending the inflorescence. The Papyrus or Paper Plant, *Cyperus papyrus*, is the best known of the umbrella sedges, *Cyperus*. The inflorescences of this group are composed of dramatic, stalked spikes subtended in most cases by whorls of long leafy bracts. Found primarily in tropical and subtropical climates, growing in or near water, the umbrella sedges range in size from 30cm to nearly 4 metres. They are popular for use in aquatic gardens, or as potted specimens indoors. Some such as *Cyperus alternifolius* 'Variegatus' have white-variegated bracts. A few sedges are colourful in fruit; seed-heads of *Carex baccans* are bright orange-red at maturity, those of *Carex atrata* are brownish black.

The foliage of sedges is even more varied in colour than that of the grasses – in tones of green, glaucous-blue, yellow, brown, and nearly red, as well as white or yellow-variegated. One of the best of the solid greens, *Carex plantaginea* has broad, pleated evergreen leaves and is ideal for shade gardens. Southern California native *Carex spissa* grows in a clump to 1.5 metres tall, with dramatic gray-green evergreen leaves. The foliage of *Carex elata* 'Bowles' Golden' is bright yellow with thin green margins, and retains its colour through most of summer. A similar and somewhat confused group of species, *Carex morrowii*, *Carex hachijoensis* and *Carex oshimensis* presents a great number of variegated forms. The whorled leaves of *Carex phyllocephala* 'Sparkler' are dark green with bold white margins. An unusual group of New Zealand native sedges including *Carex buchananii* and *Uncinia rubra* have finely tufted, evergreen foliage in various shades of copper brown and red.

In addition to clumping species for specimen and border use, the sedges include aggressively spreading species such as *Carex flacca* and *Carex muskingumensis* which make ideal groundcovers for larger areas.

In the U.S., native wetlands are increasingly threatened by development. Recent federal laws attempting to mitigate these losses are directing large projects intended to re-create wetlands in other areas. This trend has dramatically increased the demand for nursery-grown wetlands species, including a great many sedges and rushes. On another front, there is much current experimentation involving the use of resilient, drought-tolerant species such as Texas sedge, *Carex texensis* as substitutes for water-hungry lawn grasses.

The rushes, Juncaceae, are a relatively small family of 10 genera and 325 annual and perennial species. They are found only on damp or wet sites in mostly cool temperate or subarctic regions. The stems are round in cross section, with mostly basal leaves that may be flat or cylindrical. The flowers, though very small, have fairly typical perianth parts. The tiny sepals and petals are essentially alike but are in two distinct whorls. The inflorescences are terminal cymes. Only two genera are of horticultural importance.

The common rushes, *Juncus*, hail from sunny, open marshlands. They are smooth plants with cylindrical leaves. The soft rush, *Juncus effusus*, is most commonly grown ornamentally. It is valued for its stiff, dark, semi-evergreen stems and leaves and conspicuous inflorescences, which are borne below the tips of the stems. It can be quite effective as an accent or massed planting at the edges of pools or streams, and is suited to pot culture if given sufficient massed planting at the edges of pools or streams, and is suited to pot culture if given sufficient moisture. The cultivar 'Spiralis' has spiral, twisted foliage. Some of the larger species deserve more attention as ornamentals for moist garden sites, while the dwarf *J.ensifolius*, with iris-like foliage and dark, shiny flower heads is an excellent choice for the margins of even the smallest ponds. The wood-rushes, *Luzula,* are plants of rich moist woodlands, and are useful in shady woodland gardens. Their leaves are broad, flat, and hairy. Wood-rushes have attractive flower-heads that appear very early in spring. Greater wood-rush, *Luzula sylvatica*, also has attractive, relatively broad foliage and makes a good woodland ground cover. A pretty, golden-margined cultivar, 'Marginata', is available.

It seems certain the unprecedented diversity now existing in ornamental grasses will firmly and permanently establish their place in the garden palette. That diversity is clearly evinced by the genera to be found in this manual, a work intended to guide gardeners in a wide range of situations and climates, each of which offers special opportunities for gardening with grasses.

# Bamboos

Outside the cool-temperate West, bamboos are closely linked with everyday life and are used for building, scaffolding, bridges, fences, rafts, furniture, pipes, fishing rods, wang-hai canes, walking sticks, chopsticks, wickerwork, weapons, musical instruments (the shakuhachi is one), pens, paper, food (the young shoots) and medicines (e.g. tabasheer). They have also inspired fine brushwork paintings and carvings, notably on netsuke and inro. In the West, however, their value, apart from bamboo stakes, is almost entirely ornamental. Some make good screens or hedges; others do well in large pots or tubs, or may be used as bonsai. The smaller species provide excellent groundcover if suitably curbed; they may even be mown or clipped.

The first hardy introduction to the British Isles is dated to 1823 and to other parts of Europe at much the same time – one or two more tender species may have come in earlier. But their popularity was limited to a few enthusiasts and grew only gradually towards the end of the century. Two major early works on bamboos established them as a major horticultural group. *Les Bambous*, written by the Rivières in 1878, publicized work done in France and Algiers. In 1896 A.B. Freeman-Mitford (Later Lord Redesdale) of Batsford Park in Gloucestershire published *The Bamboo Garden*, a landmark in their cultivation. Freeman-Mitford had been in the Diplomatic Service in Japan, whence about 1893 he sent home quantities of many species; his exports also went to other destinations such as Wakehurst Place in Sussex and the Caledonia Nursery in Guernsey. In 1903 he could write that exactly fifty species and varieties were in cultivation in the Midlands of England.

The closing decades of the 20th century have shown a great surge of interest. Since 1979, this has been fostered by the American Bamboo Society, whose membership soon became worldwide. It was followed by the formation of the European Bamboo Society and of the British Bamboo Society. In the Far East, work on taxonomy flourishes.

CLASSIFICATION AND NOMENCLATURE.   Careful estimates indicate that there may be as many as 90 genera and 1000 species of bamboos. The scarcity or absence of flowers, and so of floral characters, has been a hindrance in classifying these plants and deciding in which genus some should be placed. In earlier days it was, understandably, often the colourful variants that tended to be introduced and named before the normal wild plant. This means that plants untypical of the species as a whole are the *type* of the species and that the forms most often encountered in the wild may have to be classed below specific rank. Examples of such 'untypical types' are *Phyllostachys nigra* and *P.sulphurea*, *Pleioblastus variegatus* and *P.auricomus*. These two factors – poor availability of floral character and types based on cultigens – have led to much confusion in bamboo nomenclature, as can be judged from the massive synonymy that a genus like *Arundinaria* will carry.

Bamboos are ancient, long-lived members of the grass family, the Gramineae; their subfamily is Bambusoideae. This differs from other grasses in having stalked leaves that join on to the leaf sheath and, in all the species we grow, woody culms rich in silica, the source of their toughness. Culms can be very short – a few centimetres – or very tall. Heights of forty metres (130ft) have been recorded from the tropics, but the limit in southern Europe and the US is 20–30m/65–98ft. Their diameter extends from 1mm to 30cm/12in. Culms attain their final stature and girth in 1–3 months; extensions of 125cm/50in. in 24 hours have been noted. Thereafter, they do no more than develop branches and, in many species, branchlets, and their full strength and vigour. This 'ceiling' has the horticultural advantage that no allowance has to be made for increased vertical growth, unlike many other woody plants. The entire final length of the culm is compressed into the burgeoning shoots. The final height is achieved by the internodes' lengthening from the base upwards, supported and sheltered by the sheaths. For industrial purposes, culms are usually harvested when they are ripe, i.e. after about four years for fully established plants, and something like this is good practice in gardens too – although too much cutting of taller species will weaken them. The thinning of older culms will admit more light and enhance the graceful appearance of established plants.

Culms of most bamboos are hollow, thin-walled or thick-walled between the solid nodes. In *Chusquea*, the culms are pith-filled and in effect solid. Nearly all have a white waxy powder below the nodes that discolours or breaks up with time. Only one branchlet is produced per node (except in *Chusquea*), but this itself will often branch so early and so low down that the fact is obscured. The number of apparent branches emerging or developing varies from few to very many and is an important diagnostic feature.

Bamboos with just plain green leaves and green-brown culms can be extremely decorative in a quiet way and need no further embellishment. Many species, however, have produced striking variations: the culms may be yellow, or gold, sometimes striped with green; they may be green mottled brown or dark purple, or glossy black. The sheaths can be colourful too, but soon fade. The leaves may show variegation, yellow or white. They are mostly in varying degree lanceolate, from 3–60cm/1–24in. long and up to 10cm/4in. wide. Those at the apex of young shoots are often larger than normal. All are evergreen. Narrow leaves may curl at the edges in wind or dry or cold weather, but soon recover. They may drop off as critical low temperatures are reached, the culms suffering as it gets colder, and finally the rhizomes. There is, however, considerable variation in cold-resistance, and much will depend, of course, on the plant's native habitat and its intended garden location. All have transverse veinlets, which show up as tessellation. Those in which the tessellation is obscure are rarely hardy. The leaves of most species are scaberulous with short bristly hairs on one or both margins. These point upwards toward the apex, and can be felt by stroking the leaf downwards. The underside is nearly always paler but frequently unevenly so, and about two-thirds glaucous and one-third green. Leaves soon curl after cutting, notably narrow ones, but can be temporarily revived for pressing etc. by soaking in water. Whole shoots can be kept from curling if the culms are cut under water in their vase, and left there.

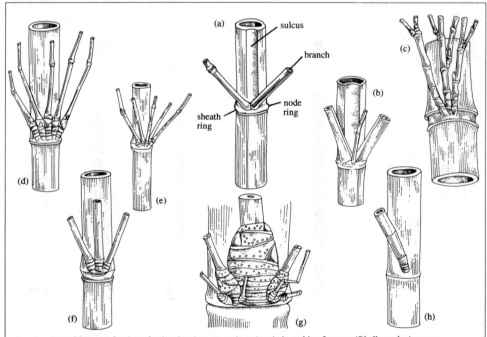

**Bamboo branching**   (a) Section of culm showing one node and main branching features (*Phyllostachys*)
(b) *Phyllostachys bambusoides*  (c) *Pseudosasa amabilis*  (d) *Fargesia nitida*  (e) *Chimonobambusa quadrangularis*
(f) *Pleioblastus gramineus*  (g) *Bambusa vulgaris*  (h) *Sasa veitchii*

**Bamboo culms 1** (a) *Phyllostachys aurea* (b) *Qiongzhuea tumidinoda* (c) *Phyllostachys nigra* (d) *Phyllostachys bambusoides* 'Holochrysa' (e) *Phyllostachys bambusoides* 'Castillonis Inversa' (f) *Phyllostachys vivax* 'Aureocaulis' (g) *Drepanostachyum khasianum* (h) *Chimonobambusa quadrangularis* (i) *Phyllostachys bambusoides* 'Castillonis'

The sheaths, which tightly encircle the young culms and leaf stalks, hold important diagnostic characters. Culm sheaths may drop off quickly, or gradually, or drop off incompletely as the culm elongates or as the branches develop, or may persist for years and end up ragged. They are usually shining inside, but vary externally in texture, indumentum, markings and colouring. At the base of the blade is a ligule which may be rudimentary, tall or short, broad or narrow, fringed or entire; by it are auricles and bristles, or just bristles. The blade of the leaf sheath is the actual green leaf. Side branches of a bamboo, above all those without a node or sheath, can rarely be confidently named.

Details of the sheaths, as of branching, often vary from bottom to top of the culms. It is therefore customary to take characters from mid-culm from mature, normally growing plants, preferably in early summer. The basal scar of the fallen culm sheath usually remains as a lower ring below the nodal protuberance, which may be prominent or hardly so. The time of emergence of culms has been used in classification and can be a help, but their incidence is not so regular away from native areas.

An awkward fact of bamboo classification is that much is made to depend on the rhizome structure, a feature not conveniently observable, or collectable and certainly not always visible above ground. The two main groups are the mainly tropical pachymorph, sympodial genera with short bulging rhizomes, typically thicker than the culms, that emerge only from their apices, producing a tight clump, and the leptomorph, monopodial genera with thin, rounded, running rhizomes usually branched, tough, very sharply pointed and capable of extending a metre or more a year. The latter have roots from their nodes, their culms are usually thicker and grow from lateral buds, well behind the apex of the rhizome, so making more open clumps. However, many of the runners stay in clumps in colder climates or non-optimum conditions (especially *Phyllostachys*), thus obscuring the basic distinction.

FLOWERING.  There is a commonly repeated and long-refuted myth concerning the flowering of bamboos – 'All plants of any given bamboo species all over the world flower at the same time (some say, in the same year) and then die' – only rarely is this true. When a given species commences to flower gregariously this flowering may be expected to spread. Guesses have been made,

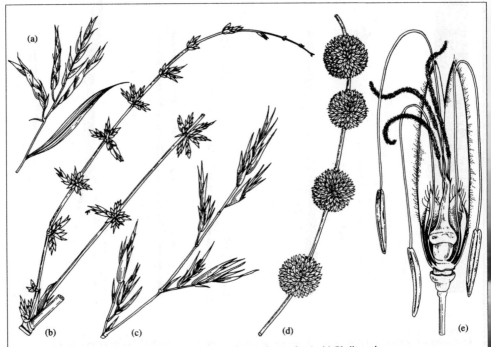

**Bamboo inflorescence types**  (a) *Pseudosasa japonica* (b) *Bambusa vulgaris* (c) *Phyllostachys aurea* (d) *Dendrocalamus strictus* (e) typical floret (*Phyllostachys*)

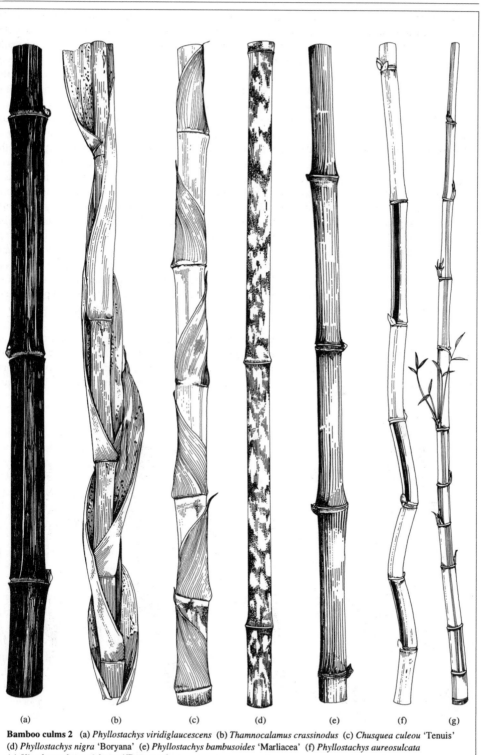

(a)     (b)     (c)     (d)     (e)     (f)     (g)

**Bamboo culms 2** (a) *Phyllostachys viridiglaucescens* (b) *Thamnocalamus crassinodus* (c) *Chusquea culeou* 'Tenuis'
(d) *Phyllostachys nigra* 'Boryana' (e) *Phyllostachys bambusoides* 'Marliacea' (f) *Phyllostachys aureosulcata*
(g) *Himalayacalamus falconeri* 'Damarapa'.

of varying ingenuity, to explain the puzzle of when flowers are produced, but the answer remains elusive. It seems that is must be controlled by genes in the rhizomes, for the culms are much shorter lived, unlike the branches of a tree or shrub. This implies an inbuilt clock that is set off once the rhizomes reach a certain age, or, dictated perhaps, by a variety of outside influences.

Records are cited for species in the Far East which go back for centuries and show apparently similar intervals between flowering bouts, sometimes as long as 120 years. Any such intervals may become blurred as species regenerate themselves over a long period. Flowering may occur in one year only or continue whenever weather allows, for twenty years or more on the same plant. It can vary from a few florets on one culm, or on all of one culm, to all culms on a single clump, to general, gregarious flowering of every clump. It can occur on one plant of a clone and not on others, although, of course, one immediate explanation for gregarious flowering is that plants of a single clone have been repeatedly repropogated and redistributed. Species have also been known to flower in one part of a country and not another, in one garden and in no other, in one country and not in another. *Chusquea* species, which flower frequently in South America, have never been known to flower in cultivation in the northern hemisphere. Dates for flowering may not be complete and some depend on possibly inaccurate identifications. Nevertheless, it can be a help in naming to know which species have been known to flower and when.

The effort of flowering diverts nutrients from the leaves, which may presage and accompany flowering by withering, with the result that the plant is deprived of chlorophyll and weakened but opened up for wind pollination (all those described in this work are wind-pollinated; some, rather softer-growing tropical species, are insect-pollinated). Culms that bear flowers tend to die, but in most cases the reserves in the plant's rhizomes survive, and sooner or later fresh green shoots appear and sufficient strength is built up for the plants to grow normally again. This may take a few years to achieve – running species are more likely to survive than clumpers. While flowering is in progress, most species are a sorry sight and might be thought to be dying, which, usually, they are not. Although the flowering culms, usually unsightly, may be excised, nothing will stop the flowering. Enriching the soil and feeding the plants can help them to recover.

DISTRIBUTION.    Bamboos grow from 50° North in Sakhalin, north of Japan, to 47° South in Chile, from sea level to 4000m/13,120ft, from very cold areas to the tropics. The region richest in species is the Far East, from Japan to the Himalayas and southward just into Australia. Africa has its own species, one of them hardy, Central and South America many, and the southeast US  one. None is native elsewhere but many have been introduced and become naturalized to the extent that their places of origin are obscured. The species with running rhizomes are those which seem most to have naturalized in Europe, especially in the British Isles.

CULTIVATION.    Within the parameters of their climatic requirements, bamboos are not difficult to grow. They prefer, but do not need, rich moist soil, but can endure waterlogging. They have even been grown in hydroculture. They do best in warm moist seasons. If it is too dry, the young shoots wither. Shelter from strong or gusty winds is desirable. Indeed, many hardy sorts do best in semi-shade and cannot endure great heat. Some of the small, running species thrive in quite deep shade. In placing specimens, bear in mind that the outer culms often splay out widely. All do better if given a mulch or other feed in late autumn. The fallen leaves and sheaths should be left if possible, for the silica they contain. The smaller species look better if they are cut back in early spring to prevent their shabby leaves spoiling the fresh display.

The running species often run too far. This may not matter in a semi-wild situation, but in many gardens, they may need to be confined by a very stout rust-proof barrier,  ideally 50cm/20in. deep, or a ditch of similar depth dug around them and kept clear. Alternatively, undesired shoots can be knocked off as they appear or the runaway rhizomes severed with a very sharp tool and disinterred. A web of such rhizomes can nevertheless be invaluable for binding the soil on slopes or river banks.

Young and newly established plants require watering until well established, which may take 2–3 years, and it will be longer still before the culms reach full size. All need water when new shoots appear and the culms are growing up.

**Bamboo foliage 1**   (a) *Indocalamus tessellatus*  (b) *Fargesia robusta*  (c) *Pseudosasa japonica*  (d) *Pleioblastus auricomus* f. *chrysophyllus*  (e) *Shibataea kumasasa*  (f) *Sasa palmata*  (g) *Pleioblastus linearis*  (h) x *Hibanobambusa tranquillans* 'Shiroshima'

**Bamboo foliage 2**  (a) *Thamnocalamus spathiflorus*  (b) *Pleioblastus viridistriatus*  (c) *Thamnocalamus crassinodus* 'Kew Beauty'  (d) *Sasa kurilensis* 'Shimofuri'  (e) *Chusquea couleou*  (f) *Pleioblastus variegatus*  (g) *Pleioblastus pygmaeus*  (h) *Phyllostachys nigra*

Given the scarcity of seed, division is the prime propagation technique: use younger, but not the youngest parts from towards the edge of the clump. Running rhizomes can be cut with 2–3 nodes; ensure that there are adequate roots as well. Clumpers call for harder treatment, a saw or even an axe, to get a suitable portion with a good bud and roots, to be tended carefully until it has recovered from the trauma. The best time to do this is just before the new shoots appear. The culms should be cut to about half their length to reduce transpiration and stress. In the hot, humid subtropics and tropics, portions of culm may be made to root. Meristem culture has also been used.

There are very few pests in temperate areas out of doors. Aphids can be a sticky nuisance, and scale insects and mites have been reported, neither of which causes serious trouble in cold conditions and can be controlled by standard means. Grey squirrels, deer and rabbits can devastate plants by eating new culm shoots that have scarcely broken the soil. They may be deterred by barricades of barbed wire or of holly and gorse twigs inserted around the base of the clump – these measures are unsightly but may be essential safeguards for the first week of a culm's life above ground, unless, that is, fencing, guns and traps are preferred. Bamboos are constitutionally robust and rarely suffer from diseases in temperate climates.

**Cyperaceae** (a) *Cyperus alternifolius* (a1) inflorescence branch (a2) spikelet (a3) flower subtended by scale
(b) *Cyperus haspan* (c) *Carex* sp. stem tip showing female inflorescences below and male above (c1) male flower with
subtending bract (c2) female flower with subtending bract (d) *Dulichium arundinaceum*

# The Families of Grasses and Grass-like Plants

**ACORACEAE** Martinov. SWEET FLAG FAMILY. Monocot. 1 genus, 2 species. Often included in Araceae. See *Acorus*.

**CYPERACEAE** Juss. SEDGE FAMILY. Monocot. 115 genera and 3600 species of perennial, usually rhizomatous herbs, rarely annuals, sometimes lianoid, shrubby and dracaenoid or vellozioid, usually with vessel-elements throughout vegetative parts; stems triangular, less often terete, usually solid; roots with root hairs (excluding *Eleocharis*), mycorrhiza absent. Leaves alternate, often 3-ranked, usually with closed sheath and long narrow blade (sometimes terete or absent), with parallel veins; ligule at junction with sheath sometimes present. Flowers small, wind-pollinated (few insect-pollinated), usually unisexual (plants monoecious, occasionally dioecious), sessile, in axils of spirally arranged or distichous bracts or scales, forming spikes or spikelets usually in secondary inflorescence, rarely solitary and terminal; perianth (1–) 6 (to numerous) free scales, bristles or absent; stamens (1–) 3 (6); ovary superior, of (2) 3 (4) fused carpels; style terminal, stigma branches equalling or not the number of carpels; ovule 1, basal, anatropous. Fruit an achene; seed free from pericarp; endosperm oily, starchy, with outer proteinaceous layer. *Eleocharis tuberosa*, Chinese water chestnut, has edible tubers. Some are used for thatching or basket-work. Papyrus of ancient Egypt was made from the pith of *Cyperus papyrus*, while *C.rotundus*, a tropical and widely naturalized tuberous species, is considered the world's most damaging weed. Cosmopolitan.

**GRAMINEAE** Juss. (Poaceae Barnhart). GRASS FAMILY. Monocot. One of the largest plant families comprising 635 genera and 9,000 species of annual, perennial and rhizomatous herbs or woody and tree-like (Bambusaceae, bamboos) but without secondary thickening; root system usually fibrous, often augmented by adventitious roots, and branch or tiller from the base to form a rosette or tussock, endo-mycorrhizae are often associated with roots; plants often spread laterally by stolons or rhizomes and vegetative propagation is very important in many members of the family; cell-walls, especially epidermis strongly silicified; stems (culms) terete, rarely flattened, jointed, usually with hollow internodes (solid in sugar cane, *Saccharum officinarum*, and some bamboos). Leaves alternate in 2 rows on opposite sides of the stem, with open sheath and elongate blade and 2 basal auricles (narrowed to a petiolar base above sheath in many bamboos); ligule adaxial at junction of lamina and sheath, usually membranous or reduced to a hairy fringe, rarely absent. Flowers usually wind-pollinated, bisexual, less often unisexual, in 1 to many-flowered spikelets, forming spicate or paniculate inflorescences; spikelets with 2 subopposite bracts (glumes) and 1 to several distichous florets on flexuous rachilla; florets enclosed in an outer scale (lemma) and a delicate membranous inner scale (palea); glumes and lemmas may be extended into bristles (awns); 2–3 (6 or more in bamboos) minute fleshy or hyaline scales (lodicules) are considered to represent a reduced perianth; stamens typically 3, rarely 1, 6 or more; anthers deeply sagittate, filaments delicate; styles usually 2, rarely 1 or 3; stigmas generally large and feathery, sometimes papillose; ovary of 2 united carpels (3 in bamboos), 1-loculed, with 1 ovule. Fruit a caryopsis (grain), 1-seeded, with a thin pericarp adhering to the seed; endosperm usually copious, starchy, with proteinaceous tissue, sometimes oily, rarely absent. The family is economically most important as it includes the cereals (barley, wheat, maize, millet, rice, oats, rye, sorghum, etc.), sugarcane, pasture and fodder crops. Many species are sources of alcohol, aromatic and edible oils, waxes, fibres, thatch, cane, edible shoots and paper pulp. Cosmopolitan.

**JUNCACEAE** Juss. Rush family. Monocot. 10 genera and 325 species of herbs, rarely annuals or pachycaul shrubs, often with starchy sympodial rhizomes; stems often not extending above ground except in flower. Leaves spirally arranged, usually basal, simple, parallel-veined, with sheath and usually flat, sometimes channelled or terete lamina; sheath often with apical auricles; leave sometimes reduced entirely to sheaths. Flowers usually bisexual (plants rarely dioecious), wind-pollinated, with no nectaries when insect-pollinated, small, solitary, in heads or cymes; perianth usually of 6 segments in 2 whorls, rarely 2+2 or 3 in one whorl, usually sepaloid, green, sometimes almost black, rarely white or yellow; stamens 3+3 or 3+0 or 2, pollen in tetrads; ovary of 3 fused carpels, 1- or 3-loculed; placentation parietal or axile (in *Luzula* 3 basal ovules); ovules 3 to numerous. Fruit a loculicidal capsule, rarely indehiscent; endosperm starchy. *Juncus* species provide cords and fibres and the split rushes are used for making baskets, mats and chair seats. Temperate and cold regions and tropical mountains, usually in damp places.

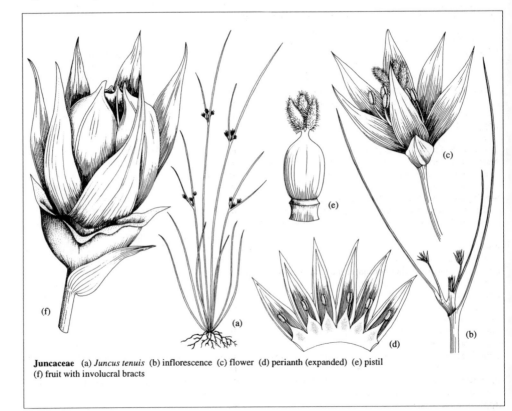

**Juncaceae**  (a) *Juncus tenuis*  (b) inflorescence  (c) flower  (d) perianth (expanded)  (e) pistil (f) fruit with involucral bracts

**POACEAE**. See Gramineae.

**TYPHACEAE** Juss. Cattail family. Monocot. 1 genus, 10 species. See *Typha*.

# Glossary

**acicular** needle-shaped, and usually rounded rather than flat in cross-section.

**acuminate** with the tip or, less commonly, the base tapering gradually to a point, usually with somewhat concave sides.

**acute** where two almost straight or slightly convex sides converge to terminate in a sharp point, the point shorter and usually broader than in an acuminate tip.

**adherent** of parts usually free or separate (i.e. petals) but clinging or held closely together. Such parts are sometimes loosely described as united or, inaccurately, as fused, which is strictly synonymous with coherent. Some authors use this word to describe the fusion of dissimilar parts.

**adpressed (appressed)** (of indumentum, leaf sheaths, etc.) used of an organ which lies flat and close to the stem or leaf to which it is attached.

**alternate** arranged in two ranks along the stem, rachis, etc., with the insertions of the two ranks not parallel but alternating.

**amphipodial** of a bamboo rhizome, essentially sympodial in habit with one culm arising from the other in clumps, but with some rhizome branches elongated, thus appearing to run, or to be monopodial in habit.

**annual** a plant which completes its entire life-cycle within the space of a year.

**anther** the pollen-bearing portion of the stamen, either sessile or attached to a filament.

**apical** borne at the apex of an organ or stem; pertaining to the apex.

**apiculate** possessing an apicule.

**apicule** a short sharp but not rigid point terminating a leaf, bract or perianth segment.

**apomictic** asexual reproduction, often through viable seeds but without fusion of gametes.

**appendage** secondary part or process attached to or developed from any larger organ.

**appressed** see adpressed.

**aquatic** a plant growing naturally in water, either entirely or partially submerged.

**arching** curved gently downwards – usually of culms, leaves and panicles, more distinctly and comprehensively so than in nodding, less so than in pendent.

**aristate** of a leaf apex abruptly terminated in an acicular continuation of the midrib. Otherwise, awned.

**articulate** jointed; possessing distinct nodes or joints, sometimes swollen at their attachment and breaking easily.

**ascending** rising or extending upwards, usually from an oblique or horizontal position, thus differing from erect.

**attenuate** (of the apex or base) tapering finely and concavely to a long drawn out point.

**auricle** an ear-like lobe or outgrowth, often at the base of an organ (i.e. a leaf), or the junction of leaf sheath and blade in some Gramineae.

**awn** a slender sharp point or bristle-like appendage found particularly on the glumes of grasses.

**awned** possessing awns.

**axil** the upper angle between an axis and any off-shoot or lateral organ arising from it, especially a leaf.

**axis** a notional point or line around or along which organs are developed or arranged, whether a stem, stalk or clump; thus the vegetative or growth axis and the floral axis describe the configuration and development of buds and shoots and flowers respectively, and any stem or point of origination on which they are found.

**barb** a hooked semi-rigid hair.

**bearded** terminating in a bristle-like hair (awned); or, more generally, possessing tufts or zones of indumentum on parts of the surface.

**bidentate** (1) (of an apex) possessing two teeth; (2) of a margin, with teeth, the teeth themselves toothed.

**biennial** lasting for two years from germination to death, generally blooming in the second year and monocarpic.

**bifid** cleft deeply, forming two points or lobes.

**bisexual** of flowers with both stamens and pistils; of plants with perfect (hermaphrodite) flowers.

**blade** the thin, expanded part of a leaf, also known as the lamina – the part of a leaf above the sheath.

**bract** a modified protective leaf associated with the inflorescence (clothing the stalk and subtending the flowers), with buds and with newly emerging shoots and stems.

**branch** a lateral growth from the main stem; in a panicle the main divisions are branches.

**branch complement** used of bamboos to denote the number of branches per node characteristic of the species.

**branchlet** a small branch or twig derived from a primary branch.

**bristle** stiff hair, or a very fine straight awn, also applied to the upper part of the awn.

**bristly** see echinate, hispid and setaceous.

**bulbous** bulb-like.

**caespitose** a habit description: tufted, with stems growing in dense clumps.

**calcareous** chalky or limy.

**callus** an abnormal or isolated thickening of tissue.

**calyx** a collective term for the sepals, whether separate or united, which form the outer whorl of the perianth or floral envelope.

**capillary** slender and hair-like; much as filiform, but even more delicate.

**capitate** (1) arranged in heads; (2) terminating in a knob or somewhat spherical tip.

**capsule** a dry, dehiscent seed vessel.

**caryopsis** (of Gramineae) a one-celled, one-seeded, superior fruit in which the pericarp and seed-wall are adherent; the grain.

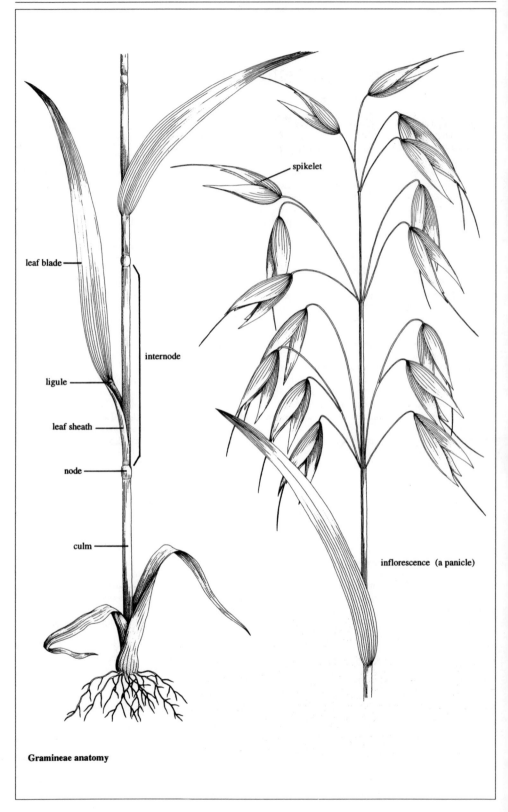

leaf blade

ligule

leaf sheath

node

culm

internode

spikelet

inflorescence (a panicle)

**Gramineae anatomy**

**cauline** attached to or arising from the stem.

**chartaceous** (of leaf and bract texture) thin and papery.

**chasmogamous** having open, not cleistogamous flowers, allowing cross-fertilization.

**chlorophyll** green colouring matter of the plant.

**chromosomes** microscopic bodies found in the nuclei at the time of division. In the vegetative cells two sets of these are normally present, their number (diploid, written 2n) usually being constant in a species.

**ciliate** bearing a marginal fringe of fine hairs (cilia).

**cleistogamous** with self-pollination occurring in the closed flower.

**collar** whitish, yellowish or purplish zone at the junction of the leaf-sheath and the blade.

**compressed** flattened, usually applied to stems and florets and qualified by 'laterally', 'dorsally' and 'ventrally'.

**concave** curving inwards, hollowed out.

**connate** united, usually applied to similar features when fused in a single structure.

**contiguous** in contact, touching but not fused.

**continuous** an uninterrupted symmetrical arrangement, sometimes used as a synonym of decurrent.

**contracted** narrowed and/or shortened.

**convolute** rolled or twisted together longitudinally, with the margins overlapping.

**cordate** heart-shaped, with a rounded base and a deep basal sinus or notch.

**coriaceous** leathery or tough but smooth and pliable in texture.

**corymb** an indeterminate flat-topped or convex inflorescence, where the outer flowers open first; cf. umbel.

**culm** the stems of Gramineae.

**cuneate** inversely triangular, wedge-shaped.

**cusp** a short, stiff, abrupt point.

**cyme** a more or less flat-topped and determinate inflorescence, the central or terminal flower opening first.

**deciduous** (1) falling off when no longer functional; (2) a plant that sheds its leaves annually, or at certain periods, as opposed to evergreen plants.

**decumbent** of a stem, lying horizontally along the ground, but with the apex ascending and almost erect.

**decurrent** where the base of a leaf blade extends down and is adnate to the sheath and the stem.

**deflexed** bent or turned abruptly downwards.

**dense** applied to flower-heads when crowded with spikelets.

**digitate** 'fingered', of a compound inflorescence, palmately arranged with the branches arising from the same point at the apex of the peduncle.

**dimorphic** occurring in two dissimilar forms or shapes, either at the same stage in a plant's life history or at different stages in a plant's development, as in juvenile and adult foliage.

**dioecious** with male and female flowers on different plants.

**diploid** plant having two sets of chromosomes in its nuclei.

**disarticulating** fracturing at the nodes, as the axis of the spikelet of many grasses.

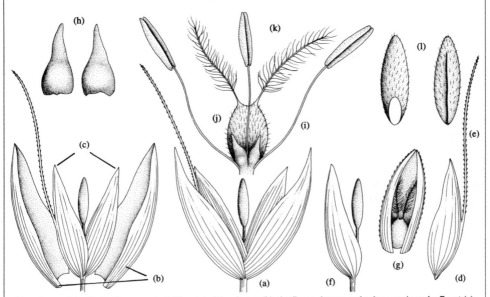

**Gramineae** (a) **spikelet** (*Avena sativa*). Divested of the **glumes** (b), the flower, lemma and palea constitute the **floret** (c). The **lemma** (d) is the outermost scale of the floret, often backed or tipped with a bristle or **awn** (e). The **palea** (f) is the innermost scale. Within this envelope lies the **flower** (g), subtended by two small scales or **lodicules** (h), it consists of three **stamens** (i), an **ovary** (j) with a single ovule and two **stigmas** (k). The ovule develops into a grain or **caryopsis** (l), with a small body or embryo at the base on its rounded side (left); the other side (right) is grooved, the groove marked by a dark line or hilum, the ovule's point of attachment to the ovary wall.

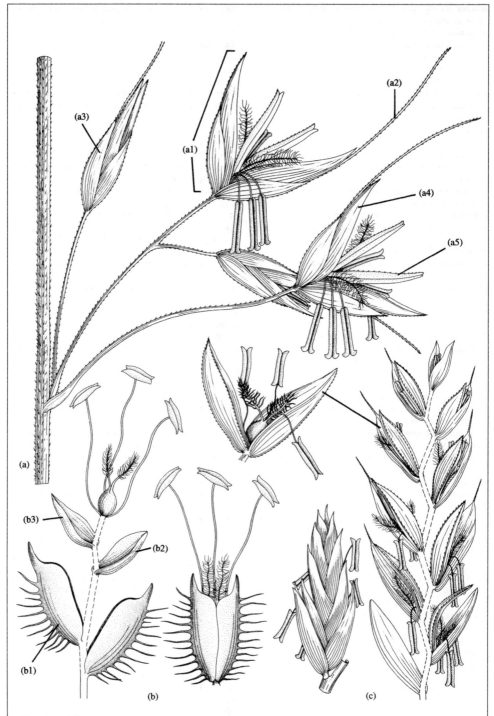

**Gramineae inflorescences**
(a) *Arrhenatherum elatius*, part of panicle showing (a1) spikelet, (a2) awn, (a3) glume, (a4) lemma, (a5) palea.
(b) *Phleum pratense*, single-flowered spikelet (right), exploded view (left) showing (b1) glume, (b2) lemma, (b3) palea.
(c) *Lolium multiflorum*, several-flowered spikelet (left), exploded view (right), single floret (centre).

**distant** widely spaced, synonymous with remote and the antithesis of proximate.

**distichous** distinctly arranged in two opposite ranks along a stem or branch.

**distinct** separate, not connate or in any way united, with similar parts; more generally meaning evident or obvious, easily distinguishable.

**ear** the spike of corn or an auricle.

**echinate** covered with many stiff hairs or bristles, or thick, blunt prickles.

**effuse** spreading widely.

**elliptic** ellipse-shaped; midway between oblong and ovate but with equally rounded or narrowed ends, with two planes of symmetry and widest across the middle.

**emarginate** (of the apex) shallowly notched, the indentation (sinus) being acute.

**embryo** the rudimentary plant within the seed.

**empty glumes** two empty bracts at the base of the spikelet.

**endemic** confined to a particular region.

**endosperm** the albumen, when it is stored inside the embryo sac.

**entire** continuous; uninterrupted by divisions or teeth or lobes, thus 'leaves entire', meaning margins not toothed or lobed.

**epidermis** the outer layer of a periderm.

**exserted** obviously projecting or extending beyond the organs or parts surrounding it; cf. included.

**extravaginal** applied to the young vegetative shoot when it grows up outside the leaf-sheath having burst through the base of the sheath; cf. intravaginal.

**fertile** producing viable seed; said of anthers containing functional pollen, of flowers with active pistils, of fruit bearing seeds. A fertile shoot is one bearing flowers, as opposed to a sterile shoot.

**filament** (1) a stalk which bears the anther at its tip, together forming a stamen; (2) a threadlike or filiform organ, hair or appendage.

**filiform** (of leaves, branches etc.) filament-like, i.e. long and very slender, usually rounded in cross-section.

**flexuous** (of an axis) zig-zag, bending or curving in alternate and opposite directions. This is a very distinctive growth pattern, to be seen in the culms of some bamboos, notably *Phyllostachys*. Confusion arises when the word is used erroneously as a synonym of *arching* or *nutant*.

**floret** a very small flower, generally part of a congested inflorescence; in grasses, a collective term embracing the lemma, palea, lodicules and enclosed flower.

**flowering glume (or lemma)** the lower of two bracts enclosing the flower.

**folded** applied to leaf-blades folded lengthwise along the midrib, with upper surface within.

**free** (strictly of dissimilar parts or organs) separate, not fused or attached to each other. 'Distinct' describes the separateness of similar organs.

**fruit** the fertilized and ripened ovary of a plant together with any adnate parts.

**fugacious** falling off or withering rapidly; transitory.

**fusiform** spindle-shaped, swollen in the middle and tapering to both ends.

**fusoid** somewhat fusiform.

**geniculate** bent abruptly, in the form of a flexed knee.

**gibbous** swollen on one side or at the base.

**glabrescent** (1) nearly glabrous, or becoming glabrous with age; (2) minutely and invisibly pubescent.

**glabrous** smooth, hairless.

**glaucous** coated with a fine waxy bloom, whitish, or blue-green or grey and easily rubbed off.

**globose** spherical, sometimes used to mean near-spherical.

**glume** a small, dry, membranous bract found in the inflorescences of Gramineae and Cyperaceae and usually disposed in two ranks.

**grain** the fruit of Gramineae.

**gynoecium** the female element of a flower comprising the pistil or pistils.

**habit** the characteristics of a plant's appearance, concerning shape and growth and plant type.

**habitat** the area or type of locality in which a plant grows naturally.

**herbaceous** (1) pertaining to herbs, i.e. lacking persistent aerial parts (as in the idea of a herbaceous perennial or the herbaceous border) or lacking woody parts; (2) softly leafy, pertaining to leaves; (3) thin, soft and bright green in texture and colour, as in the leaf blades of some grasses.

**hermaphrodite** bisexual, having both pistils and stamens in the same flower.

**hexaploid** having six sets of chromosomes in its nuclei.

**hilum** the scar on a seed at the point at which the funicle was attached.

**hirsute** with long hairs, usually rather coarse and distinct.

**hispid** with stiff, bristly hairs, not so distinct or sharp as when setose.

**husk** the outer layer of certain fruits.

**hyaline** transparent, translucent, usually applied to the margins of leaves and bracts.

**hybrid** a plant produced by the cross-breeding of two or more genetically dissimilar parents; the parents generally belong to distinct taxa. A natural hybrid is one arising in nature, a spontaneous hybrid one arising without the direct intervention of man but, usually, in gardens, an artificial hybrid is a cross deliberately made by man. Specific hybrids (interspecific hybrids) are those between species belonging to the same genus, intergeneric hybrids those between taxa belonging to different genera – the number of genera involved is usually denoted by a prefix, e.g. bigeneric, trigeneric, etc.

**hypogynous** borne beneath the ovary, generally on the receptacle; said of the calyx, corolla and stamens of a superior ovary.

**imbricate** (of organs such as leaves or bracts) overlapping; more strictly applied to such organs when closely overlapping in a regular pattern, sometimes encircling the axis.

**included** 'enclosed within', as in a grass floret by its glume; neither projecting nor exserted.

**incurved** bending inwards from without.

**indurate** hardened and toughened.

**inflexed** bent inwards towards the main axis.

**inflorescence** the arrangement of flowers and their accessory parts on an axis.

**innovation** basal vegetative shoot of a grass.

**internode** the portion of stem between two nodes.

**interrupted** not continuous; the disturbance of an otherwise continuous or symmetrical arrangement.

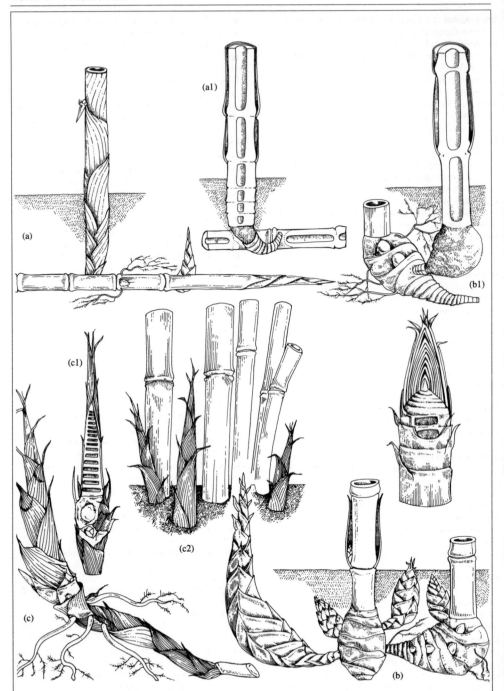

**Bamboo morphology** – rhizomes and culms   (a) Monopodial growth: running or leptomorphic rhizome, narrower than the culms, with long internodes not or scarcely branching and continuously growing straight leader (al) section. (b) Sympodial growth: a clump-forming or pachymorphic rhizome, stouter than the culms with short thick internodes, branching freely, with no apparent leader and culms arising one from the other (bl) section.   (c) Amphipodial growth: the rhizome is essentially monopodial, running with a strong leader, this however, ascends to form a culm and sympodial branches are produced from its base (in the cross section (cl), the buds can be seen at the base of the culm shoot). The overall effect is clumping (c2)

**intravaginal** of a young vegetative shoot that grows up within the enveloping sheath; cf. extravaginal.

**involucre** a single, highly conspicuous bract, bract pair, whorl or whorls of small bracts or leaves subtending a flower or inflorescence.

**involute** rolled inward, toward the uppermost side (see revolute).

**joint** used for the node of the culm; also applied to the internodes of the spike-axis and spikelet-axis; such axes are said to be jointed.

**keel** (1) a prominent ridge, like the keel of a boat, running longitudinally down the centre of the undersurface of a leaf, bract or glume; (2) the two lower united petals of a papilionaceous flower.

**lamina** see blade.

**lanceolate** lance-shaped, narrowly ovate, but 3–6 times as long as broad and with the broadest point below the middle, tapering to a spear-like apex.

**lateral** on or to the side of an axis or organ.

**lax** loose, open, as in the branches of a light and airy panicle.

**lemma** the lower and stouter of the two glumes which immediately enclose the floret in most Gramineae; also referred to as the flowering glume.

**leptomorphic** of the rhizomes of some bamboos, essentially monopodial in habit, thus little-branched with a strong, continuously growing main leader and culms arising along its length. Such bamboos are said to be running rather than clump-forming and can be invasive. cf. pachymorphic, sympodial.

**ligule** the thin, scarious, sometimes hairy projection from the top of the leaf sheath in Gramineae.

**linear** very slender and elongated (as are most grass leaves), the margins parallel or virtually so.

**lobe** a division, such as the apical teeth or segments of a lemma, then said to be lobed.

**lodicule** one of two, or sometimes three minute extrastaminal scales, adpressed to the base of the ovary in most Gramineae; they swell at flowering-time to cause the divergence of the lemma and palea (the two bracts immediately enclosing the floret).

**lower palea** lemma.

**membranous** thin-textured, soft and flexible.

**monoecious** with bisexual flowers or both staminate and pistillate flowers present on the same plant; cf. dioecious.

**monopodial** of a stem or rhizome in which growth continues indefinitely from the apical or terminal bud, and which generally exhibits little or no secondary branching.

**mucro** an abrupt, sharp, terminal spur, spine or tip.

**mucronate** possessing a mucro.

**muticous** lacking an awn or mucro.

**navicular** shaped like a deeply keeled boat.

**nerve** applied to slender veins or ribs of leaves, glumes, lemmas and paleas; e.g. the paleas are usually two-nerved.

**nodding** inclined from the vertical, hanging or arching slightly downwards, said of such panicles and spikelets.

**node** the point on an axis where one or more leaves, shoots, whorls, branches or flowers are attached.

**oblanceolate** as with lanceolate, but with the broadest part above the middle, and tapering to the base rather than the apex.

**oblique** (1) (of the base) with sides of unequal angles and dimensions; (2) (of direction) extending laterally, the margin upwards and the apex pointed horizontally.

**oblong** essentially linear, about 2–3 times as long as broad, with virtually parallel sides terminating obtusely at both ends.

**obovate** ovate, but broadest above rather than below the middle, thus narrowest toward the base rather than the apex.

**obtuse** (of apex or base) terminating gradually in a blunt or rounded end.

**open** loose, as in a panicle with widely spaced spikelets.

**opposite** two organs at the same level, or at the same or parallel nodes, on opposite sides of the axis.

**orbicular** perfectly circular, or nearly so.

**oval** elliptic.

**ovary** the basal part of a pistil, containing the ovules; a simple ovary is derived from a single carpel, a compound ovary from two or more; a superior ovary is borne above the point of attachment of perianth and stamens, an inferior ovary is borne beneath, and a subinferior, semi-inferior or half-inferior ovary is intermediate between these two positions. **ovate** egg-shaped in outline, rounded at both ends but broadest below the middle, and 1.5–2 times as long as it is broad (see elliptic).

**ovoid** egg-shaped, ovate but 3-dimensional.

**ovule** the body in the ovary which bears the megasporangium, developing into a seed after pollination.

**pachymorphic** of the rhizomes of some bamboos, freely branching with short, usually swollen internodes and culms arising one from the other, thus clump-forming. Virtually synonymous, in the case of bamboos, with sympodial; cf. leptomorphic; monopodial.

**palea** specifically the upper and generally slimmer of the two glumes enclosing the floret in most Gramineae; more generally a small, dry bract or scale.

**palmate** with three or more free lobes or similar segments originating in a palm-like manner from the same basal point.

**panicle** an indeterminate branched inflorescence, the branches generally racemose or corymbose.

**pectinate** pinnately divided or arranged, the segments being many and close together, like the teeth of a comb.

**pedicel** the stalk supporting an individual flower or fruit.

**peduncle** the stalk of an inflorescence.

**pendent** hanging downwards, more markedly than arching or nodding, but not as a result of the weight of the part in question, or the weakness of its attachment or support. cf. pendulous.

**pendulous** hanging downwards, but as a result of the weakness of the support, as in a slender-stalked heavy fruit or a densely flowered weeping raceme.

**penicillate** brush-shaped, or like a tuft of hairs.

**pentamerous** with parts in groups of five, or multiples of five; often written 5-merous.

**pentangular** five-angled.

**perennate** to survive from year to year; to overwinter.

**perennial** a plant lasting longer than two years.

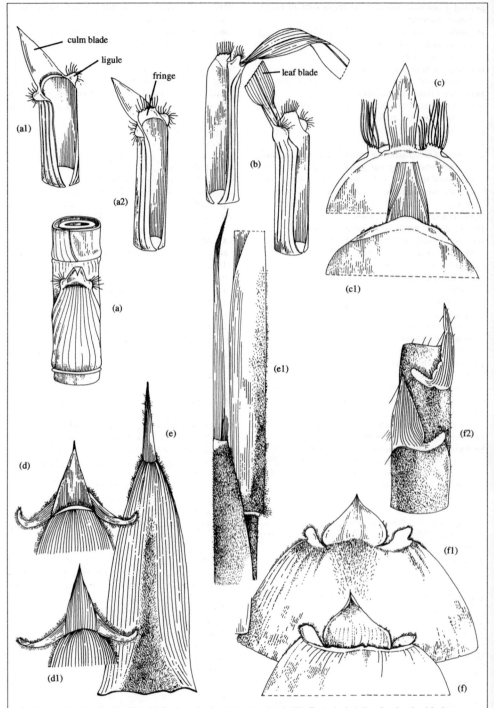

**Bamboo leaf and culm sheaths**  (a) Section of culm with culm sheath (*Phyllostachys*)  (a1) culm sheath with short ligule and fringe  (a2) culm sheath with long ligule and fringe  (b) leaf sheaths  (c) *Phyllostachys nigra*, apex of culm sheath dorsal view showing ciliate auricles  (c1) ventral view showing short ligule  (d) *Phyllostachys nidularia*, apex of culm sheath  (d1) dorsal view  (e) *Pseudosasa amabilis*, culm sheath  (e1) attached to culm  (f) *Bambusa vulgaris*, culm sheath  (f1) dorsal view  (f2) attached to culm.

**perfect** a bisexual flower or an organ with all its constituent members.

**perianth** the collective term for the floral envelopes, the corolla and calyx, especially when the two are not clearly differentiated.

**pericarp** the wall of a ripened ovary or fruit, sometimes differentiated into exocarp, mesocarp and endocarp.

**periderm** a three-layered tissue of skin.

**persistent** of an organ, neither falling off nor withering.

**petiolate, petioled** furnished with a petiole.

**petiole** the leaf stalk.

**pilose** covered with diffuse, soft, slender hairs.

**pistil** one of the female reproductive organs of a flower, together making up the gynostemium, and usually composed of ovary, style and stigma. A simple pistil consists of one carpel, a compound pistil of two or more.

**plano-convex** shallowly curved.

**plicate** folded lengthwise, pleated, as a closed fan.

**plumose** broadly speaking, feather-like; when applied to indumentum, with long, fine hairs, which themselves have fine secondary hairs.

**pollen** the microspores, spores or grains producing male gametes and borne by an anther in a flowering plant.

**pollination** the mechanical or physical transfer of pollen from an anther to a receptive stigma.

**polymorphic** occurring in more than two distinct forms, possessing variety in some morphological feature.

**polyploid** with more than the regular two sets of chromosomes.

**procumbent** trailing loosely or lying flat along the surface of the ground, without rooting.

**proliferous** bearing vegetative buds or plantlets at the tips of the spikelets; see also viviparous.

**prostrate** lying flat on the ground.

**protandrous** describes a flower in which pollen is released from the mature anthers before the stigma of the same flower becomes receptive; in incomplete protandry the shedding of pollen continues after the stigma becomes receptive, enabling self-pollination to take place.

**protogynous** of a flower in which the stigma becomes receptive before pollen is shed from the anthers of the flower.

**pruinose** thickly frosted with white, rather than blue-grey bloom, sometimes on a dark ground colour, e.g. purple or black.

**puberulent, puberulous** minutely pubescent; covered with minute, soft hairs.

**puberulose** see puberulent.

**pubescent** generally hairy; more specifically, covered with short, fine, soft hairs.

**pungent** ending in a rigid and sharp long point.

**raceme** an indeterminate, unbranched and, usually, elongate inflorescence composed of stalked flowers or spikelets.

**racemose** of flowers borne in a raceme; of an inflorescence that is a raceme.

**rachilla** the secondary axis of a branched inflorescence.

**rachis** (rachises, rachides) the axis of a compound inflorescence, as an extension of the peduncle.

**recurved** curved backwards.

**reflexed** abruptly deflexed at more than a 90° angle.

**revolute** of margins, rolled under, i.e. toward the dorsal surface; cf. involute.

**rhachilla (rachilla)** term for main axis of a spikelet.

**rhizomatous** producing or possessing rhizomes; rhizome-like.

**rhizome** a specialized stem, slender or swollen, branching or simple, subterranean or lying close to the soil surface that produces roots and aerial parts (stems, leaves, inflorescences) along its length and at its apex.

**rib** main or prominent nerves on leaves; leaves with such nerves are said to be ribbed.

**rostrum** beak.

**runner** see stolon.

**scabrid** a little scabrous, or rough.

**scabrous** rough, harsh to the touch because of minute projections, scales, tiny teeth or bristles.

**scale** miniature leaf without blade, found at the base of stems and on rhizomes.

**scariose** see scarious.

**scarious** thin, dry and shrivelled; consisting of more or less translucent tissue.

**secund** where all the parts are borne along one side of the axis, or appear to be arranged in this way because of twists in their stalks or their weight.

**sessile** stalkless, usually of a leaf lacking a petiole.

**setaceous (setose)** either bearing bristles, or bristle-like.

**sheath** a tubular structure surrounding an organ or part; most often, the basal part of a leaf surrounding the stem in Gramineae and other monocot families, either as a tube, or with overlapping edges and therefore tube-like and distinct from the leaf blade.

**simple** not compound; not divided into secondary units, thus in one part or unbranched.

**smooth** lacking minute rough points and hairs.

**solitary** used either of a single flower or spikelet which constitutes a whole inflorescence, or a single flower in an axil (but perhaps in a much larger overall inflorescence).

**spathaceous** furnished with a spathe.

**spathe** a conspicuous leaf or bract subtending an inflorescence, or clothing culms.

**spicate** spike-like, or borne in a spikelike inflorescence.

**spike** an indeterminate inflorescence bearing sessile flowering units on an unbranched axis.

**spikelet** a small spike, forming one spicate part of a compound inflorescence, especially the flower head of Gramineae, composed of flowers and their bracts or glumes.

**spike-like** resembling a spike, applied to compressed, narrow racemes and panicles.

**spreading** directed outwards, e.g. the culms, or the branches of panicles.

**stem** (or culm) main axis of plant, bearing leaves and flowers.

**stamen** the male floral organ, bearing an anther, generally on a filament, and producing pollen.

**sterile** (of sex organs) not producing viable seed, non-functional, barren; (of shoots, branches etc.) not bearing flowers; (of plants) without functional sex organs, or not producing fruit.

**stigma** the apical unit of a pistil which receives the pollen and normally differs in texture from the rest of the style.

**stolon** a prostrate or trailing stem, taking root and giving rise to plantlets at its apex and sometimes at nodes.

**stoloniferous** producing stolons.

**style** the elongated and narrow part of the pistil between the ovary and stigma; absent if the stigma is sessile.

**subtend** (verb) of a bract, bracteole, spathe, leaf, etc., to be inserted directly below a different organ or structure, often sheathing or enclosing it.

**subterete** almost circular in section.

**subulate** awl-shaped, tapering from a narrow base to a fine, sharp point.

**sympodial** a form of growth in which the terminal bud dies or terminates in an inflorescence, and growth is continued by successive secondary axes growing from lateral buds; cf. monopodial.

**tepal** a unit of an undifferentiated perianth, which cannot be distinguished as a sepal or petal.

**terete** cylindrical, and circular in cross-section.

**tessellation** marked with a net-like pattern of veins; such leaves are said to be tessellate, an important character in the identification of bamboos and in judging their hardiness.

**terminal** at the tip or apex of a stem, the summit of a growth axis.

**tetraploid** having four basic sets of chromosomes, instead of the usual two.

**tomentose** with densely woolly, short, rigid hairs, perceptible to the touch.

**triploid** having three basic sets of chromosomes.

**truncate** where the organ is abruptly terminated as if cut cleanly straight across, perpendicular to the midrib.

**tufted** with a compact basal cluster of vegetative shoots, leaves and/or stems.

**tumid** see turgid.

**turgid** more or less swollen or inflated, sometimes by fluid contents.

**turion** a detached winter-bud, often thick and fleshy, by which aquatic plants perennate; also an adventitious shoot or sucker.

**umbel** a flat-topped inflorescence like a corymb, but with all the flowered pedicels (rays) arising from the same point at the apex of the main axis; in a compound umbel, all the peduncles supporting each secondary umbel (umbellule) arise from a common point on their primary ray.

**umbellate** borne in or furnished with an umbel; resembling an umbel.

**unisexual** either staminate or pistillate; with flowers of one sex only.

**upper palea** see palea.

**utricle** a round, inflated sheath or appendage, as can be found in *Coix lacryma-jobi*, also a bladder-like, one- to two-seeded, indehiscent fruit with a loose, thin pericarp.

**valve** one of the parts into which a dehiscent fruit or capsule splits at maturity.

**valvule** palea.

**variegated** marked irregularly with various colours.

**villous** with shaggy pubescence.

**viscid** covered in a sticky or gelatinous exudation.

**viviparous** with seeds which germinate, or buds or bulbs which become plantlets while still attached to the parent plant.

# Grasses in the Garden

## SPRING/SUMMER FOLIAGE COLOUR

GREY-GREEN
*Arundo donax*
*Koeleria macrantha*
*Saccharum ravennae*
*Sesleria heufleriana*

GREY-BLUE
*Bothriochloa saccharoides*
*Elymus hispidus*
*Leymus condensatus* 'Canyon Prince'
*Leymus racemosus*
*Helictotrichon sempervirens*
*Koeleria glauca*
*Koeleria vallesiana*
*Panicum virgatum* 'Heavy Metal'
*Sorghastrum nutans* 'Sioux Blue'

BLUE-GREY
*Elymus magellanicus*
*Festuca cinerea*
*Festuca glauca* and cultivars
*Festuca ovina* ssp. *coxii*
*Festuca valesiaca* 'Silbersee'
*Leymus arenarius*
*Muhlenbergia lindheimeri*
*Poa colensoi*

BLUE-GREEN
*Leymus secalinus*
*Muhlenbergia mexicana*
*Muhlenbergia pubescens*
*Poa glauca*
*Schizachyrium scoparium* 'Aldous'
*Sesleria caerulea*

COPPER TO BRONZE
*Carex berggrenii*
*Carex buchananii*
*Carex comans* 'Bronze'
*Carex dipsacea*
*Carex flagellifera*
*Carex petriei*
*Carex testacea*

*Carex uncinifolia*
*Uncinia egmontiana*
*Uncinia rubra*
*Uncinia uncinata*

PURPLE-BRONZE
*Acorus calamus* 'Purpureus'
*Schoenus pauciflorus*

RED/RED-BURGUNDY
*Imperata cylindrica* 'Red Baron'
*Pennisetum setaceum* 'Rubrum'
*Pennisetum setaceum* 'Cupreum Compactum'
*Pennisetum* 'Burgundy Giant'

SOLID GOLD TO BRIGHT YELLOW-GREEN
*Acorus gramineus* 'Oborozuki'
*Milium effusum* 'Aureum'
*Deschampsia flexuosa* 'Aurea'
*Luzula sylvatica* 'Aurea'
*Luzula sylvatica* 'Woodsman'

YELLOW-STRIPED VARIEGATION
*Alopecurus pratensis* 'Variegatus'
*Carex brunnea* 'Variegata'
*Carex oshimensis* 'Aureavariegata'
*Phragmites australis* 'Variegatus'
*Spartina pectinata* 'Aureomarginata'

YELLOW-GREEN-STRIPED VARIEGATION
*Carex elata* cultivars
*Carex firma* 'Variegata'
*Miscanthus sinensis* 'Goldfeder'

WHITE-BANDED VARIEGATION
*Juncus effusus* 'Zebrinus' & 'Vittatus'
*Schoenoplectus lacustris* 'Zebrinus'

YELLOW-BANDED VARIEGATION
*Juncus effusus* 'Aureostriatus'
*Miscanthus sinensis* 'Strictus'
*Miscanthus sinensis* 'Zebrinus'

CREAM-YELLOW-STRIPED VARIEGATION
*Acorus calamus* 'Variegatus'
*Acorus gramineus* 'Variegatus'
*Carex morrowii* 'Fisher's Form'
*Carex oshimensis* 'Evergold'
*Glyceria maxima* 'Variegata'
*Hakonechloa macra* 'Aureola'
*Molinia caerulea* ssp. *caerulea* 'Variegata'
*Phragmites australis* 'Striatopictus'

WHITE-STRIPED VARIEGATION
*Acorus gramineus* 'Albovariegatus'
*Arrhenatherum elatius* ssp. *bulbosum*
  'Variegatum'
*Arundo donax* 'Variegata'
*Carex conica* cultivars
*Carex morrowii* 'Variegata'
*Carex oshimensis* 'Variegata'
*Carex phyllocephala* 'Sparkler'
*Carex siderosticha* 'Variegata'
*Cyperus albostriatus* 'Variegatus'
*Cyperus alternifolius* 'Variegatus'
*Holcus mollis* 'Variegatus'
*Luzula sylvatica* 'Marginata'

*Miscanthus sinensis* 'Cabaret'
*Miscanthus sinensis* 'Cosmopolitan'
*Miscanthus sinensis* 'Morning Light'
*Miscanthus sinensis* 'Silberpfeil'
*Miscanthus sinensis* 'Variegatus'
*Schoenoplectus lacustris* 'Albescens'
*Miscanthus tinctorius* 'Nanus Variegatus'
*Phalaris arundinacea* 'Picta'

PINK AND WHITE-STRIPED VARIEGATION
*Glyceria maxima* 'Variegata'
*Oplismenus hirtellus* 'Variegatus'
*Phalaris arundinacea* 'Feesey'
*Zea mays* 'Quadricolor'

CREAM AND WHITE-STRIPED VARIEGATION
*Acorus calamus* 'Variegatus'
*Acorus gramineus* 'Ogon'
*Calamagrostis* × *acutiflora* 'Overdam'

LEAVES WOOLLY
*Alopecurus lanatus*
*Holcus lanatus*
*Luzula* (several)

## AUTUMN FOLIAGE COLOUR

ORANGE
*Andropogon elliottii*
*Andropogon glomeratus*
*Andropogon virginicus*
*Hakonechloa macra*
*Sorghastrum nutans*
*Sporobolus heterolepis*

ORANGE-RED
*Andropogon gerardii*
*Saccharum contortum*
*Miscanthus sinensis* 'Purpurascens'
*Schoenus pauciflorus*

RED-ORANGE
*Saccharum ravennae*
*Schizachyrium scoparium*
*Themeda japonica*

BRONZE TO ORANGE-RED
*Andropogon ternarius*
*Apera spica-venti*
*Bouteloua curtipendula*
*Elymus virginicus*

*Miscanthus sinensis* 'Graziella'
*Saccharum brevibarbe*

BRONZE TO DEEP-RED
*Panicum virgatum* 'Rotstrahlbruch'
*Spodiopogon sibiricus*
*Stipa arundinacea* 'Autumn Tints'
*Tridens flavus*

WINE-RED
*Imperata cylindrica* 'Red Baron'
*Panicum virgatum* 'Hänse Herms'

VARIEGATION SUFFUSED WITH WINE-RED
*Hakonechloa macra* 'Aureola'

VARIEGATION SUFFUSED WITH PINK
*Miscanthus sinensis* 'Cabaret'
*Miscanthus sinensis* 'Cosmopolitan'
*Phalaris arundinacea* 'Feesey'

BRIGHT YELLOW
*Chasmanthium latifolium*

GOLDEN YELLOW

*Molinia caerulea* ssp. *arundinacea*  and
  cultivars
*Panicum clandestinum*
*Panicum virgatum*

*Pennisetum alopecuroides*
*Pennisetum alopecuroides* 'Hameln'
*Phragmites australis*
*Stipa arundinacea* 'Gold Hue'

## SUITABLE FOR OUTDOOR CONTAINER DISPLAY

*Acorus gramineus* – variegated cultivars
*Alopecurus pratensis* 'Variegatus'
*Arundo donax*
*Arundo donax* 'Variegata'
*Calamagrostis* × *acutiflora* 'Karl Foerster'
*Calamagrostis brachytricha*
*Carex phyllocephala* 'Sparkler'
*Chasmanthium latifolium*
*Cortaderia selloana* – all cultivars
*Festuca amethystina* – all cultivars

*Festuca cinerea*
*Festuca glauca* – all cultivars
*Festuca ovina* ssp. *coxii*
*Glyceria maxima* 'Variegata'
*Helictotrichon sempervirens*
*Imperata cylindrica* 'Red Baron'
*Miscanthus sinensis* – most cultivars
*Miscanthus tinctorius* 'Nanus Variegatus'
*Molinia caerulea* ssp. *caerulea* 'Variegata'
*Pennisetum* – most species and cultivars

## FOR GLASSHOUSE AND CONSERVATORY

*Acorus gramineus* –variegated cultivars
*Carex brunnea* 'Variegata'
*Coix lacryma-jobi*
*Cymbopogon* – all species
*Cyperus* – many species
*Gynerium sagittatum*
*Hyparrhenia* – both species
*Isolepis cernua*
*Isolepis setacea*
*Neyraudia reynaudiana*
*Oplismenus* – all

*Oryza sativa*  and cultivars
*Pennisetum setaceum* cultivars
*Rhynchospora nervosa*
*Saccharum officinarum*
*Setaria palmifolia*
*Setaria plicatilis*
*Sorghum bicolor*
*Stenotaphrum secundatum* 'Variegatum'
*Themeda triandra*
*Thysanolaena maxima*
*Zizania latifolia*

## SCENT

*Acorus calamus* (rhizomes aromatic)
*Anthoxanthum odoratum* (foliage with a fresh,
  hay scent )
*Cymbopogon citratus* (strong lemon scent)

*Cymbopogon nardus* (lemon scent)
*Sporobolus heterolepis* (inflorescence scented)
*Vetiveria zizanoides* (roots perfumed)

## FORM

*Ampelodesmos mauritanicus* (very large, reedy)
*Arundinella nepalensis* (somewhat tender reed
  with plumes of flower)
*Arundo donax* (giant reed)
*Calamagrostis* × *acutiflora* 'Karl Foerster'
  (medium to tall, strongly erect, pearly
  plumes)

*Carex comans* 'Frosted Curls' (leaves very
  narrow, densely curly)
*Carex muskingumensis* (slender stems of
  narrow leaves)
*Carex umbrosa* 'The Beatles' (low, neat mop)
*Carex pilulifera* 'Tinney's Princess' (very
  dwarf, yellow-striped)

*Carex plantaginea* (broad, ribbed leaves)

*Carex siderosticha* (broad leaves)

*Chionochloa* – most species (large, arching tussocks)

*Cortaderia* – all species and cultivars (large mounds of fine foliage, tall, silvery plumes)

*Cyperus papyrus* (very large and architectural, tender)

*Eragrostsis trichodes* 'Bend' (strongly arching panicles)

*Festuca* (many form dense, low tussocks of needle-like leaves, notably *F.f cinerea,eskia, glauca, rubra, valesiaca*)

*Gynerium sagittatum* (large reed with showy plumes,tender)

*Helictotrichon sempervirens* (strongly architectural - arching flower spikes over grey tussocks)

*Juncus decipiens* 'Curly Wurly' (stems tightly coiled)

*Juncus effusus* 'Spiralis' (stems spiralled)

*Molinia caerulea* ssp. *arundinacea* and cultivars (very dramatic with tall,fine flower heads)

*Spodiopogon sibiricus* (leaves held at right angles to culms forming rounded clumps)

*Stipa calamagrostis* (windswept spikes held high over foliage mounds)

*Stipa gigantea* (very tall shimmering panicles of long-awned spikelets)

## FLOWERS

*Aegilops* – most species (small, bearded spikes, dries well)

*Agrostis* – many species (light, delicate inflorescences, dries well)

*Aira caryophyllea* (delicately, branched, fine panicles, silver-grey)

*Aira elegantissima* (very finely branched, silver-purple, dries well)

*Andropogon gerardii* (purple-tinged, forked racemes)

*Andropogon ternarius* (silver-grey-plumed)

*Andropogon virginicus* (silvery, luminous spikelets)

*Apera interrupta* (small, delicate panicles)

*Apera spica-venti* (broader and finer than the last)

*Arundo donax* (very large, grey-white plumes, dries well)

*Avena barbata* (long-awned spikelets on weeping branches)

*Avena sterilis* (light and graceful flowerheads)

*Bothriochloa caucasica* (purple-tinted)

*Bouteloua curtipendula* (erect spikes with mosquito-like spikelets)

*Bouteloua gracilis* (as the last, but with spike bent and spikelets hanging, dries well)

*Briza* – all species ( delicate panicles of papery, puffed-up spikelets, very popular for drying)

*Bromus* – most species (delicate panicles, dries well)

*Calamagrostis* × *acutiflora* 'Karl Foerster' (strongly erect,translucent plumes)

*Calamagrostis brachytricha* (loose, silvery plumes)

*Carex atrata* (black flowerheads)

*Carex baccans* (red fruit)

*Carex fraseri* (white pompons)

*Carex pendula* (tall, arching)

*Carex pseudocyperus* (arching)

*Chasmanthium latifolium* (arching panicles of flattened, spangle-like spikelets, dries well)

*Chionochloa* – most species (tall plumes)

*Coix lacryma-jobi* (bead-like utricles on arching stems)

*Cortaderia selloana* and all cultivars (very large, silvery plumes, dries well)

*Cynodon* – all species (small, 'fingered' flower spikes)

*Cyperus papyrus* (large, mop-like heads)

*Deschampsia* – both species and cultivars (delicate, shimmering panicles )

*Digitaria sanguinalis* (slender, fingered panicles)

*Echinochloa crus-galli* (long-awned spikes, sometimes purple, dries well)

*Eleusine coracana* (unusual, fingered panicles, dries well)

*Eragrostis* – most species (large, fine panicles, often attractively tinted)

*Eriophorum* – most species (white flowerheads like silky cotton wool)

*Hordeum* – both species (feathery, long-awned flowerheads, dries well)

*Hystrix patula* (bottlebrush-like flower spikes, dries well)

*Lagurus ovatus* (soft, hare's tail inflorescences, very popular for drying)

*Lamarckia aurea* (golden, downswept plumes, dries well)

*Luzula nivea* and cultivars (small, fine white flower clusters)

*Melica ciliata* (small, soft, luminous pearly spikes)

*Miscanthus* – all species and cultivars (very beautiful panicles, typically with a fountain of branches swept in one direction, pearly white, dries well)

*Molinia caerulea* ssp.*arundinacea* all cultivars (very fine, translucent plumes)

*Muhlenbergia filipes* (cloud-like, purple panicles)

*Muhlenbergia rigens* (slender, parchment spikes lasting well into winter)

*Panicum virgatum* and all cultivars – (open, cloud-like panicles)

*Paspalum* – all species (slender, fingered flower spikes)

*Pennisetum* all species and cultivars (long-bristled, feathery spikes)

*Phalaris canariensis* (rounded, compact spikes, dries well)

*Phragmites australis* (pearly plumes, dries well)

*Polypogon monspeliensis* (fine, silky spikes)

*Rhychelytrum repens* (showy panicles with pink, red, purple and white hairs)

*Rhynchospora nervosa* (like a small papyrus with white flower bracts)

*Saccharum contortum* (large plumes, tinted red at first, paler when dried)

*Saccharum ravennae* (large, silvery plumes, dries well)

*Schizachyrium scoparium* (small, translucent, beautiful en masse in winter, dries well)

*Sorghastrum nutans* and all cultivars (coppery, long-awned spikes with yellow anthers, dries well)

*Sorghum* – both species (large, drooping dense panicles, dries well)

*Spodiopogon sibiricus* (narrow, white-hairy panicles)

*Sporobolus heterolepis* (delicate, open panicles held high above foliage)

*Stipa* – see under FORM

*Trichloris crinita* (dense, feathery spikes, usually treated as half-hardy annual)

*Tridens flavus* (airy, purple panicles)

*Typha* – all (velvety, brown 'cat tails', dries well if properly treated)

*Uniola paniculata* (erect to nodding panicles of spangle-like spikelets, dries well)

*Zea mays* – ornamental cultivars (cobs in mixed shades of yellow, red, white, blue)

*Zizania* – both species (showy pyramidal panicles)

## SHADE-TOLERANT

*Alopecurus pratensis* 'Variegatus'

*Arrhenatherum elatius* ssp.*bulbosum* 'Variegatum'

*Bolboschoenus maritimus*

*Briza media*

*Calamagrostis × acutiflora* 'Karl Foerster'

*Calamagrostis brachytricha*

*Calamagrostis epigejos*

*Carex fraseri*

*Chasmanthium latifolium*

*Deschampsia cespitosa* and all cultivars

*Deschampsia flexuosa* and all cultivars

*Saccharum alopecuroides*

*Hakonechloa macra* and cultivars

*Hystrix patula*

*Luzula* – all

*Melica ciliata*

*Melica uniflora*

*Milium effusum*

*Miscanthus sinensis* 'Purpurascens'

*Molinia caerulea* subsp.*caerulea* and all cultivars

*Pennisetum alopecuroides* 'Moudry'

*Phalaris arundinacea* and all cultivars

*Scirpus*

*Sesleria autumnalis*

*Sesleria heufleriana*

*Spodiopogon sibiricus*

## MOISTURE-TOLERANT

Acorus gramineus and cultivars
Alopecurus pratensis 'Variegatus'
Andropogon glomeratus
Arundo donax
Arundo donax 'Variegata'
Bolboschoenus maritimus
Carex – almost all
Cyperus – all
Deschampsia cespitosa and all cultivars
Eriophorum – all
Glyceria maxima 'Variegata'
Juncus – all
Isolepis – all
Luzula – all

Miscanthus sacchariflorus
Miscanthus sinensis (most cultivars)
Miscanthus 'Giganteus'
Molinia caerulea ssp.arundinacea and all
  cultivars
Panicum virgatum 'Hänse Herms'
Phalaris arundinacea and all cultivars
Phragmites australis and all cultivars
Saccharum contortum
Saccharum giganteum
Schoenus
Scirpoides
Scirpus
Spartina pectinata 'Aureomarginata'

## AQUATICS AND MARGINALS

Acorus calamus
Cyperus – most species (many tender)
Dulichium arundinaceum
Eleocharis – all (some tender)
Eriophorum – all
Glyceria – all
Gynerium sagittatum (tender)
Juncus – many
Oplismenus hirtellus and cultivars (tender)

Oryza sativa (tender)
Phragmites australis
Pycreus filicinus
Rhynchospora nervosa (tender)
Schoenoplectus
Scirpus
Spartina – both species
Typha – all
Zizania – both species (only Z.aquatica hardy)

## SOIL AND SAND-BINDERS

Agropyron cristatum
Ammophila arenaria
Ammophila breviligulata
Bolboschoenus maritimus
Buchloe dactyloides
Cynodon dactylon
Eragrostis curvula
Leymus arenarius

Mibora minima
Oplismenus – all (tender)
Phragmites australis
Spartina alternifolia
Spartina pectinata
Stenotaphrum secundatum (tender)
Stipagrostis pennata
Uniola paniculata

## DROUGHT-TOLERANT

Andropogon elliotii
Andropogon gerardii
Andropogon ternarius
Andropogon virginicus
Bouteloua curtipendula
Bouteloua gracilis
Calamagrostis × acutiflora 'Karl Foerster'
Eragrostis curvula
Leymus racemosus
Festuca amethystina 'Superba'
Festuca cinerea
Festuca glauca

Festuca mairei
Helictotrichon sempervirens
Leymus condensatus
Miscanthus sinensis 'Gracillimus'
Miscanthus sinensis 'Graziella'
Miscanthus sinensis 'Malepartus'
Muhlenbergia – all
Panicum virgatum and all cultivars
Schizachyrium scoparium
Sorghastrum nutans and all cultivars
Sporobolus heterolepis

# Genera Listed by Family

Illustrated genera are marked with an asterisk (*).

**Araceae (Acoraceae)**
Acorus

**Cyperaceae**
Bolboschoenus
Carex*
Cladium
Cyperus*
Dulichium*
Eleocharis
Eriophorum*
Isolepis
Pycreus
Rhynchospora*
Schoenoplectus*
Schoenus
Scirpoides
Scirpus*
Uncinia

**Gramineae (Poaceae)**
(grasses)
Aegilops
Agropyron
Agrostis*
Aira*
Alopecurus*
Ammophila*
Ampelodesmos
Andropogon*
Anthoxanthum
Apera*
Arrhenatherum*
Arundinella
Arundo*
Avena
Bothriochloa
Bouteloua*
Brachiaria
Brachypodium*

Briza*
Bromus
Buchloe
Calamagrostis*
Cenchrus
Chasmanthium*
Chionochloa
Chloris
Coix*
Cortaderia*
Ctenium
Cymbopogon
Cynodon*
Cynosurus*
Dactylis*
Danthonia
Deschampsia*
Dichanthium
Digitaria*
Echinochloa
Ehrharta
Eleusine
Elymus
Eragrostis*
Eriochloa
Festuca*
Glyceria*
Gynerium
Hakonechloa*
Helictotrichon
Hilaria
Holcus*
Hordeum
Hyparrhenia
Hystrix*
Imperata
Koeleria*
Lagurus*
Lamarckia

Leymus*
Lolium
Melica*
Mibora*
Milium*
Miscanthus*
Molinia*
Muhlenbergia
Neyraudia
Oplismenus
Oryza*
Oryzopsis
Panicum
Paspalum
Pennisetum*
Phalaris*
Pharus
Phleum*
Phragmites*
Poa*
Polypogon*
Rhynchelytrum
Rostraria
Saccharum*
Schizachyrium*
Sesleria
Setaria*
Sorghastrum*
Sorghum
Spartina*
Spodiopogon
Sporobolus
Stenotaphrum
Stipa*
Stipagrostis
Themeda
Thysanolaena
Trichloris
Tricholaena

Tridens
Uniola
Vetiveria
Zea*
Zizania

(bamboos)
Arundinaria
Bambusa*
Chimonobambusa*
Chusquea*
Dendrocalamus*
Drepanostachyum*
Fargesia*
× Hibanobambusa*
Himalayacalamus*
Indocalamus*
Otatea
Phyllostachys*
Pleioblastus*
Pogonatherum*
Pseudosasa*
Qiongzhuea*
Sasa*
Sasaella*
Sasamorpha*
Semiarundinaria
Shibataea*
Sinobambusa*
Thamnocalamus*
Yushania*

**Juncaceae**
Juncus*
Luzula*

**Typhaceae**
Typha*

# Temperature Conversion

$$°C = 5/9 \ (°F - 32) \qquad °F = 9/5 \ °C + 32$$

Celsius

| −18° | −10 | 0 | 10 | 20 | 30 | 40 |

Fahrenheit

| 0° | 10 | 20 | 32 | 40 | 50 | 60 | 70 | 80 | 90 | 100 |

---

# Conversions of Measurements

| Length | | | |
|---|---|---|---|
| | 1 millimetre | = | 0.0394 inch |
| | 1 centimetre | = | 0.3937 inch |
| | 1 metre | = | 1.0936 yards |
| | 1 kilometre | = | 0.6214 miles |

---

# Range of Average Annual Minimum Temperature
# for each Climatic Zone

| Zone | °F | °C |
|---|---|---|
| 1 | < −50 | < −45.5 |
| 2 | −50 to −40 | −45.5 to −40.1 |
| 3 | −40 to −30 | −40.0 to −34.5 |
| 4 | −30 to −20 | −34.4 to −28.8 |
| 5 | −20 to −10 | −28.8 to −23.4 |
| 6 | −10 to 0 | −23.3 to −17.8 |
| 7 | 0 to +10 | −17.7 to −12.3 |
| 8 | +10 to +20 | −12.2 to −6.7 |
| 9 | +20 to +30 | −6.6 to −1.2 |
| 10 | +30 to +40 | −1.1 to +4.4 |
| 11 | > +40 | > +4.4 |

**Acorus** L. (Latin name derived from Greek *akoron*, applied to both *Acorus calamus*, the Sweet Flag, and to the Yellow Flag, *Iris pseudacorus*.) Araceae (Acoraceae). 2 species of perennial herbs found in wetlands. Rhizomes horizontal, sympodial, aromatic. Leaves equitant, unstalked but basally sheathing, iris- or grass-like. Spathe and peduncle scarcely distinguishable, appearing overall like a more rigid, erect leaf; spadix sessile, borne at or above midpoint of spathe/peduncle and held at 45°, cylindrical or club-shaped; flowers bisexual, densely crowded, perianth 6-parted. Old World, N America. Genetic conformation and distribution are apparently linked: diploids predominate in N America and Siberia, sterile triploids in Europe and N India, tetraploids in E Asia.

CULTIVATION   Found in a range of shallow freshwater habitats, *Acorus* species are well suited to the bog garden and use as marginal and submerged aquatics grown primarily for their sheaves of attractive and aromatic foliage; the spadix is not particularly ornamental, and is usually produced only when the plant is grown in water.   The variegated selections of both species listed below are handsome additions to any damp border, bog garden or pond margin. *A.calamus* 'Variegatus', with boldly striped flag iris-like leaves, provides a bright relief for the darker forms of *Astilbe* × *chinensis*, *Gentiana asclepiadea* and *Ligularia*. *A.gramineus*, with lower fans of fine grassy foliage, is an excellent ground cover for the foreground of bog gardens. Its variegated forms are slow to increase and spread. They have been used for container cultivation, often indoors, and for bonsai treatment. With their fascinating phytogeography and long history as sources of fragrant oils, medicines, rushes (for flooring), insecticides and liquor flavourings, even the typical forms deserve a place at pond and lake margins, to which they will bring a natural grace. Grow in a permanently moist, rich loamy soil or in shallow water, about 25–30cm/10–12 in. deep, in an open position in sun. Remove faded foliage in autumn. Propagate by division every third or fourth year if clumps become congested; also from seed sown when still very fresh.

*A.calamus* L. SWEET FLAG; SWEET CALAMUS; MYRTLE FLAG; CALAMUS; FLAGROOT.
Rhizome thick, tinted pink. Leaves to 150 × 3cm, resembling a large flag-iris, rather rigid, bright green, with a distinct midrib and, in emerging leaves, sporadic zones of lateral wrinkling and puckering. Peduncle narrower than leaves, strongly 2–3-ridged; spadix to 10 × 1.2cm, yellow, becoming green with development of fruit. Asia, SE US, widely naturalized throughout the N Hemisphere and most notably in Europe, where it was introduced via Turkey in the mid-16th century. 'Purpureus': leaves, stems and inflorescence tinted purple-bronze. 'Variegatus': leaves striped cream and yellow. var. **angustifolius** (Schott) Engl. Rhizome with many air spaces; leaves to 1.5cm wide, grass-like, thin-textured, the midrib especially pronounced. SE Asia. Z3.

*A.gramineus* Ait.
Leaves 8–50 × 0.25–0.8cm, glossy and sedge-like, tapering very finely, lacking a prominent midrib and disposed in a marked fan at the lead-edge of the short, creeping rhizome, this often tilted slightly forward. Spathe 6–12.5cm; spadix 4.5–8cm. E Asia. 'Albovariegatus' ('Argenteostriatus'): dwarf, leaves striped white. 'Pusillus' (*A.gramineus* var. *pusillus* (Sieb.) Engl.): habit dwarf and compact, the leaves seldom exceeding 8cm, the inflorescence never seen. 'Oborozuki': leaves vibrant yellow. 'Ogon' ('Wogon'): leaves variegated chartreuse and cream. 'Variegatus' ('Aureovariegatus') (*Carex variegatus* hort.): leaves striped cream and yellow. 'Yodonoyuki': leaves variegated with paler green. Z5.

**Aegilops** L. (From the Classical Greek name *aigilops*, for a kind of bearded grass.) GOAT GRASS. Gramineae. Some 21 species of annual grasses. Culms slender, erect or ascending. Leaves linear, flat. Inflorescence rigid, narrow, spicate or racemose, lanceolate to ovoid; spikelets 2–8-flowered, solitary at nodes, cylindric to ovoid, all hermaphrodite or upper spikelets male or sterile; glumes oblong to ovate, equal, convex, tough, flexible; palea 2-ribbed; awns 2–5, often conspicuous; lemmas convex. Summer. Mediterranean, Asia Minor, N Africa. Z9.

CULTIVATION   Annual grasses from dry warm habitats, their inflorescences are usually attractively bearded with long conspicuous awns, and are ideal for drying. *A.ovata*, the Annual Goat Grass, is sown as an annual outdoors in spring in a sunny, sheltered position. It flowers from late spring to midsummer.

*A.kotschyi* Boiss. GOAT GRASS.
Culms 15–25cm, branched from base. Leaves glabrous or slightly pubescent beneath. Spikelets 2–3cm, narrowly lanceolate or linear-lanceolate. Afghanistan, Egypt, N Africa.

*A.ovata* L. ANNUAL GOAT GRASS.
To 30CM. Culms several, decumbent, becoming erect, slender, nodes prominent. Leaves to 10 × 0.6cm, pubescent, flat; sheaths rounded beneath, somewhat inflated. Inflorescence ovoid to ellipsoid, to 2.5cm; spikelets in fascicles of 2–4, upper 2 spikelets sterile; glumes awned, awns to 5 per glume, spreading, rough, to 3.5cm.

*A.triuncalis* L.
To 45cm. Leaves pubescent. Inflorescence linear to narrow-lanceolate; spikes bearded, constricted between spikelets; spikelets to 7; glumes awned, awns to 3 per glume, erect to spreading, glumes of terminal spikelet longer than those of lateral spikelet.

*A.ventricosa* Tausch.
To 45cm. Culms numerous. Leaves sparsely pubescent. Inflorescence constricted between spikelets; spikelets 5–10 per inflorescence; glumes awnless, or awns short, 3-toothed, in terminal florets; lemmas of sterile flowers long-awned, of fertile flowers reduced or absent.

**Agropyron** Gaertn. (From Greek *agros*, field, and *puros*, wheat.) (*Agropyrum*) WHEATGRASS; DOG GRASS. Gramineae. Some 40 species of perennial grasses. Rhizomes often creeping. Culms usually erect, nodes prominent. Leaves usually flat, slender; sheaths rounded beneath with short auricles. Inflorescence spicate, pectinate, usually erect, green or tinged purple; spikelets 3- to many-flowered, sessile, placed sideways, singly and alternately on the opposite sides of a continuous rachis; glumes equal, firm, variously nerved, acute or awned; lemmas firm, 5–7-nerved, acute or more or less awned from the apex; palea ciliate on the keels, roughly equal to the lemma. Temperate and cool regions.

CULTIVATION   The ornamental members of this genus removed to *Leymus* and *Elymus*, there remains a number of pernicious weed species, spreading by means of creeping rootstock and by seed; they are controlled by assiduous removal of roots or by black polythene mulch. *A.cristatum*, a drought-resistant perennial grass, is, however, widely used for pasture and hay in the Great Plains, and sometimes for erosion control. *A.repens*, the notorious Couch Grass, is less welcome in gardens than even Ground Elder or Bindweed.

**A.cristatum** (L.) Gaertn. (*A.pectiniforme* Roem. & Schult.). CRESTED WHEATGRASS; FAIRWAY CRESTED WHEATGRASS.
To 40cm. Spikes to 6cm, ovate-oblong, distinctly pectinate; spikelets crowded, divergent from rachis, horizontal to ascending; glumes to 5mm, glumes and lemma short-awned. Eurasia; introduced N America (Great Plains).

**A.elongatum** (Host) Scribn. & J.G. Sm.) Rydb. See *Elymus elongatus*.
**A.glaucum** hort. See *Elymus hispidus*.
**A.inerme** (Scribn. & J.G. Sm.) Rydb. See *Elymus spicatus*.
**A.intermedium** (Host) P. Beauv. See *Elymus hispidus*.
**A.littorale** Nash. See *Schizachyrium scoparium* var. *littorale*.
**A.magellanicum** Desv. See *Elymus magellanicus*.
**A.pauciflorum** (Schweinf.) A. Hitchc. See *Elymus trachycaulos*.

**A.pectiniforme** Roem. & Schult. See *A.cristatum*.
**A.pseudoagropyron** (Griseb.) Franch. See *Leymus chinensis*.
**A.pubiflorum** (Steud.) Parodi. See *Elymus magellanicus*.

**A.repens** (L.) Beauv. (*Elymus repens* (L.) Gould). COUCH; TWITCH; QUICK; SCUTCH; QUACK GRASS.
30–120cm, spreading rapidly and invasively by creeping, wiry rhizomes. Spikes to 20cm, slender, loose or compact; spikelets alternating in two rows, more or less adpressed to axis; glumes 7–12mm, blunt or pointed (long-awned in var. *aristatum* Baumg.) Eurasia. Widespread weed of temperate climates.

**A.sibiricum** (L.) Beauv. See *Elymus sibiricus*.
**A.spicatum** (Pursh) Scribn. & J.G. Sm. See *Elymus spicatus*.
**A.tenerum** Vasey. See *Elymus trachycaulos*.

**Agrostis** L. (From Greek *agrostis*, a kind of grass.) BENT GRASS. Gramineae. Some 120 species of annual or perennial grasses to 1.5m, stoloniferous or tufted. Leaves flat to revolute, linear or filiform, bristly; ligules membranous. Panicles usually much-branched, green, occasionally tinged mauve; spikelets numerous, whorled, single-flowered, laterally slightly flattened; glumes as long as spikelets, equal in length, acute to awned, coarse, 1-veined; awn sharply bent, borne on outer surface; palea minute, often reduced. Cosmopolitan (especially N Hemisphere).

CULTIVATION *A.tenuis*, Browntop Bent, *A.castellana*, Highland Bent, and *A.stolonifera* are amongst the most commonly used species in cool-season turf grass mixtures. The ornamental species described below occur in predominantly dry, open and rocky habitats and are valued for the light delicate and airy inflorescences which last well when cut and dried. Sow seed *in situ* in spring or autumn, in any well drained soil in sun.

**A.alba** L. See *Poa nemoralis*.
**A.algeriensis** hort. See *Aira elegantissima*.
**A.arundinacea** L. See *Calamagrostis arundinacea*.

**A.canina** L. VELVET BENT; VELVET BENT GRASS; BROWN BENT; RHODE ISLAND BENT.
Perennial. Culms to 60cm, tufted, slender, erect. Leaves to 6.5 × 0.3cm, basal leaves narrow, to 1.5mm diam., culm leaves broader, flat, to 8 × 2.5cm, usually scabrous; ligule to 3mm, acute. Panicles to 11 × 5cm; branches spreading, scabrous, 3–6-clustered; spikelets 2mm; awn geniculate, exserted. Summer. Europe. 'Silver Needles': leaves very small and fine, edged silver-white; sometimes offered as *Festuca rubra* 'Silver Needles'.

**A.capillaris** hort. non L. See *A.nebulosa*.
**A.elegans** Thore. See *A.tenerrima*.

**A.nebulosa** Boiss. & Reut. (*A.capillaris* hort. non L.). CLOUD GRASS.
Annual to 35cm. Culms very slender, glossy dark green,

nodes swollen, brown-tinted. Leaves to 10 × 1cm, few, linear, flat, rough; ligule obtuse, to 5mm. Panicles loose, broadly spreading with long branches, oblong in outline, to 15cm; spikelets to 2mm, awns absent. Summer. Morocco, Iberia. Z7.

**A.pulchella** (R. Br.) Roth. See *Sporobolus pulchellus*.

**A.setacea** Curtis. BRISTLE-LEAVED BENT GRASS.
Perennial, densely tufted, hummock-forming, slightly glaucous. Panicle purple, close, oblong; branches and flower stalks rough; outer flowering glume with an awn twice its length. Europe.

**A.spica-venti** L. See *Apera spica-venti*.

**A.tenerrima** Trin. (*A.elegans* Thore).
Annual to 30cm. Culms tufted, flimsy. Leaves linear, to 10cm × 0.3cm; ligules to 2mm. Panicles branched, loose, to 15cm; spikelets to 1mm, awns absent. Summer. W Mediterranean. Z6.

**Agrostis**  (a) *A.canina*  (b) *A.setacea*

**Aira**   (a) *A.praecox*   (b) *A.caryophyllea*

**Aira** L. (Greek name for a similar grass.) HAIR GRASS. Gramineae. Some 9 species of annual grasses to 40cm high. Culms slender, flimsy. Leaves to 5cm, filiform or narrow-linear, often inrolled, grooved; ligules membranous. Panicles lax, finely branched, to 10cm; spikelets small, laterally compressed, each with 2 flowers, usually awned, shiny; glumes of equal length, papery, 1–3-veined; lemma 5-veined, with 2 minute teeth at apex and an abruptly bent awn arising from the back, longer than palea. Mediterranean, Asia, Africa, Mauritius. Z6.

CULTIVATION Found in dry, open habitats, *A.praecox* characteristically on sandy, acidic soils, *Aira* species are valued for their light delicate inflorescences, useful in dried arrangements; those of *A.elegantissima* are especially fine, carrying diffuse silvery spikelets on delicate, purple-tinted branches. Sow seed *in situ* in spring in any well drained soil in sun or part shade.

*A.capillaris* Host. See *A.elegantissima*.

**A.caryophyllea** L. SILVER HAIR GRASS.
To 30cm. Culms tufted, glabrous. Leaves 3–4, to 50 × 1mm, grooved, finely scabrous; sheaths rough; ligule to 5mm, acute. Panicles to 10cm, loose, with widely spreading branches; spikelets 2.5–3mm, tinged silver-grey or purple, on pedicels to 1cm; glumes acute; lemmas with exserted awns. Summer. Europe, N & W Asia, N Africa, Tropical Africa (mts).

*A.cespitosa* L. See *Deschampsia cespitosa*.
*A.elegans* auct. See *A.elegantissima*.

**A.elegantissima** Schur (*A.capillaris* Host; *A.elegans* auct.; *Agrostis algeriensis* hort.). HAIR GRASS.
To 30cm. Forming loose tufts; culms flimsy. Leaves to 50 × 1mm; ligule to 5mm. Panicles very diffuse, 2.5–10 × 2.5–10cm; branches spreading, loosely divided, branchlets delicate; spikelets to 2.5–3mm, long- and slender-stalked, with 1–2, more or less exserted awns. Spring–summer. Mediterranean.

*A.flexuosa* L. See *Deschampsia flexuosa*.

**A.praecox** L. EARLY HAIR GRASS.
To 20cm. Culms solitary or tufted. Leaves to 50 × 0.5mm, sheaths somewhat inflated, smooth; ligule lanceolate, to 3mm. Panicles spike-like, narrow-obovate, to 5cm; spikelets to 3mm; glumes acute; lemmas with shortly exserted awns. Late spring–early summer. N, C & W Europe.

**A.provincialis** Jordan.
Similar to *A.caryophyllea*. To 40cm. Culms solitary or forming clumps. Leaves grooved longitudinally, sheaths scabrous. Panicles very diffuse; branches spreading; spikelets to 3mm with 1 awn; glumes acute. Summer. S France.

*A.pulchella* Link. See *A.tenorei*.

**A.tenorei** Guss. (*A.pulchella* Link).
10–40cm. Culms slender, solitary or forming tufts. Panicles diffuse; branches spreading; spikelets to 1mm, without awns; glumes obtuse. Spring. S Europe.

**Alopecurus** L. (From Greek *alopekouros*, fox-tail, a name used by Pliny (21.17.61).) FOXTAIL GRASS. Gramineae. 25 species of annual or perennial grasses to 1.5m, similar to *Phleum*. Leaves flat; ligules usually obtuse. Inflorescence a spike-like, cylindric, soft panicle, typically resembling a fox-brush; spikelets arranged in a dense spiral, 1-flowered, laterally compressed, falling after maturity; glumes more or less equal in length, often joined at base, 3-veined, keeled, with small hairs on keel, not awned; lemma membranous, 5-veined, obtuse, margins usually joined at base, awned on dorsal surface; awn 2–3× length of spikelet. North temperate regions, temperate S America. Z5.

CULTIVATION A genus including notorious agricultural weeds (*A.myosuroides*), and meadow and pasture species (*A.pratensis*), with some grown for ornament, the dried flowers used in arrangements. The variegated forms of *A.pratensis* make attractive brightly coloured groundcover, especially if clipped over in early summer, before blooming; they are hardy, undemanding and easily grown in sun or light part shade in any moderately fertile, well drained soil. The woolly *A.lanatus* is perhaps the most attractive of the genus, requiring a sharply drained soil in rather dry and airy conditions. Also suitable for troughs and the alpine house. Propagate by division.

**A.*alpinus*** Sm.
Low-growing, lax-caespitose perennial. Culms with 2–3 internodes. Leaf sheaths somewhat dilated; lower leaves narrow, upper leaves broader. Inflorescence spicate, ovoid or short-cylindric, to 2cm × 9mm, densely pubescent, grey; spikelets ovoid, to 4.5mm; glumes somewhat divergent, with long, dense hairs, especially on keel; lemma acute, glabrous; awn straight, not exserted or exserted to 1.5mm. Summer. Scotland, Arctic Europe.

*A.armenus* K. Koch. See *A.arundinaceus*.

**A.*arundinaceus*** Poir. (*A.ventricosus* Pers., non Huds.; *A.nigricans* Hornem.; *A.pratensis* var. *armenus* K. Koch; *A.armenus* (K. Koch) Grossh.; *A.arundinaceus* ssp. *armenus* (K. Koch) Tzvelev).
Perennial to 105cm. Rhizome creeping; culms usually erect, 2-noded. Leaf sheaths inflated, glabrous to somewhat pubescent; ligule obtuse-truncate, to 5mm; leaves linear, to 4.5cm, acuminate, scabrid above and on margins. Panicle broadly cylindric, to × 1.5cm, green to green-purple; spikelets urceolate, to 7mm; glumes lanceolate, acute, divergent at apex, ciliate on keels; lemma ovate, to 6mm, oblique-truncate; awn to 7.5mm, subdorsal, usually slender and somewhat curved. Temperate Eurasia.

*A.arundinaceus* ssp. *armenus* (K. Koch) Tzvelev. See *A.arundinaceus*.

**A.*lanatus*** Sibth. & Sm. WOOLLY FOXTAIL GRASS.
Caespitose perennial to 30cm. Rootstock cylindric, thick, black; culms erect or somewhat curved, slightly geniculate at the single node, slender, glabrous above, white-tomentose below. Leaf sheaths very close to base of stem, white-tomentose, inflated; ligule acute, to 2.5mm; leaves linear, usually convolute, to 5.5 × 0.3cm, thick, mucronate or obtuse, densely white-tomentose. Panicles ovoid-globose, to 1.5 × 1.3cm; spikelets to 6mm; glumes lanceolate, aristate, somewhat connate at base, densely hispid-pubescent; lemma oblique-truncate, to 3.5mm, somewhat ciliate; awn to 11mm, geniculate, tortuous in lower half. E Mediterranean.

*A.nigricans* Hornem. See *A.arundinaceus*.

**A.*pratensis*** L. COMMON FOXTAIL GRASS; FOXTAIL GRASS; LAMB'S TAIL GRASS; MEADOW FOXTAIL.
Perennial to 1.2m. Culms smooth, upright, clump-forming. Leaves flat, scabrous, to 8cm; sheath somewhat inflated, smooth; ligule membranous, obtuse, to 2mm. Panicle cylindric, dense, to 10 × 1cm, pale green, often tinged purple; spikelets to 7mm, each with 3–8 flowers; glumes to 7mm, acute, minutely fringed; keel pubescent; lemma acute, awn exserted. Summer. Eurasia, NE Africa. 'Glaucus': leaves glaucous blue. 'Variegatus' ('Aureovariegatus'): leaves boldly striped and edged gold. A sport of 'Variegatus' with all-gold leaves (*A.p.* 'Aureus') has been reported. Z4.

*A.pratensis* var. *armenus* K. Koch. See *A.arundinaceus*.
*A.ventricosus* Pers. non Huds. See *A.arundinaceus*.

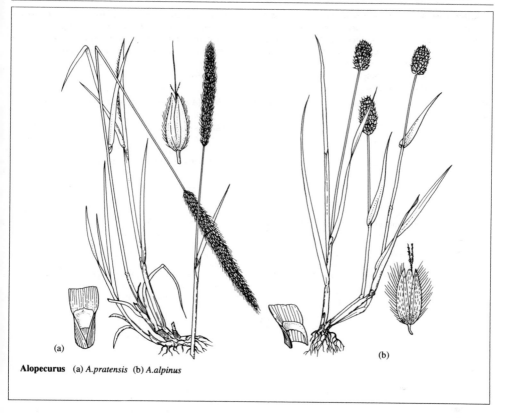

**Alopecurus**  (a) *A.pratensis*  (b) *A.alpinus*

**Ammophila** Host. (From Greek *ammos*, sand, and *philos*, loving, referring to its habitat.) MARRAM GRASS; BEACH GRASS; MEL GRASS. Gramineae. 2 species of coarse, perennial grasses to 1.3m. Rhizomes long, spreading, hard, scaly. Leaves rolled into a tube, unrolling in high humidity, pungent, to 60 × 0.5cm, coarsely pubescent on veins, otherwise glabrous, grey-green. Inflorescence a dense, spike-like panicle, pale yellow to straw-coloured; spikelets single-flowered, somewhat flattened; glumes equal; lemma without awns, with basal ring of fine white hairs; palea almost as long as lemma. Europe, N Africa, N America. Z5.

CULTIVATION  Though not obtrusively ornamental, *Ammophila* species are invaluable in erosion control and in wetlands restoration, particularly in the stabilization of coastal dunes (their natural habitat), inserted at 60cm/2ft spacings. Propagate by division.

*A.arenaria* (L.) Link.
Ligules to 3cm, bifid, flimsy. Panicle to 25cm. Summer.
Coastal Europe, N Africa.

*A.breviligulata* Fern.
Culm clothed in old leaf sheaths, thinner than in *A.arenaria*. Ligule to 3mm, rigid. Panicle to 30cm, narrower than in *A.arenaria*. Coastal NE America.

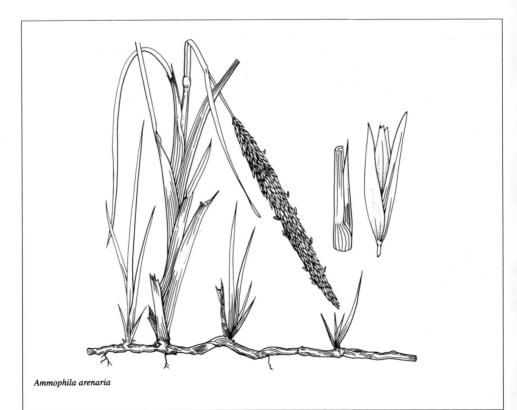

*Ammophila arenaria*

**Ampelodesmos** Link. MAURITANIA VINE REED. Gramineae. 1 species, a robust, perennial grass forming large clumps to 1m diam., resembling *Cortaderia*. Culms to 3m, glabrous, pale green. Leaf sheaths glabrous, striate; ligule to 1.5cm, ciliate at base; leaf blades elongate, to 100 × 0.8cm, wiry, curved basally, strongly veined, scabrous, subulate-acuminate. Panicle to 50cm, branches slender, pendulous, apparently one-sided, scabrous; spikelets crowded, 2–5-flowered, to 1.5cm, yellow or green-yellow and suffused purple, densely pilose on lemma and rachilla joints; lower glume to 1cm, obscurely 5-veined, upper glume to 12mm, 5-veined, scabridulous; lemma linear-lanceolate, to 1.5cm, bidentate, awn to 4mm, subterminal. Mediterranean. Z8.

CULTIVATION    A striking large, reedy specimen for a sheltered sunny spot in mild climates. Treatment as for *Arundo*.

**A.mauritanicus** (Poir.) T. Dur. & Schinz (*Arundo mauritanica* Poir.; *Arundo tenax* Vahl; *Arundo ampelodesmos* Cyr.; *Arundo festucoides* Desf.; *Arundo bicolor* Desf., non Poir.).

**Andropogon** L. Gramineae. (From Greek *aner*, man and *pogon*, beard; referring to the silky hairs on the spikelets of some species.) Some 100 species of annual or perennial grasses to 2.5m. Rhizomes creeping, short or absent. Culms solid, occasionally branched toward summit, sometimes flattened, with extrafloral nectaries. Leaves flat, sheathed; sheaths often villous. Inflorescence compound, racemes paired or palmately arranged, flimsy, on a jointed pubescent rachis, usually sheathed with spathes; spikelets paired at rachis joints, unequal, one sessile, bisexual, often awned, the other stalked, sterile or male, without awns, with a fine apical hair, occasionally reduced; fertile spikelet with 2 florets, glumes of both florets equal, upper floret bisexual; lemma 2-toothed, awned; awn glabrous, sharply bent; palea translucent, shorter than lemma or absent; lower floret reduced to a translucent lemma, sterile; sterile spikelet sometimes composed of 1–2 vestigial glumes, sometimes only present as a pedicel. Warm temperate and tropical regions.

CULTIVATION    Many are valued for their noble inflorescences and handsome foliage: in autumn, the leaves of several species turn a vivid orange-red, becoming tawny as the winter hardens. *A.gerardii* produces strongly upright glaucous clumps to 1.5m/5ft, turning a pale bronze-purple in autumn. The tallest US native in this genus, it has been described as the 'king of grasses'. Grow all in light, porous, sandy soils in full sun. Propagate by division or seed.

*A.annulatus* Forssk. See *Dichanthium annulatum*.
*A.argenteus* Sw. See *Bothriochloa saccharoides*.

**A.capillipes** Nash. To 80cm.
Culms erect, tufted, rhizomatous, branching from upper nodes, conspicuously glaucous. Leaf sheaths congested at base, distichous, chalky white, yellow-green or somewhat purple; basal leaves elongate, to 5mm diam., flat or folded; ligule obsolete. Inflorescence of 2+ racemes, subtended by purple-bronze inflated spathes to 3cm, reflexed or arching on capillary peduncles; sessile spikelet 3mm, villous; awn straight, to 1cm; pedicels villous; sterile floret absent. SE US Florida. Z8.

*A.caricosus* L. See *Dichanthium caricosum*.
*A.caucasicus* Trin. See *Bothriochloa caucasica*.
*A.citratus* DC. ex Nees. See *Cymbopogon citratus*.

**A.elliottii** Chapm. ELLIOTT'S BROOM SEDGE; ELLIOTT'S BEARD GRASS.
To 1m. Culms tufted, upper node strongly bearded. Leaf sheaths somewhat pilose. Leaves to 40×0.5cm, flat, elongate, later curled or tortuous. Racemes softly sericeous, primary racemes long-exserted from sheaths, secondary racemes paired, rarely exserted; sessile spikelet, to 5mm, awn geniculate, to 2.5mm; pedicelled spikelets rudimentary, white-pubescent. Autumn. E US. Z5. Differs from *A.virginicus* in having the flowering portion of the stem mostly aggregated at the top and surrounded by broad sheaths – the sheaths turn a striking orange in winter.

*A.furcatus* Muhlenb. See *A.gerardii*.

**A.gerardii** Vitm. (*A.furcatus* Muhlenb.). BIG BLUESTEM.
Perennial to 2m, forming large clumps, producing short stolons. Rhizomes short or absent. Leaves sheathed; basal sheaths and leaves glaucous and villous; blades flat, to 1cm wide, margin rough. Inflorescence composed of 3–6 purple-tinged racemes to 10cm, terminal on far-exserted peduncle; rachis straight, joints pubescent; bisexual spikelet to 1cm; lower glume sulcate, hispid; awn twisted at base, bent, to 2cm; stalked spikelet male, not reduced. Summer. N America (Canada to Mexico). Z4.

**A.glomeratus** (Walter) BSP (*A.macrourus* Michx.). BUSHY BEARDGRASS.
Perennial. Culms erect, forming tussocks, laterally flattened. Leaf sheaths crowded at stem base, keeled. Inflorescence a dense, feathery, club-shaped panicle. Summer. N. America, C America, W Indies. Z6.

*A.halapensis* (L.) Brot. See *Sorghum halapense*.

**A.hallii** Hackel.
As for *A.gerardii*, except rhizome spreading. Leaves pale green or glaucous grey-green. Racemes shorter, hairy; hairs straw-coloured; bisexual spikelet awned or awnless, awn to 5mm. Summer. N America. Z5.

*A.intermedius* Retz. See *Bothriochloa bladhii*.
*A.ischaemum* L. See *Bothriochloa ischaemum*.
*A.littoralis* Nash. See *Schizachyrium scoparium* var. *littorale*.

**A.longiberbis** Hackel.
To 80cm. Culms wiry, branching above. Leaf sheaths grey-villous, subglabrous; leaves elongate, to 5mm diam., ultimately arching, ligule truncate to 1mm. Inflorescence elongate, with paired, distant racemes; subtending spathes to 6cm, narrow; ultimate peduncles paired, capillary; racemes flexuous; sessile spikelet 3mm with 12mm awn; pedicels tufted, pilose. Autumn. SE US. Z6.

*A.macrourus* Michx. See *A.glomeratus*.
*A.muricatus* Retz. See *Vetiveria zizanoides*.
*A.nardus* L. See *Cymbopogon nardus*.
*A.nodosus* (Willem.) Nash. See *Dichanthium aristatum*.
*A.schoenanthus* L. See *Cymbopogon schoenanthus*.
*A.scoparius* Michx. See *Schizachyrium scoparium*.
*A.sibiricus* (Trin.) Steud. See *Spodiopogon sibiricus*.
*A.sorghum* (L.) Brot. See *Sorghum bicolor*.

**Andropogon**   (a) *A. gerardii*  (b) *A. elliottii*  (c) *A. glomeratus*  (d) *A. virginicus*

**A.ternarius** Michx. SPLITBEARD BLUESTEM.
Perennial to 120cm. Culms tufted, erect, branching above; branches usually long, slender, erect. Leaves to 20 × 0.4cm, purple-glaucous, glabrous or lax-villous beneath. Inflorescence lax, elongate; racemes to 6cm, to 12-jointed, silver-cream, or grey-plumose, usually on long-exserted peduncles from slender, inconspicuous spathes; sessile spikelets to 7mm, glabrous; awn to 2cm, tortuous below. Spring–autumn. SE US. Z6. Differs from *A.virginicus* in the spikelets, which stand out from the main stems on long, fine stalks; when dry they become silvery and luminescent.

**A.virginicus** L. BEARD GRASS; BROOM SEDGE. To 120cm. Culms tufted, slender or rather thick, branching above. Leaf sheaths glabrous or sparsely pilose; leaves to 0.5cm diam., pubescent above, green, glaucous or somewhat purple. Inflorescence simple, paniculate or corymbiform; spathes to 6cm; racemes later exserted from spathe; rachis slender, flexuous, long-sericeous; sessile spikelets to 4mm, awn straight, 2cm; pedicels long-villous. Autumn. E US. Z5. Leaves turn orange in autumn, this colour persisting through the winter. The silvery spikelets have a luminous quality.

**A.zizanoides** (L.) Urban. See *Vetiveria zizanoides*.

**Anthoxanthum** L. (From Greek *anthos*, flower, and *xanthos*, yellow, referring to colour of mature spikelets.) VERNAL GRASS. Gramineae. 15 species of annual or perennial grasses to 60cm. Leaves flat, not revolute, aromatic when crushed. Flowers in a dense, spike-like panicle; spikelets somewhat laterally compressed; florets 3 per spikelet, 2 sterile, attached to terminal bisexual floret at base; glumes 2, of unequal length, papery; upper glume 3-veined; lower glume 1-veined; lemmas of sterile floret papery, bifid, occasionally with 2 obtuse lobes at tip, 3-veined, hairy, bearing bent dorsal awn; lemmas of fertile floret shorter than those of sterile florets, 5–7-veined, stiff, without awns. Eurasia, E Africa, N & S America. Zone 7.

CULTIVATION *Anthoxanthum* species are not noted for their great ornamental value but both species described have a characteristic fresh coumarin fragrance when bruised or cut, and may be useful in forming the base for garlands of more attractive dried flowers. *A.gracile* is the more commonly cultivated (*A.odoratum* is widespread in grassland, especially on hill pasture and heath). Sow seed *in situ* in spring in well drained soil in sun.

**A.gracile** Biv.
As for *A.odoratum*, except annual. Culms to 30cm, forming loose tussocks. Leaf blades arching, downy above; ligule jagged. Panicles ovate, silver-grey ; spikelets to 12mm. Spring. Mediterranean.

**A.odoratum** Lagasca. SWEET VERNAL GRASS.
Perennial to 60cm, most parts sweetly scented. Culms clustered. Leaves glabrous, to 1cm wide, becoming straw-coloured. Flowers in panicle to 8cm; spikelets to 1cm; glumes unequal, rough-textured; sterile florets with lemmas pubescent of equal length; lower lemma with awn from just below tip; upper lemma with awn from base – awns geniculate; fertile flowers with lemma to 3mm, smooth, brown, lustrous. Spring. Eurasia.

*Anthoxanthum odoratum*

**Apera** Adans. (Name invented by Michel Adanson (1727–1806).) SILKY BENT. Gramineae.

3 species of annual grasses to 1m. Leaves flat or revolute, blades narrowly linear, to 25 × 0.3–1cm, somewhat scabrous, glabrous; sheaths scabrous to smooth, tinged purple; ligules papery, oblong, obtuse to acute. Flowers in a panicle to 25cm, often tinged purple; spikelets laterally flattened, 1-flowered, to 4mm; glumes unequal, upper 3-ribbed, lower 1-ribbed; lemma 5-ribbed, equalling upper glume, awned at apex; awn to 1cm, not bent. Caryopsis narrow-ellipsoid, tightly enclosed in palea and lemma. Summer. Europe, N Asia. Z6.

CULTIVATION   Occurring on sandy soils in warm sunny situations, given an approximation of these conditions in gardens, *Apera* species may self sow freely. They are grown for their delicately ornamental inflorescences and *A.spica-venti* also for the chestnut-red colours of the leaves in autumn. Sow seed *in situ* in spring.

*A.interrupta* (L.) P. Beauv. DENSE SILKY BENT; WIND GRASS. Differs from *A.spica-venti* in its contracted, dense or interrupted panicles and very small anthers. To 70cm. Leaves to 4mm wide; sheaths smooth; ligules acute. Panicles narrow-ovate, to 25cm, tinged purple, branches distant, interrupted, spreading; spikelets to 3mm; awns to 1cm.

*A.spica-venti* (L.) P. Beauv. (*Agrostis spica-venti* L.) LOOSE SILKY BENT.
To 1m. Culms flimsy. Leaves flat, 7–25cm × 5–10mm, scabrous above, smooth beneath; sheaths scabrous, tinged purple; ligules obtuse, to 1cm. Panicles ovate to oblong, densely branched, to 25cm, tinged purple; spikelets to 3mm, awns to 1cm.

**Apera**  (a) *A.interrupta*  (b) *A.spica-venti*

**Arrhenatherum** P. Beauv. (From Greek *arren*, male, and *ather*, awn, referring to the bristled staminate flowers.) OAT GRASS. Gramineae. 6 species of perennial grasses to 1.5m. Culms erect or somewhat spreading; basal nodes occasionally swollen into bulbous or pear-shaped structures. Leaves flat, to 40cm; sheaths usually glabrous, smooth; ligules translucent, papery. Panicles loose to dense, narrow, to 30cm; spikelets 2-flowered, to 1cm, somewhat flattened laterally, 3-flowered, 1 floret vestigial; lower floret usually male, rarely hermaphrodite with an abruptly bent, twisted awn arising from base of lemma, upper flower hermaphrodite with a short awn at lemma apex or rarely without awns; glumes acuminate, membranous; lower glume 1-ribbed; upper glume longer, 3-ribbed; lemmas convex, acute, 5–7-ribbed. Europe, N Africa, N & W Asia. Z6.

CULTIVATION  *A.elatius*, found commonly on roadside verges, rough grassland and in hay meadows, has light and attractively tinted racemes of flower, sometimes dried for use in naturalistic flower arrangements. It is potentially invasive and seldom considered ornamental. *A.e.* ssp. *bulbosum* is occasionally grown for the interest of its habit of growth, forming many small 'bulbs' that root readily on contact with the ground. The only true ornamental, the brightly variegated *A.e.* ssp. *bulbosum* 'Variegatum', one of the purest white-variegated grasses, requires regular division and replanting to maintain vigour. It performs best in climates that have lower humidity and cool summer nights: in the humid summers of the Central and Eastern US. for example, it goes into near dormancy. Grow in any moderately fertile soil in sun or part shade.

*A.avenaceum* P. Beauv. See *A.elatius*.

*A.bulbosum* (Willd.) C. Presl. See *A.elatius* ssp. *bulbosum*.

*A.elatius* (L.) Presl & C. Presl (*A.avenaceum* P. Beauv.; *Avena elatior* L.). FALSE OAT; FRENCH RYE.

To 1.5m. Culms forming tussocks, glabrous, occasionally with hairy nodes. Leaves scabrous, to 40 × 1cm, usually glabrous, coarse; sheaths smooth, or hispidulous; ligule to 3mm, obtuse. Racemes lanceolate to oblong, often purple-tinged, lustrous, to 30cm; spikelets to 1cm; lower glume to 5mm, upper glume to 1cm; lemma to 1cm, acute, glabrous to hairy at base, lower lemma awned from middle of dorsal surface, awn to 2cm, abruptly bent, twisted. Summer. Europe. ssp. **bulbosum** (Willd.) Schübl. & Martens (*A.bulbosum* (Willd.) C. Presl; *A.tuberosum* (Gilib.) F.W. Schultz). Basal internodes swollen into a chain of bulbous organs to 1cm diam. Europe. 'Variegatum': not vigorous, to 30cm; leaves erect, narrow, boldly striped and edged pure white.

*A.tuberosum* (Gilib.) F.W. Schultz. See *A.elatium* ssp. *bulbosum*.

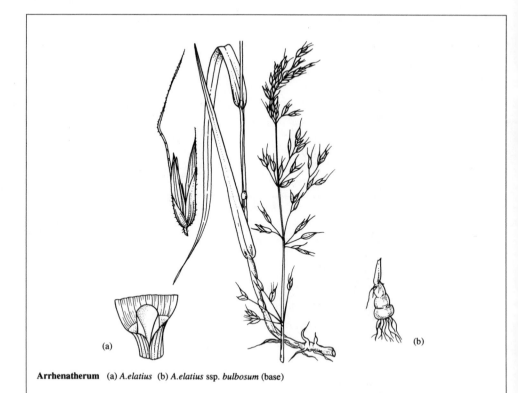

(a)                                                    (b)

**Arrhenatherum**  (a) *A.elatius* (b) *A.elatius* ssp. *bulbosum* (base)

**Arundinaria** Michx. (From Latin *arundo*, a reed.) Gramineae. 1 species, a bamboo; rhizomes running. Culms 2–10m × 2–7cm, thin-walled but stout, terete, erect; sheaths rather persistent, glabrescent, usually with coarse dark, scabrous bristles and, often, auricles; new shoots branching in their second year from the upper part of the culm; branches many, erect. Leaves 10–30 × 2.5–4cm, tessellate, margins scabrous; sheaths bristly. This generic name has been widely used in the past for a multiplicity of species now usually included in other genera. In this work see also *Chimonobambusa, Drepanostachyum, Fargesia, Himalaycalamus, Indocalamus, Pleioblastus, Pseudosasa, Sasaella, Sasamorpha, Semiarundinaria, Thamnocalamus* and *Yushania*. Some authorities still include some Asiatic species in *Arundinaria*; the genus is, however, most naturally restricted to the one species native and endemic to N America.

CULTIVATION See BAMBOOS.

*A.amabilis* McClure. See *Pseudosasa amabilis.*
*A.anceps* Mitford. See *Yushania anceps.*
*A.argenteostriata* (Reg.) Vilm. See *Pleioblastus argenteostriatus.*
*A.aristata* Gamble. See *Thamnocalamus aristatus.*
*A.atropurpurea* Nak. See *Sasaella masumuneana.*
*A.auricoma* Mitford. See *Pleioblastus auricomus.*
*A.borealis* (Hackel) Mak. See *Sasamorpha borealis.*
*A.chino* Franch. & Savat. See *Pleioblastus chino.*
*A.chino* var. *argenteostriata* f. *elegantissima* Mak. See *Pleioblastus chino* f. *gracilis.*
*A.disticha* (Mitford) Pfitz. See *Pleioblastus pygmaeus* var. *distichus.*
*A.falcata* Nees. See *Drepanostachyum falcatum.*
*A.falconeri* (Hook. ex Munro) Benth. & Hook.f. See *Himalayacalamus falconeri.*
*A.fastuosa* (Marliac ex Mitford) Houz. See *Semiarundinaria fastuosa.*
*A.fortunei* (Van Houtte) Nak. See *Pleioblastus variegatus.*
*A.gauntlettii* auct. See *Pleioblastus humilis* var. *pumilus.*

**A.gigantea** (Walter) Muhlenb. (*A.macrosperma* Michx.; *A.tecta* (Walter) Muhlenb.). GIANT REED; CANE REED; SWITCH CANE.
SE US. Z6.

*A.graminea* (Bean) Mak. See *Pleioblastus gramineus.*
*A.hookeriana* Munro. *See Himalayacalamus hookerianus.*
*A.humilis* Mitford. See *Pleioblastus humilis.*
*A.japonica* Sieb. & Zucc. ex Steud. See *Pseudosasa japonica.*
*A.jaunsarensis* Gamble. See *Yushania anceps.*
*A.latifolia* Keng. See *Indocalamus latifolius.*
*A.linearis* Hackel. See *Pleioblastus linearis.*

*A.macrosperma* Michx. See *A.gigantea.*
*A.maling* Gamble. See *Yushania maling.*
*A.marmorea* (Mitford) Mak. See *Chimonobambusa marmorea.*
*A.murielae* Gamble. See *Fargesia murielae.*
*A.narihira* Mak. See *Semiarundinaria fastuosa.*
*A.niitikayamensis* Lawson, non Hayata. See *Yushania anceps* 'Pitt White'.
*A.nitida* Mitford. See *Fargesia nitida.*
*A.pumila* Mitford. See *Pleioblastus humilis* var. *pumilus.*
*A.pygmaea* (Miq.) Mitford. See *Pleioblastus pygmaeus.*
*A.quadrangularis* (Fenzi) Mak. See *Chimonobambusa quadrangularis.*
*A.racemosa* auct. non Munro. See *Yushania maling.*
*A.ragamowskii* (Nichols.) Pfitz. See *Indocalamus tessellatus.*
*A.simonii* (Carr.) A. & C. Riv. See *Pleioblastus simonii.*
*A.simonii* var. *albostriata* Bean. See *Pleioblastus simonii* f. *variegatus.*
*A.simonii* var. *striata* Mitford. See *Pleioblastus simonii* f. *variegatus.*
*A.spathacea* (Franch.) D. McClintock. See *Fargesia murielae.*
*A.spathiflora* Trin. See *Thamnocalamus spathiflorus.*
*A.tecta* (Walter) Muhlenb. See *A.gigantea.*
*A.tessellata* (Nees) Munro. See *Thamnocalamus tessellatus.*
*A.vagans* Gamble. See *Sasaella ramosa.*
*A.variegata* ( Sieb. ex Miq.) Mak. See *Pleioblastus variegatus.*
*A.variegata* var. *akebono* Mak. See *Pleioblastus akebono.*
*A.viridistriata* (Reg.) Mak. ex Nak. See *Pleioblastus auricomus.*

**Arundinella** Raddi. (Diminutive of *arundo*, Latin for reed.) Gramineae. Some 47 species of annual or perennial grasses. Culms erect. Leaves linear, flat or rolled, rarely flaccid and lanceolate; ligule short, scarious, ciliate. Panicle oblong, open or contracted; spikelets in pairs, yellow-green or purple-tinged, lower floret male or sterile; glumes membranous; lower lemma ovate-elliptic, 3–7-nerved, palea hyaline; upper lemma coriaceous, 1–7-nerved, emarginate or bilobed, awned. Pantropical. Z9.

CULTIVATION An attractive reed with plume-like panicles,. it requires a damp, fertile soil in full sun in a sheltered situation. It may be damaged or cut back to the ground by frosts. In subtropical or tropical gardens, it tends to become invasive.

*A.ecklonii* Nees. See *A.nepalensis.*

**A.nepalensis** Trin. (*A.ecklonii* Nees).
Tufted perennial to 180cm, erect. Leaves linear or convolute, to 30 × 1cm, glabrous to hirsute. Panicle oblong or contracted, to 40cm, branches densely spiculate, spikelets lanceolate, to 6mm; lower glume ovate, to 5mm, acuminate; lower lemma obtuse. Tropics.

**Arundo** L. (From Latin *arundo*, reed.) GIANT REED. Gramineae. 3 species of large, rhizomatous, perennial grasses to 6m. Leaves alternate, equally spaced on thick, reed-like clums, deflexed, broad-linear, flat, slightly scabrous, bases sheathing. Florets bisexual, in large terminal, much branched, dense, feathery panicles; spikelets with 2–7 florets each, laterally flattened; glumes equal, papery, 3–5-ribbed; lemma 3–5-ribbed, acute, downy in lower half of dorsal surface, awned, coarsely but minutely toothed. Tropical and subtropical Old World. Z7.

CULTIVATION  Occurring by riversides and in ditches, *A.donax* is a giant reed grown for its magnificent glaucous leaves, which arch gracefully from stout culms; although invasive in its native zones, in temperate regions growth is sufficiently restricted for this seldom to be the case. The best foliage effects are obtained by cutting down annually to the base, in late autumn. On uncut plants, the large plume-like panicles are particularly fine when dried for floral arrangements, although in cool temperate climates these are usually only produced following long hot summers. *A.plinii* will tolerate temperatures as low as −10°C/14°F. *A.donax* and *A.d.*'Macrophylla' withstand several degrees of frost (between −5°C and −10°C/23–14°F). In general the variegated forms are considered tender, making extremely handsome specimens for borders and large tubs in the cool glasshouse and conservatory. Grow in full sun in a fertile, well drained but moist soil with shelter from winds which may bruise or shred the leaves. Propagate by seed, cultivars by division.

*A.ampelodesmos* Cyr. See *Ampelodesmos mauritanicus.*
*A.bicolor* Desf., non Poir. See *Ampelodesmos mauritanicus.*
*A.conspicua* Forst.f. See *Chionochloa conspicua.*

**A.donax** L.
Rhizome knobbly, thick. Culms to 6m, stout, robust, forming clumps. Leaves to 60 × 6cm, slightly scabrous, grey-green. Florets in panicles to 60cm, tinged with red, becoming light grey-white; spikelets to 1.5cm; lemma to 1cm. Autumn. Mediterranean. 'Macrophylla': stems sometimes tinted mauve; leaves larger, to 9cm wide, glaucous, grey-green to blue-green. 'Variegata' (var. *versicolor* (Mill.) Stokes; 'Versicolor'): to 3m; leaves striped and edged off-white. 'Variegata Superba': stems to 1m; leaves wide, to 30×6.5cm, striped and edged off-white, closer together on culms.

*A.festucoides* Desf. See *Ampelodesmos mauritanicus.*
*A.madagascariensis* Kunth. See *Neyraudia arundinacea.*
*A.mauritanica* Poir. See *Ampelodesmos mauritanicus.*
*A.phragmites* L. See *Phragmites australis.*

**A.plinii** Turra (*A.pliniana* Turra).
Resembles a smaller, finer *A.donax*. Culms to 2m, slender; leaves to 30 × 0.7cm, pungent, projecting stiffly at 45–90° from culm; spikelets smaller than in *A.donax*, 2-flowered. Mediterranean.

*A.richardii* Endl. See *Cortaderia richardii.*
*A.selloana* Schult. See *Cortaderia selloana.*
*A.tenax* Vahl. See *Ampelodesmos mauritanicus.*

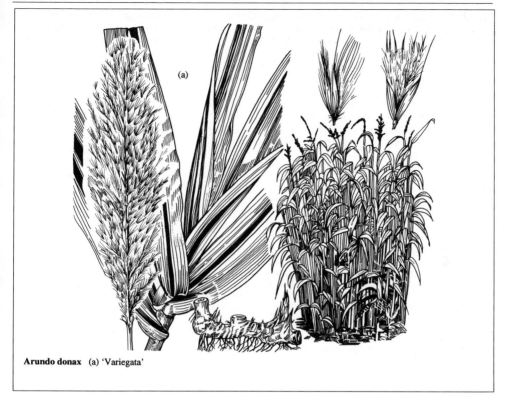

**Arundo donax** (a) 'Variegata'

**Avena** L. (Latin name for oat.) OATS. Gramineae. Some 10–15 species of annual grasses to 1.5m. Culms solitary or tufted, upright. Leaves flat, linear; sheaths slightly hairy to smooth; ligule obtuse, papery. Florets few, in very loose panicles to 30cm, spikelets 1–7-flowered, scattered, pendulous on peduncles to 5cm, flattened laterally, spreading; glumes membranous, lanceolate, with 7–9 ribs, not awned; lemmas 5–9-ribbed, hairy at base, emarginate to 2-toothed at apex, awn to 7.5cm, arising from middle of dorsal surface, twisted and bent. Eurasia, N Africa. Z5.

CULTIVATION Occurring on dry wasteland, cultivated ground and in meadows, especially on heavy soils, *A.sterilis* is particularly valued in the garden for the light graceful flowerheads, used dried in flower arrangements. The long-awned spikelets of *A.barbata*, hung on weeping branches, are a perfect foil for smaller, bolder bedding plants. Grow in any moderately fertile soil in sun, and collect flowerheads before seeding. Sow seed *in situ* in spring or autumn.

**A.barbata** Pott ex Link. SLENDER WILD OATS.
To 1m. Culms solitary to fasciculate, geniculate to erect. Leaf sheaths densely pilose; ligule to 6mm, obtuse; leaves to 25 × 2cm, acute, pilose to glabrous. Panicle lax, to 36 × 15cm, subsecund; branches to 18cm; spikelets narrow, to 2.5cm, with 2–3 florets, on slender, curved, capillary pedicels; lemma 7-veined, to 19mm, stiffly red-pubescent, teeth terminating in slender setae to 5mm, dorsally awned; awn geniculate, to 4cm, tortuous below. Spring–summer. W & C Europe (Mediterranean area), Caucasia, Asia (except Siberia).

**A.elatior** L. See *Arrhenatherum elatius*.
**A.sempervirens** Vill. See *Helictotrichon sempervirens*.

**A.sterilis** L. ANIMATED OAT.
To 1m. Culms upright or somewhat spreading, 1 or several in a tuft. Leaves to 30 × 1cm, rough; ligule to 5mm. Panicle to 30 × 20cm, lax, widely branched, erect to nodding; spikelets to 4cm, with 3–5 florets, green ripening corn yellow; lemmas narrowly lanceolate; awn to 7.5cm. Summer. Atlantic Is., Mediterranean, C Asia.

**Bambusa** Schreb. (From the Malay vernacular.) Gramineae. Some 120 species of tender clumping bamboos, the name formerly much more widely used than now. Culms hollow (except for one variant), usually glabrous, mostly terete, sometimes distorted; sheaths somewhat persistent, thick, sometimes with auricles and bristles, blade broad, triangular; branches many, mostly unequal, re-branching with age. Leaves small or medium-sized, rarely showing any tessellation, margins scabrous. Tropical & subtropical Asia.

CULTIVATION    Tall clumping bamboos of long-standing economic and horticultural importance throughout SE Asia. *B.multiplex* is used for hedging throughout SE Asia. *B.vulgaris* provides light timber, paper pulp and edible young shoots. Plant in a humus-rich soil. Site in full sun or dappled shade in humid, warm and damp conditions. *Bambusa* will not tolerate prolonged exposure to temperatures below −5°C/23°F. Propagate by division.

*B.argentea* hort. See *B.multiplex*.

*B.glaucescens* (Willd.) Sieb. ex Munro. See *B.multiplex*.

***B.multiplex*** (Lour.) Rausch. (*B.glaucescens* (Willd.) Sieb. ex Munro; *B.nana* Roxb.; *B.argentea* hort.; *Leleba multiplex* (Lour.) Nak.). HEDGE BAMBOO.
A variable species. Culms 3–15m×1–4.5cm, slender, arching, glabrescent; sheaths sometimes slow to fall, usually glabrous and lacking auricles and bristles, ligule ciliate, blade glabrescent; nodes rather prominent; branches very numerous, developing in the second year. Leaves 2.5–15×0.5–1.5cm, in 2 rows, somewhat silvery beneath; leaf sheaths almost glabrous, with no, or poor, auricles and bristles. S China; much cultivated in warmer climates, and for bonsai. Variants include 'Alphonse Karr': culms and branches striped orange-yellow and green, tinged pink when young, their sheaths similarly striped. 'Fernleaf' ('Wang Tsai'): smaller, the culms thicker-walled yet more slender than the type with leaves numerous small, 2.5–4.5cm×4–7mm. 'Golden Goddess': small; culms golden; leaves larger. 'Riviereorum' (CHINESE GODDESS BAMBOO): culms delicate, somewhat sinuous but solid; leaves 1.6–3×0.3–0.8cm. 'Silver Stripe': the largest variant: culms, culm sheaths and leaves variously striped white or yellow. Z9.

*B.nana* Roxb. See *B.multiplex*.

***B.oldhamii*** Munro (*Dendrocalamopsis oldhamii* (Munro) Keng f.; *Leleba oldhamii* (Munro) Nak.; *Sinocalamus oldhamii* (Munro) McClure). OLDHAM BAMBOO.
Habit thick, robust. Culms 6–15m×3–13cm, erect, glabrous, with white powder below the nodes; sheaths glabrescent with few or small auricles and bristles, ligule short, blades puberulent above; branches rather short, nodes not prominent. Leaves 15–30×3–6cm, tough and broad, denticulate; sheaths usually lacking auricles and bristles. China, Taiwan. Much planted in the warmer parts of the US. Z9

***B.ventricosa*** McClure. BUDDHA'S BELLY BAMBOO.
Culms 8–20m × 3–5.5cm, thin-walled, glabrous; sheaths glabrous, usually with auricles and bristles, ligule short; nodes not prominent. Leaves 10–20×1.5–3cm, sheaths with small auricles and bristles. S China. Grown in pots or in poorer soil, it produces culms to only 0.5–2m with obese internodal swellings; these disappear in optimum growing conditions. Z9.

***B.vulgaris*** Schräd. ex Wendl. (*Leleba vulgaris* (Schräd. ex Wendl.) Nak.). COMMON BAMBOO.
Extremely vigorous. Culms 5–25m × 5–25cm, thin-walled, glabrescent; sheaths dark brown-pubescent especially above, with auricles and bristles, ligule short; nodes prominent. Leaves 10–25 ×1.8–4cm, near-glabrous; sheaths with auricles and bristles. So long and widely cultivated that its native area is uncertain. 'Vittata': culms, sheaths and, often, leaves striped green and yellow. Z10.

**Bambusa** (a) *B.vulgaris* (b) *B.vulgaris* 'Vittata' (c) *B.ventricosa* (d) *B.multiplex*

**Bolboschoenus** Asch. ex Palla. (From Greek *bolbos*, a bulb, and *schoinos*, a rush.) Cyperaceae. Around 8 species of annual or perennial, grass-like herbs. Stems tuberous and swollen at base, 3-angled. Leaves grasslike, thin, usually equalling or exceeding stem. Inflorescence terminal, usually simple, subtended and exceeded by leaf-like bracts; flowers hermaphrodite, very small, subtended by a scale-like bract (glume) spirally arranged; sepals and petals represented by 6 bristles or absent. E Asia, NE America. Close to and sometimes included in *Scirpus*.

CULTIVATION   An attractive sedge suited to naturalistic plantings in a range of soils and sites from damp woodland to sandy slopes. Treatment as for *Scirpus*.

*B.maritimus* (L.) Palla (*Scirpus maritimus* L.).
Rhizome creeping, branching, scaly. Stem solitary, 3-angled, leafy below. Leaves 10–35 × 0.3–0.7cm, exceeding stems, tapering to tip; sheaths often membranous, lower sheaths lacking blades. Bracts twice inflorescence length, rough; spikelets 1–10, sessile or stalked, terete, 1–3 × 0.5–0.6cm, pointed; glumes 3–4mm, sparsely hairy, tipped with 3–6 white bristles to 2mm. Summer. Cosmopolitan. Z6.

**Bothriochloa** Kuntze. (From Greek *bothrion*, a shallow pit, and *chloe*, grass: in some species the lower glume of sessile spikelets is pitted.) Gramineae. Some 28 species of perennial grasses to 1.5m. Characters as for *Andropogon* except racemes ascending to erect; rachis joints with a translucent groove; one pair of spikelets may be unisexual; glumes sometimes pocked with small indentations; upper lemma of fertile spikelet entire, tapering into awn. Summer. Cosmopolitan (warm temperate to tropical regions).

CULTIVATION   Although *B.caucasica* exhibits a sprawling and untidy habit which may make placement in carefully manicured parts of the garden difficult, it is notable for its intensely purple-pink-tinted panicles and drought-tolerance. *B.saccharoides* is a fine overall silver-grey and excellent for silver gardens and for providing pale contrasts. Given a hot sunny position in well drained soil, most species are hardy where temperatures drop to −20°C/−4°F, or possibly below (reported to survive −14°F in Pennsylvania). Alternatively, they may be treated as annuals, sown in spring in a warm, sunny situation. Propagate by seed, sown *in situ* in spring, or by division.

*B.bladhii* (Retz.) S.T. Blake (*B.intermedia* (R. Br.) Camus; *Andropogon intermedius* Retz).
Racemes to 7cm; rachis with simple branches or, rarely, divided branches at base; fertile spikelets to 4.5m; basal glumes of fertile spikelets glabrous, as long as basal lemma. Indochina, Malaysia. Z7.

*B.caucasica* (Trin.) C. Hubb. (*Andropogon caucasicus* Trin.).
PURPLE BEARD GRASS; CAUCASIAN BLUE STEM.
As for *B.bladhii* except racemes loose, open, purple; spikelets to 3mm, fertile spikelets with basal glume twice as long as lower lemma. Russia, India. Z5.

*B.intermedia* (R. Br.) Camus. See *B.bladhii*.

*B.ischaemum* (L.) Keng (*Andropogon ischaemum* L.).
YELLOW BLUE STEM.
Culms upright, forming tussocks to 1m. Inflorescence a terminal, palmate cluster of 3–15 panicles to 7cm; spikelets to 4.5mm, narrow, tapering to obtuse base and acute apex, papery; lemmas awned; awn to 1.5cm. S Europe. Z5.

*B.saccharoides* (Sw.) Rydb. (*Andropogon argenteus* Sw.)
SILVER BEARD GRASS.
Culms forming tussocks to 120cm. Leaves glabrous, blue-green, to 5mm across. Panicle much-branched, oblong, to 15cm, composed of numerous upright racemes to 5cm, grey-downy; spikelets to 4mm; bisexual spikelets with awned lemmas; awns sharply bent, to 1.5cm. N America to Brazil. Z6.

**Bouteloua** Lagasca. (Named for the brothers Boutelou, Claudio (1774–1842) and Esteban (1776–1813), Spanish botanists.) GRAMA GRASS. Gramineae. 39 species of annual or perennial grasses to 80cm. Stoloniferous or rhizomatous; culms in clusters or forming dense clumps, stiff, slender. Leaves mostly basal; blades flat or folded; sheaths rounded; ligules usually a ring of hairs. Panicle 1- to many-branched but one-sided with the branches aligned in one direction and short; branches spicate, close or spreading; rachis angular to flattened, occasionally terminating in a reduced floret; spikelets 1 to many, sessile, in 2 rows on one side of rachis, with 1 lower, fertile floret and 1–3 staminate or sterile florets above; glumes unequal, 1-ribbed, unawned to short-awned; lemma of lower, sexual floret 3-nerved, awnless or nerves extending as awns; lemma of imperfect floret 3-awned. US, W Indies, C & S America.

CULTIVATION   One of the predominant genera of short grassland species of the Great Plains of North America, *Bouteloua* species are suitable for the herbaceous border or for larger rock gardens, valued for the densely tussock-forming habit and for the curiously shaped inflorescence where the flower spikes are held horizontally on wire-thin stems with the appearance of mosquitoes. The flower heads are suitable for drying. Easily grown in full sun on any well drained garden soil, although performance is better on near-neutral or lime-free soils. Propagate by seed sown *in situ* or in pots under glass in spring, or by division.

**B.curtipendula** (Michx.) Torr. SIDEOATS GRAMA.
Differs from *B.gracilis* in its larger, coarser habit, its erect, not obliquely angled inflorescence, and the remote, not densely packed spikelets. Perennial to 80cm. Rhizome scaly; culms erect, in tussocks. Leaf blades flat to somewhat involute, linear, to 0.5cm wide, scabrous or slightly hairy; ligule a ring of cilia. Panicle 30–80-branched, to 25cm, spikelets 1–12 per branch, compact or spreading, to 1cm, tinged purple; glumes unequal, glabrous or scabrous; fertile lemma glabrous, acute or 3-toothed at apex, often mucronate, 3-ribbed; lemma of imperfect florets reduced. Summer. America (Canada to Argentina). Z6.

**B.gracilis** (HBK) Griffiths (*B.oligostachya* Torr. ex A. Gray). BLUE GRAMA; MOSQUITO GRASS.
Perennial to 60cm. Rhizome short, stout; culms in clumps, erect, glabrous to minutely hairy. Leaf blades linear, flat or slightly involute, to 0.3cm wide, scabrous; sheaths glabrous or sparsely hairy; ligule a short fringe. Panicles to 5cm, 1–4-branched, dense, strongly 1-sided with the branches on the underside of a straight, ascending then horizontal rachis, the florets exserted and hanging below the spikelets; rachis with a terminal, vestigial spikelet; spikelets 40–90 per branch, densely arranged, to 0.5cm; glumes scabrous or glabrous; lemmas to 0.5cm, 3-awned, awns to 0.3cm. Summer. S & W US, Mexico. Z5.

*B.oligostachya* Torr. ex A. Gray. See *B.gracilis*.

**Brachiaria** (Trin.) Griseb. Gramineae. Some 90 species of annual or perennial grasses. Leaves linear to lanceolate, ligule a line of cilia. Inflorescence of compound racemes on a common axis; spikelets solitary or in pairs, ovate to oblong, obtuse to acute; lower glume sheathing, not exceeding spikelets; lower floret male or sterile; upper lemma coriaceous, obtuse or acute, margins inrolled. Tropics and subtropics. Z10.

CULTIVATION   As for *Panicum*.

**B.eruciformis** (Sm.) Griseb.
Loosely tufted annual to 60cm, slender. Leaves linear to narrowly lanceolate, to 15 × 0.6cm, glabrous or pubescent. Inflorescence of 3–14 racemes on an axis to 8cm; racemes to 2.5cm; spikelets solitary, elliptic, to 3mm, pubescent, subacute; lower glume reduced to a scale; upper lemma caducous, smooth, lustrous, obtuse. Tropics to Subtropics.

**B.mutica** (Forssk.) Stapf.
Sprawling perennial; culms to 125cm, prostrate, rooting at nodes. Leaves to 30 × 1.5cm. Inflorescence of 5–20 racemes to 10cm on an axis to 20cm; spikelets usually in pairs on a winged rachis, elliptic, to 4mm, acute; lower glume to 1mm; upper lemma rugulose, obtuse, often mucronate. Tropical Africa.

**B.ramosa** (L.) Stapf (*Panicum ramosum* L.).
Loosely tufted annual to 70cm. Leaves broadly linear, to 25 × 1.5cm. Inflorescence of 3–15 racemes on an axis to 10cm; racemes to 8cm, simple or branching at base; spikelets elliptic to broadly elliptic, to 4mm, glabrous or pubescent, acute; lower glume to 1mm; upper glume and lower lemma membranous, upper lemma rugose, subacute to acute. Tropical Africa, Middle East, Tropical Asia.

**B.subquadripara** (Trin.) Hitchc. (*Panicum subquadriparum* Trin.).
Creeping annual to 50cm from prostrate base. Leaves broadly linear to narrowly lanceolate, to 20 × 1cm. Inflorescence of 3–5 racemes on an axis to 10cm; racemes to 6cm, spikelets solitary on a winged axis, narrowly elliptic, 3–4mm, glabrous, acute; upper lemma rugulose, acute. Tropical Asia, Australasia.

**Brachypodium** P. Beauv. (From Greek *brachys*, short, and *pous*, foot, referring to the very short pedicels.) Gramineae. 17 species of annual or perennial grasses to 1.2m. Rhizomes branching or absent, sometimes becoming hard and thickened at base. Culms upright, somewhat flimsy, forming tussocks. Leaves flat or revolute, lax or stiff, often hairy; sheaths glabrous to coarse, hispidulous; ligules papery. Florets in arched or erect spike-like racemes; spikelets short-pedicellate, adaxially compressed, 5–25-flowered; glumes unequal, shorter than lemmas; lemmas stiff, 7-ribbed, tapering into a short point or straight awn at apex. Summer. Temperate N hemisphere. Z5.

CULTIVATION    *B.sylvaticum*, a native of woodland and shady hedgebanks, and *B.pinnatum* of dry calcareous grassland, are not in the first rank of ornamental grasses but are occasionally grown for their flower spikes in situations approximating to those in habitat. Propagate by seed or division.

**B.pinnatum** (L.) P. Beauv. TOR GRASS.
Perennial to 1.2m. Rhizome narrow, scaly. Culms stiff, forming tussocks. Leaves stiff, linear, to 45 × 0.5cm, glabrous, scabrous, yellow-green to light green. Raceme erect, spike-like, to 25cm; spikelets 3–15, to 25-flowered, to 4cm; glumes acute, convex, upper 6-ribbed, lower 7-ribbed; lemmas with awn to 5mm. N hemisphere.

**B.sylvaticum** (Huds.) P. Beauv. SLENDER FALSE BROME; WOOD FALSE BROME.
Perennial to 90cm. Culms upright, forming tussocks. Leaves linear-lanceolate to 35 × 1cm, scabrous, pubescent, green to yellow-green, sheaths hairy. Racemes loose, erect to arching, spike-like, to 20cm; spikelets stalked, 8–16-flowered, to 4.5cm; glumes acute, lower 7-ribbed, upper 9-ribbed; lemmas with awn to 1cm. Europe, temperate Asia, N Africa.

**Brachypodium**  (a) *B.pinnatum*  (b) *B.sylvaticum*

**Briza** L. (Greek name for a grass, probably rye.) QUAKING GRASS. Gramineae. 12 species of annual or perennnial grasses to 1m. Leaves flat; ligules translucent, papery. Panicles erect, lax with slender branches; spikelets 4–20-flowered, on filiform pedicels, ovate, compressed; glumes ovate, 3–9-ribbed, margins membranous, translucent; lemmas alternate, spreading in one plane, closely imbricate, ovate, very convex on exterior (appearing inflated), cordate at base, the whole resembling a rattlesnake's tail suspended from a slender stalk, ultimately straw-coloured or pearly and papery. Summer. Cosmopolitan in (mainly N) temperate regions. Z5.

CULTIVATION Found in a range of natural grasslands, but commonly on chalk hills and in old pasture on light and heavy, moist or dry soils. *Briza* species are amongst those grasses most often grown for drying and dyeing, although the natural buff and parchment shades of the very graceful flower heads are extremely attractive. For drying, pick as soon as fully developed but before anthesis. Because of the gentle rustling sound they make when disturbed by breeze, quaking grasses are of particular value in gardens designed for people with visual handicaps. Grow in moderately fertile, well drained but retentive soils in full sun. Propagate by seed, *B.media* also by division.

*B.geniculata* Thunb. See *Eragrostis obtusa.*

*B.gracilis* hort. See *B.minor.*

*B.major* C. Presl. See *B.maxima.*

**B.maxima** L. (*B.major* C. Presl). GREAT QUAKING GRASS; PUFFED WHEAT.
Annual to 60cm. Culms solitary or tufted. Leaves glabrous; blades linear, long-acuminate, to 20 × 1cm, margins scabrous; sheaths smooth; ligules oblong, obtuse. Panicles loose, sparsely branched, nodding, to 10cm; spikelets 1–12, 1–3 per branch, cordate or ovate to oblong, to 2.5×2cm, 7–20-flowered, tinged red-brown, light grey or purple; lemmas usually glabrous, 7–9-ribbed, to 1cm. Mediterranean. 'Rubra': lemmas purple-red edged white.

**B.media** L. COMMON QUAKING GRASS; DIDDER; COW QUAKES; DILLIES; DODDERING DICKIES; LADY'S HAIR GRASS; MAIDENHAIR GRASS; PEARL GRASS; RATTLE GRASS; SHIVERING GRASS; TOTTER; TREMBLING GRASS.
Perennial to 1m. Rhizome spreading; Culms upright forming tussocks, some infertile. Leaves glabrous; blades to 15 × 0.5cm, margins scabrous; sheaths smooth; ligule truncate, to 2mm, translucent, papery. Panicles erect, spreading, to 18cm, pyramidal in outline; branches only bearing spikelets in upper half; spikelets ovate to broadly deltoid, to 1cm, 4–12-flowered, often tinged purple at first; glumes 3–5-ribbed; lemmas marked purple. Eurasia. 'Flore Albo': spikelets white. 'Flore Viride': spikelets green.

*B.minima* hort. ex Nichols. See *B.minor.*

**B.minor** L. (*B.minima* hort. ex Nichols.). LESSER QUAKING GRASS.
Annual to 60cm. Culms flimsy, erect, forming loose tussocks. Leaves glabrous; blades linear-lanceolate, to 15 × 1cm, scabrous; sheaths smooth; ligules obtuse, to 5mm. Panicle broadly deltoid, to 20cm; branches ascending; spikelets 4–8-flowered, rounded to deltoid-ovate, to 5mm, light green to purple-tinged, lustrous; glumes 3–5-ribbed. Eurasia. 'Minima': smaller in all respects.

**Briza** (a) *B.media* (b) *B.maxima* (c) *B.minor*

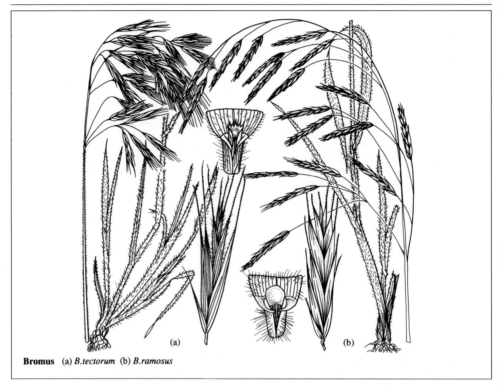

**Bromus** (a) *B.tectorum* (b) *B.ramosus*

**Bromus** L. (From Greek *bromos*, oat.) BROME; CHESS. Gramineae. Some 100 species of annual, biennial or perennial grasses Culms flimsy to rigid, usually erect, solitary or tufted. Leaves flat or revolute; sheaths tubular, usually pubescent; ligules translucent, papery. Panicles erect or nodding, spreading or contracted; spikelets 1- to many-flowered, to 10cm; glumes of unequal length, acute, 1–9-ribbed; lemmas 5–13-ribbed, usually awned; awn apical, straight or recurved. Summer. Temperate regions (usually Northern). Z5.

CULTIVATION  Woodland and meadow grasses grown for their often showy, usually delicate and elegant panicles which, with the possible exception of *B.madritensis*, dry well if picked before the florets open. Grow in any fertile soil in sun. Propagate by seed sown in autumn or spring, perennials by division.

**B.anomalus** Rupr.
Perennial to 1.6m.  Leaves flat. Panicles to 30cm; spikelets laterally compressed, to 3cm; lemma with a central, longitudinal rib, developed as awn. Mexico.

**B.arvensis** L. FIELD BROME.
Annual to 1.1m. Leaves sparsely pubescent; upper sheaths sometimes glabrous. Panicle to 25×20cm; spikelets 4–10-flowered, narrow-lanceolate to oblong, to 2.5cm, borne on long, flimsy branchlets; lemmas to 6mm, spreading at maturity, awn to 1cm. Europe, temperate Asia.

**B.asper** Murray. See *B.ramosus.*

**B.briziformis** Fisch. & Mey.
Annual to 60cm. Leaves downy; sheaths splitting; blades to 15cm. Panicles loose, broadly pyramidal, to 25 × 20cm, sometimes tinged purple; branches somewhat pendent, bearing 1–2 spikelets each; spikelets lanceolate to ovate, slightly flattened laterally, to 2×0.5cm, with 4–10 florets; glumes obtuse, lower 3–5-ribbed, upper 5–9-ribbed; lemma somewhat quadrilateral, obtuse, to 1.5×1cm, without awns. Europe, temperate Asia.

**B.canadensis** Michx. (*B.ciliatus* auct. non L.). FRINGED BROME.
Perennial to 120cm, non-rhizomatous. Leaf sheaths glabrous or short-pilose on lower leaves; leaves lax, to 8mm, glabrous to pilosulous. Panicles to 25.5cm, open, with slender, pendulous branches; spikelets to 15cm, 3–9-flowered, somewhat compressed, occasionally tinged bronze or purple; first glume

1-nerved, second trinerved; lemma to 12.5mm, pubescent on margin and dorsally below; awn to 6.5mm. US, Mexico.

**B.ciliatus** auct. non L. See *B.canadensis.*

**B.danthoniae** Trin.
Annual to 45cm. Leaves downy, blades to 10cm. Panicles narrow, erect, often purple-tinged; spikelets 1 to few, lanceolate or oblong, to 5cm, pubescent or glabrous, shiny; lemmas to 1cm, 3-awned from just below apex; central awn flattened in lower halves, becoming recurved and twisted, tinged purple or maroon. SW & C Asia.

**B.erectus** Huds. UPRIGHT BROME.
Perennial to 1.2m. Culms tufted, erect, not rhizomatous. Leaves sparsely hairy; lower leaves revolute, upper flat; blades erect, to 30×0.5cm. Panicles narrow, erect, to 25cm, branches ascending; spikelets 5–10-flowered, red-purple; glumes acuminate, unequal; lemmas to 1.5cm, scabrous; awn to 5mm. Europe, SW Asia, NW Africa.

**B.fibrosus** Häckel.
Perennial, tufted. Leaves long, arching, with a soft, semi-glaucous tinge. E Europe.

**B.inermis** Leysser.
Perennial, with long creeping rhizomes. Culms erect, to 1m; leaf sheaths usually glabrous; leaves flat, to 35×0.8cm, usually glabrous, scabrous, especially on margins. Panicle contracted or diffuse, erect or slightly nodding, to 20 × 10cm; lower branches to 10cm, bearing 1–2 spikelets;

**Bromus**    (a) *B.erectus*  (b) *B.madritensis*  (c) *B.arvensis*

spikelets narrow-oblong, to $3 \times 0.5$cm, pale green to grey-purple; lower glumes to 9mm, upper glumes to 12mm; lemma narrow-oblong, to 12mm, sometimes obtuse, glabrous or pubescent beneath; awn to 1.5mm, sometimes absent. Europe, temperate Asia.

**B.japonicus** Thunb.
Annual or biennial to 80cm. Leaves hairy; blades to 15cm; ligule to 2mm. Panicles spreading, broadly pyramidal, nodding, to 20cm; branches pendent; spikelets 7–10-flowered, lanceolate to oblong, to 4cm, often tinged purple; lower glume acute, 3-ribbed, upper glume obtuse, 5-ribbed; lemmas lanceolate, obtuse, to 5mm, smooth to 5mm wide; awn to 1cm, often twisted, divergent. Mediterranean, temperate Asia.

**B.lanceolatus** Roth. See *B.macrostachys*.
**B.lanuginosus** Poir. See *B.macrostachys* var. *lanuginosus*.

**B.macrostachys** Desf. (*B.lanceolatus* Roth). BROME GRASS.
Annual to 60cm. Leaves and sheaths pubescent; ligule to 3mm. Panicles erect, dense, narrow, contracted, to 20cm, often tinged purple; spikelets sparse, to 3cm, 8–20-flowered, hairy or glabrous; lemmas to 1.5cm, awned; awn apical, bent and twisted, to 1.5cm, purple at base. Mediterranean. var. **lanuginosus** (Poir.) Coss. & Dur. (*B.lanceolatus* var. *lanuginosus* (Poir.) Dinsm.). Spikelets covered in pale grey woolly hairs. Cultivated more often than the type.

**B.madritensis** L. COMPACT BROME; STIFF BROME; WALL BROME.
Annual to 60cm. Leaf blades flat, to 20cm; sheaths hairy, particularly at culm base, sometimes glabrous and smooth; ligules to 5mm. Panicles upright, obtusely conic, dense or loose, to 15 x 6cm, often tinged purple; spikelets oblong, laterally flattened, becoming outspread, cuneate, to 6cm, 6–13-flowered; lemmas toothed, to 2cm, with slightly curved awns to 2cm. Mediterranean.

**B.ramosus** Huds. (*B. asper* Murray). HAIRY BROME; WOOD BROME.
Perennial to 1.5m. Leaf sheaths auriculate, stiffly retrorse-pubescent; ligule to 6mm; leaves flat, to $60 \times 1.5$cm, pendulous, glabrous or thinly pubescent. Panicles very lax, to $40 \times 25$cm, pendulous; branches rough, mostly paired, patent or pendulous; spikelets 1–9 on each branch, linear-oblong, to $4 \times 0.6$cm, pendent, green or green-purple, short-pubescent, 4–12-flowered; lower glume subulate, to 8mm, upper glume lanceolate, to 11mm, rarely very short-awned; lemma lanceolate, to 14mm, short-pubescent, at least below; awn straight, to 7mm. Europe, N Africa, SW Asia.

**B.squarrosus** L. ROUGH BROME.
Annual or biennial to 60cm. Culms single or tufted, erect or ascending. Leaves with long hairs. Panicles spreading, near-secund, racemose; spikelets laterally flattened, ovate to elliptic, to 20-flowered, to 7 x 0.5cm; lemmas to 1cm, emarginate; awn recurved, to 1cm. Mediterranean.

**B.tectorum** L. DROOPING BROME.
Annual to 90cm. Culms single or tufted, glabrous to sparsely hairy. Leaves hairy; blades flat, to 16cm ligules to 5mm, jagged. Panicles dense, nodding, to 18cm; spikelets 1–4 per branch, 4–8-flowered, cuneate, laterally flattened, to 2cm, tinged purple, glistening; lemmas to 1.5cm; awns to 2cm. Mediterranean.

**Buchloe** Engelm. (A shortened form of *Bubalochloe* – a rendition of the vernacular 'buffalo grass'.) BUFFALO GRASS. Gramineae. 1 species, a slow-spreading, stoloniferous, tufted grass. Leaves grey-green, sparsely hairy, to 0.3cm across. Male inflorescences composed of 2–3 racemes on a common axis to 20cm; female inflorescences sessile clusters enclosed in surrounding leaves; male spikelets 2-flowered, sessile, in 2 rows on one side of the rachis; glumes unequal, 1-nerved, acute, lemma 5-nerved; female spikelets 4–5 in a short spike; lower glume narrow, mucronate, upper glume broadening at middle, acuminately 3-lobed. N America. Z4.

CULTIVATION    As for *Panicum*. In arid regions, it has been used extensively as an alternative to the more usual lawn grasses.

**B.dactyloides** (Nutt.) Engelm.

**Calamagrostis** Adans. (From Greek *kalamos*, a reed, and *agrostis*, a kind of grass.) REED GRASS. Gramineae. Some 250 species of reedy rhizomatous perennial grasses. Culms flimsy to robust, usually glabrous. Leaves long, linear, flat or channelled; ligules papery. Inflorescence paniculate, compressed, dense; spikelets narrow lanceolate, 1-flowered; glumes subequal, longer than floret, 1-ribbed, acute to acuminate; lemma papery, 5-ribbed, surrounded by fine basal hairs, apex with 2 lobes, or denticulate; awns to twice length of lemma, straight or sharply bent. Summer. Temperate N hemisphere. Z7.

CULTIVATION   Natives of marshland fen and wet woodland, *Calamagrostis* species are graceful reed-like grasses which thrive on permanently damp, fertile, even heavy soils. Their feathery inflorescences are attractive *in situ* and valued for drying. Many can become invasive by seeding, notably *C.arundinacea*, and they may be used to best effect in the wild and woodland garden; regular division and replanting will help confine to allotted space.

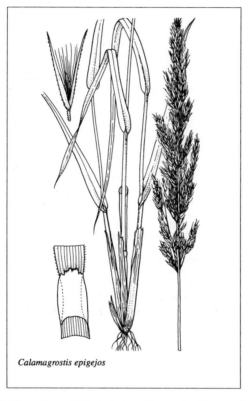

*Calamagrostis epigejos*

*C.× acutiflora* (Schräd.) DC. (*C.arundinacea × C.epigejos*.) (*C.* 'Hortorum'). FEATHER REEDGRASS.
Tight, tufted, stiffly erect to 2m. Foliage lax, arching, slightly glossy, midrib prominently ridged. Panicle loose, soft and silvery bronze to purple-brown. 'Karl Foerster' ('Stricta'): erect, clump-forming, to 2m; inflorescence held upright, red-bronze, later buff. 'Overdam': mound-forming, to 1m; leaves arching, striped white. Z4.

*C.arundinacea* (L.) Roth (*Agrostis arundinacea* L.).
Tufted perennial to 125cm, with creeping rhizomes. Culms unbranched, glabrous, scabrous below inflorescence. Leaf sheaths usually scaberulous; ligule to 3mm; leaves linear, to 7mm diam., long-acuminate, scabrous, with fine raised pubescent above. Panicle contracted, to 25cm; branches very scabrous; spikelets 5mm; glumes lanceolate, subequal; lemma elongate-lanceolate, cleft at apex; awn geniculate, to 7mm. Summer. Scandinavia, Europe to Asia Minor.

*C.brachytricha* Steud. (*C.arundinacea* var. *brachytricha*; *C.varia* Turcz.; *C.sylvatica* Maxim.; *Achnatherum brachytrichum*). KOREAN FEATHER REEDGRASS.
Perennial to 1m, densely tufted, with short-creeping rhizomes. Culms glabrous, unbranched. Leaf sheaths short-pilose, rarely glabrous, scabrous; ligule to 3mm; leaves linear, to 1cm diam., long-acuminate. Panicle contracted, to 18cm; branches very scabrous; spikelets lanceolate, 5mm, somewhat purple; glumes lanceolate, equal, scaberulous; lemma elongate-lanceolate, 4mm, serrate at apex; awn geniculate, to 7mm. Summer–autumn. C to E Asia. Z4.

*C.canescens* (Weber) Roth (*C.lanceolata* Roth). PURPLE SMALL REED.
To 1.2m. Culms tufted, graceful, erect. Leaves to 45 × 0.6cm, short pubescent above, scabrous; ligules to 5mm. Inflorescence to 25 × 10cm, lanceolate to oblong, tinged purple, green or yellow; spikelets lanceolate, to 6mm. W, N, C Europe. 'Variegata': leaves variegated white.

*C.epigejos* (L.) Roth. WOOD SMALL-REED; BUSH GRASS; FEATHER REED GRASS.
To 1.9m; culms erect, tufted, internodes 2–4. Leaves to 60 × 1cm, scabrous; ligules to 13mm, splitting. Panicle erect, lanceolate to oblong, to 30 × 6cm, purple, brown or green; spikelets to 6mm, rachilla sometimes extended, finely pubescent; glumes longer than floret, linear-lanceolate, acute to acuminate; lemma leathery to membranous, 5-nerved; awn sometimes absent, callus pubescent, hairs to twice length of lemma. Eurasia.

**C.foliosa** Kearn.
Tufted, erect, 1.5–3m. Leaves many, crowded towards base, blades involute, firm, smooth, almost as long as the culm. Panicles pale, dense, 5–12cm; glumes 1cm, acuminate; lemma 5–7mm, acuminate, apex with 4 setaceous teeth, awn from near the base, bent, 8mm long above the bend, callus-hairs 3mm long. Spring–summer. Western N America.

C.'Hortorum'. See *C.× acutiflora*.
C.lanceolata Roth. See *C.canescens*.

**C.nutkaensis** (Presl) Steud.
Rhizomes short. Culms stout, 1–1.5m tall. Leaf blade flat, 6–12mm wide, later involute, scabrous. Panicle usually tinted purple, narrow, rather loose, 15–30cm, with ascending branches; glumes 5–7mm long, acuminate; lemma 4mm long, indistinctly nerved, with stout awn from near base, slightly geniculate, as long as lemma; callus-hairs short. Spring–summer. Western N America.

C.'Stricta'. See *C.× acutiflora* 'Karl Foerster'.
C.sylvatica Maxim. See *C.brachytricha*.
C.varia Turcz. See *C.brachytricha*.

**Carex** L. (From Greek *keiro*, to cut, alluding to the leaf margins which are often sharp.) SEDGE. Cyperaceae. Around 1000 species of grass-like, rhizomatous, perennial herbs. Rhizomes creeping, radiating or clumped, scaly. Sterile shoots very short, fertile shoots long, often triangular in section, solid. Leaves grass-like, usually basal, in 3 ranks, leaf bases sheathing stem, with or without ligules, leaf blades linear-lanceolate to filiform, flat, keeled or folded in section. Inflorescence a panicle, composed usually of unisexual spikes; flowers very small, sepals and petals absent; male flowers consisting of 3 stamens arising from a small receptacle subtended by a bract (glume), female flowers consisting of an ovary encased in a bottled-shaped sac (utricle) subtended by a glume, stigmas 2–3, projecting from utricle, axis of inflorescence sometimes continuing from within utricle. Fruit a biconvex or 3-sided nut, brown, yellow or purple. Cosmopolitan, especially temperate and Arctic regions.

CULTIVATION The Sedges are a large genus, offering considerable diversity of form and use. Many inhabit damp woodlands, bogs, moors, ditches and water margins and readily adapt to similar situations in the garden, where they form an unobtrusive green backdrop, occasionally relieved by attractive inflorescences (as in the drooping, tall panicles of *C.pendula* and *C.pseudocyperus*), or striking fruiting heads (red in *C.baccans*, black in *C.atrata*, mace-like in *C.grayi*). Several *Carex* species deserve closer attention – notably the New Zealanders, *Cc.buchananii, comans, flagellifera, petriei* and *uncifolia*. These form dense tussocks or fountains of very fine foliage in intriguing tones of orange-pink, bronze and copper, a perfect foil for bolder, blue- or yellow-leaved grasses. They require neutral or acidic, moist but well drained soils in full sun with some protection from harsh winds and prolonged frosts. Hard winters may cause the foliage to turn a pale buff, even to curl and die off at the tips. This too may be judged ornamental (especially in the elfin-knot perruque of *C.* 'Frosted Curls') and may be left until replaced by new growth in spring.

Among the most popular sedges are those with variegated leaves, including *C.conica* 'Variegata', *C.elata* 'Aurea', 'Bowles Golden' and 'Knightshayes', *C.morowii* 'Variegata' and *C.oshimensis* 'Evergold'. All are colourful foliage additions to borders, pond margins and woodland or shrubby fringes on damp soils. *C.fraseri* (*Cymophyllus fraseri*) is a beautiful US native, remarkable for its broad leaves and dainty, white, ball-like inflorescences held erect in early spring. It is best seen *en masse* in damp, acidic woodland gardens or near the water's edge. Other broad-leaved sedges include the plantain-like *C.plantaginea*, worth growing for curiosity's sake, and *C.siderosticha*, which makes a tussock of smooth foliage almost redolent of some Liliaceae (e.g. *Liriope*) and especially worthwhile in its white-margined cultivar 'Variegata'.

Two further unsedge-like sedges deserve special mention – *C.muskingumensis*, the Palm Sedge, and *C.phyllocephala*. These handsome perennials create dense stands of erect stems clothed with slender, strap-like leaves recalling a very small, fine *Dracaena*. Few hardy grass-allies are so exotic in appearance and so ideal for creating a strong visual impact, particularly in modernist landscapes. *C.* 'Sparkler', a white-edged cultivar of *C.phyllocephala*, is an exceptional plant, meriting container cultivation in the conservatory or on sheltered terraces.

Increase all by division in spring or (species) from ripe seed.

**C.acuta** L. SLENDER-TUFTED SEDGE.
30–120cm. Rhizome creeping great distances. Stem 3-angled above, nearly round below. Leaves 30–140×0.3–0.7cm, glaucous, tip pendulous, sheaths dark red-brown, margin rough. Bracts long, leaf-like, much exceeding inflorescence; male spikes 1–4, 20–40mm, female spikes 2–4, 10–40mm, female glumes blunt, inrolled, utricle 2–3.5mm, green, short-beaked, stigmas 2. Europe. The variegated cultivars are assignable to *C.elata*. Z3.

**C.acutiformis** Ehrh. SWAMP SEDGE. To 1m.
Leaves 8mm wide, grey-green. Spike to ×0.8cm, somewhat drooping. E Europe. Z3.

**C.alba** Scop.
Laxly tufted, 10–40cm. Rhizome slender, creeping. Stem smooth, blunt, 3-angled. Leaves filiform, 20 × 0.1–0.2cm. Female spikes 1–3, round to oblong, bracts glume-like, glume ovate to nearly round, white, utricle 3.5–4mm, dark brown, short-beaked, stigmas 3. C & S Europe to NE Russia. Z7.

**C.albula** Allan. BLONDE SEDGE.
Differs from *C.comans* in its almost white leaves. Tightly tufted, to 35cm. Stems smooth, sometimes deeply grooved. Leaves arching, to 20 × 0.1cm, grooved, scabrid; sheaths chestnut to maroon. Spikes small, 3–6, remote. New Zealand. Z7.

**C.arenaria** L. SAND SEDGE. 10–90cm.
Rhizome creeping great distances. Leaves 60 × 0.15–0.3cm, dark green, blade flat, recurved, or keeled, tapering to a 3-angled tip, lower sheaths dark brown, dead leaves persistent, ligule 3–5mm. Inflorescence dense, to 8cm; bracts bristle-pointed; upper spike male, central spikes bisexual, lower spikes female; female glumes 5–7mm, red-brown, pointed, utricle 4–5.5mm, broad-winged, pale green-brown, stigmas 2. N & W Europe. Z7.

**C.atrata** L. BLACK ALPINE SEDGE; JET SEDGE; BLACK SEDGE.
Clump-forming; rhizome short. Culms to 60cm; basal sheaths dark purple-brown. Leaves shorter than or equal to stems, flat

or with outward-rolled margins, pale blue-green. Spikes 2–7, to 1–3.5cm, rounded to oblong; female glumes acute, purple-black; utricle 3–5mm, broadly ovoid to obovoid, short-beaked. N Europe, C & S Europe. Z3.

**C.baccans** Nees ex Wight. CRIMSON SEEDED SEDGE.
60–120cm, forming loose tufts. Stems triangular in section. Leaves 60 × 0.5–1.5cm, deep green, flat, leathery, long-acuminate; lower sheaths 10–16mm, brown-purple, fibrous. Inflorescence a panicle, acute to obtuse, bracts exceeding spicate branches; female glumes 2.5–3.5mm, utricle 3.5–4mm, inflated, becoming red-orange and berry-like, somewhat 3-angled, stigmas 3. India, Sri Lanka to S China. Z8.

**C.baldensis** L.
Tuft-forming, 10–40cm. Stems smooth. Leaves 2–4mm wide; lower sheaths rust-brown. Inflorescence 1.5–2.5cm; bracts 2–5mm, horizontal; spikes 3–9, male above, female below; terminal spike entirely male; female glume white to pale yellow, utricle 4–5mm, white, becoming dark brown, not beaked, stigmas 3. E Alps. Z7.

**C.bebbii** Olney (*C.tribuloides* var. *bebbii* Bail.). BEBB'S SEDGE.
Stems 20–40cm, erect, acutely triangular and roughened above, rather slender, in dense clumps. Leaves 2.5–5.7cm wide, shorter than the culm; lower one or two bracts usually developed but inconspicuous; spikes usually 5–10, brown-tinged, blunt, densely many-flowered, subglobose to broadly ovoid, 5.6–11.3×3.8–8cm wide, aggregated into an oblong or linear-oblong head 17–35×10–15cm. Summer. N America.

**C.berggrenii** Petrie.
Broad and low, tufted, small. Stems to 3cm, stout, spreading, green or red-brown. Leaves to 5cm, metallic grey-brown, linear, obtuse at tip, entire or serrate above. Spikes 2–3, to 6mm, subsessile, terminal male, the remainder female; glumes elliptic, dark red-brown, very short-beaked; styles 2 or 3. New Zealand. Bronze, glaucous and a narrow-leaved form are also in cultivation. Z7.

**C.brunnea** Thunb.
30–90cm, tufted. Stems slender. Leaves 20–60×0.5cm, yellow to bright green, robust; lower sheaths brown. Spikes 10–30×2–3mm, cylindrical, nodding; female glumes ovate, pointed, red-brown, utricle 2.5–3mm, exceeding glume, short-beaked, stigmas 2. Summer. Japan, Australia. Z8. 'Variegata': small, leaves broadly edged yellow. A good house- or glasshouse plant.

**C.buchananii** Berggr. (*C.tenax* Berggr.). LEATHERLEAF SEDGE.
Densely tufted, 10–75cm. Rhizome ascending, red-brown. Leaves to 40–75 × 0.1cm, semi-terete, tapering to a flat-pointed, often twisted or spiralling tip, leaning outwards, pink-orange or red-green below to copper or bronze above; lower sheaths dark red-purple. Bracts exceeding inflorescence; male spikes 1–2, sessile, female spikes 4–5, long-stalked, 5–30× 3–4mm; female glumes bristle-tipped, white, membranous, utricle 2.5–3mm, cream-white, darker above, beak 2-lobed, stigmas 2. New Zealand. Z7.

**C.caryophyllea** Latour. (*C.praecox* Jacq.; *C.verna* Chaix). SPRING SEDGE.
Laxly tufted, 2–30cm. Rhizome short, creeping. Leaves 20 × 0.1–0.2cm, upper surface rough, recurved, tip 3-angled; lower sheaths becoming fibrous; ligule 1–2mm, blunt. Inflorescence 2–3cm; lower bracts leaf-like; male spike 10–15×2–3mm, terminal, female spikes 1–3, 5–15 × 3–5mm; pedicels 3–10mm; female glumes 2–2.5mm, red-brown, midrib green, utricle 2–3mm, short-hairy, green, stigmas 3. Europe inc. Scandinavia. Z7.

**C.comans** Berggr. (*C.pulchella* Berggr.). NEW ZEALAND HAIRY SEDGE.
Densely tufted, 6–40cm. Stem terete, glabrous, drooping in fruit. Leaves around 40 × 0.05–0.15cm, drooping to form a pale green mophead, margin rough; lower sheaths dull purple-brown. Male spikes 1–2, terminal, slender, female spikes 5–7,

5–25 × 3–4mm, becoming longer-stalked lower down the stem; female glumes shorter than utricle, notched at apex, light brown, utricle 2.5–3.5×1mm, red-brown, beak 2-lobed; stigmas 3. New Zealand. Z7. 'Bronze': leaves deeply bronze-coloured all year. 'Frosted Curls': compact, leaves grey-green-bronze, slender, densely curled and tangled.

**C.conica** Boott. (*C.excisa* Boott.; *C.digama* Nak.).
Tufted, 20–50cm. Stem 3-angled, smooth. Leaves 2–4mm wide, dark green, rigid, flat, rough. Bract blades short, sheaths brown-purple, inflated; male spike terminal, erect, female spikes erect, 3–4, lateral, widely spaced, 1–2.5cm, purple-brown, bristle-tipped, utricle beak recurved, light green, 2.5–3mm, just exceeding glume. Summer. Japan, S Korea. 'Snowline': small, neat; leaves arching, deep green edged white. 'Variegata' ('Hime-Kan-suge'; 'Marginata'): leaves variegated white. Z7.

**C.curvula** All.
Densely tufted, 2–40cm. Stems smooth, blunt, 3-angled. Leaves 1–2.5mm wide, rough, grooved or planoconvex in section. Inflorescence 1–3cm, irregular-oblong; lower bract spike-tipped, glume-like; spikes 2–6, oblong-ovoid, male above, female below; female glumes pale yellow to red-brown, utricle 5–8mm, 3-angled in section, yellow to green-brown at base, long-beaked; stigmas 3. S & C Europe. Z8.

*C.digama* Nak. See *C.conica*.

**C.digitata** L.
Clump-forming; rhizome short, ascending. Stems to 40cm, leafless, decumbent to erect, smooth or inconspicuously scabrid above. Leaves 2–5mm wide, shorter than or exceeding stems, flat, green to dark green. Male spike 8–15 × 1–1.5mm, subsessile; female spikes 2–3, 10–25×2–3mm, linear; female glumes as long as utricles, obovate, red-brown; utricles to 4mm, yellow-brown, puberulent, beak obscure, conical. Europe. Z7.

**C.dipsacea** Berggr.
Densely tufted. Stems 30–60cm, erect, slender, smooth. Leaves 2mm wide, exceeding stems, bronzy olive-green, numerous, flat, keeled, striate, with sharply scabrid margins. Spikes 4–7, many-flowered, pale or dark brown to black, terminal male, remainder mostly female, to 2.5cm, sessile; bracts long and leafy; glumes orbicular-ovate, obtuse, pale or dark chestnut-brown; utricles elliptic-ovoid, densely packed, short-beaked. New Zealand. Z7.

**C.dissita** Sol. & Boott.
Densely tufted. Stems 30–75cm, slender, smooth, leafy. Leaves 6mm wide, flat, grass-like, deeply grooved, dark green. Spikes 5–8, to 2.5cm, erect or nodding, short, cylindric, pale-brown, terminal male, the remainder female or male at the base only; bracts long, leafy; glumes broadly ovate, dark chestnut-brown with paler margins; utricle as long as glumes, ovoid, turgid; beak with 2 often widely divergent teeth. New Zealand. Z7.

**C.elata** All. TUFTED SEDGE.
Densely tussock-forming, 25–120cm. Rhizome short, erect. Leaves 40–100×0.1–0.6cm, glaucous, folded in section, tip flat; lower sheaths lacking a lamina, brown-yellow, becoming fibrous; ligule 5–10mm, pointed. Inflorescence short; bracts leaf-like; male spikes 1–3, 20–50mm, female spikes 2–3, overlapping, 15–40mm; female glumes 3–4mm, blunt, utricle 3–4mm, green, short-beaked; stigmas 2. Europe inc. Scandinavia. 'Aurea' (*C.acuta* 'Aureovariegata'): leaves with yellow margins. 'Bowles Golden': leaves golden-yellow, with green margins. 'Knightshayes': leaves yellow. Z7.

*C.excisa* Boott. See *C.conica*.

**C.firma** Host.
Densely tufted, mat-forming, to 20cm. Stem obscurely 3-angled. Leaves rigid, pointed, 10×0.4cm, blue-green; lower sheaths yellow-dark brown. Male spike terminal, 5–10mm, female spikes 1–3, 5–10×3–4mm; female glumes shorter than utricle, utricles 3.5–4.5×1.5mm, glabrous; stigmas 3.

**Carex**   (a) *C.oshimensis* 'Evergold'   (b) *C.pseudocyperus*   (c) *C.* 'Frosted Curls' (2 detached stems)   (d) *C.buchananii*

C Europe (mts). 'Variegata': leaves striped creamy yellow. Z7.

**C.flacca** Schreb. non Carey ex Boon.
Stems 20–50cm. Leaves shorter than stems, 3–6mm wide, glaucous beneath, dark green above, apex acute. Male spikes 2–3, to 40 × 3mm; female spikes 1–5, 12–30×3–7mm; female glumes acute or acuminate, purple-black to pale red-brown; utricles 2–4mm, broadly ellipsoid to obovate, green-brown to purple-black, rounded at apex, beak very short. Europe. Z5.

**C.flagellifera** Colenso.
Close to *C.buchananii*, but with longer stems and coarser leaves of a pale bronze to tawny brown arching to form a low dome. New Zealand. Z7.

**C.fortunei** hort. See *C.morrowii*.

**C.fraseri** Andrews (*Cymophyllus fraseri* (Andr.) Mackenzie ex Britt.; *Carex fraseriana* Ker-Gawl.; *Cymophyllus fraserianus* (Ker-Gawl.) Kartesz & Gandhi). FRASER'S SEDGE.
Tufted, to 35cm. Leaves 15–60×2.5–5cm, stiff, smooth, evergreen, broad, serrate. Spike bisexual, globose, bright white, solitary, male above, female below; female glumes white, exceeded by utricle, utricle straw-coloured, ovoid, membranous. Spring. W US (Virginia). Z7.

**C.gaudichaudiana** Kunth.
Loosely tufted, 30–60cm. Rhizome creeping great distances. Stems 6–23cm, 3-angled, subglabrous. Leaves grass-like, very numerous, 30–60×0.1–0.25cm, plicate; lower sheaths light brown. Male spike sessile, terminal, female spikes 2–4, lateral, 5–20mm, occasionally with male flowers at tip; glumes dark brown, shorter than utricles, utricle 2–4×1–2mm, beak very short; stigmas 2. New Zealand, New Guinea. Z9.

**C.gracilis** Curtis. See *C.acuta*.

**C.grayi** Carey. GRAY'S SEDGE; MACE SEDGE.
Clump-forming, 30–100cm. Leaves broad, flat, 6–11mm wide, pale green, margin rough. Bracts leaf-like, exceeding inflorescence; male spike terminal, thin, female spikes 1–2, 12–30-flowered, round, resembling a mace; utricle inflated, firm, glabrous, 15–20mm in fruit, spreading from a single point, short-beaked. Summer. Eastern N America. Z7.

**C.hachijoensis** Akiyama.
Evergreen, tufted; similar to *C.morrowii* but with narrower, softer leaves and beak of utricle short and barely notched. Japan. This species is commonly confused in cultivation with *C.morowii* and *C.oshimensis*. Two variegated cultivars, 'Variegata' and 'Evergold', assignable respectively to these last two species are often included here. Z5.

**C.humilis** Leysser.
Low, clump-forming, gradually forming a dense semi-evergreen mat. Leaves 3–5cm, narrow, involute. Inflorescence small, exceeded by leaves. Z4.

**C.hystricina** Muhl. PORCUPINE SEDGE.
Stems tufted, 15–100cm tall, shorter or longer than the leaves; blades flat, 2–10mm wide. Spikelets 2–5, forming an inflated, semi-translucent head, the terminal male, linear, 1–5cm long, female spikelets approximate or separate, the lower nodding on long stalks, oblong, 1–6cm long, bracts leaf-like; female scales strongly awned from a small obovate or oblanceolate base, the base narrower and shorter than the perigynia; perigynia tapering into a slender beak about 2mm long. US.

**C.intumescens** Rudge.
30–100cm. Leaves 3–8mm wide, dark green, soft, rolling inwards when dry. Bracts leaf-like; male spike narrow, on a long peduncle, female spikes 1–3, ovoid to round on short peduncles; female glumes lanceolate, half length of utricle, utricle thin, green, 10–20×5–8mm, spreading, inflated, beaked. Summer. Eastern N America. Z7.

**C.kaloides** Petrie.
Tussock-forming. Stems 30–90cm, slender, smooth, drooping at tips. Leaves 2mm wide, shorter than stems, erect, flat or involute, grass-like, ochre to pale rust or orange-brown, mar-

gins scabrid above. Inflorescence loose, linear, to 12.5cm; spikes 8mm, numerous, few-flowered, with male flowers usually uppermost; bracts very long, leaf-like; glumes ovate-lanceolate, tapering upwards into a long beak with serrate margins. New Zealand. Z7.

**C.maxima** Scop. See *C.pendula*.

**C.montana** L. MOUNTAIN SEDGE.
10–40cm. Rhizome thick, creeping. Stems slender, flexuous, 3- or 6- angled, sometimes leafless. Leaves 10–35 × 0.15–0.2cm, flat, gradually tapering to the tip, soft, becoming glabrous; sheaths persistent, becoming fibrous, red; ligule 1mm, blunt. Inflorescence small, 1–2cm; bracts glume-like; male spike terminal, 10–20mm, female spikes 1–4, few-flowered, 6–10mm; female glumes 3–5, red-black, midrib pale, utricle 3–4mm, pear-shaped, stalked; stigmas 3. Europe to C Russia. Z7.

**C.morrowii** Boott. (*C.fortunei* hort.).
Tufted, 20–40cm. Leaves evergreen, 8mm wide, stiff, thick, tapering to the tip, deep green, margin rough; lower sheaths dark brown, becoming fibrous. Male spike terminal, female spikes 2–4, lateral, slender, 2–5cm, on long, erect pedicels; female glume brown, utricle 3–3.5mm, with a long, bifid beak; stigmas 3. Summer. C & S Japan. 'Fisher's Form': large; leaves variegated cream. 'Variegata': leaves narrowly striped white near margin. Z8. See note under *C.hachijoensis*.

**C.muskingumensis** Schweinf. PALM SEDGE.
Stems erect, tufted, 35–100cm. Leaves very numerous, crowded on stems, 3–7mm wide, pale green; sheaths loose, green; fertile shoots sparsely leafy. Spikes 5–12, adpressed to stem, 1.5–2.5cm; female glumes brown, half length of utricle, utricle thin, long-beaked, narrow-lanceolate, 7–10 × 2.5mm. Summer. Western N America. Z7. Called the Palm Sedge because of its exotic appearance – tall stems clothed with narrow, arching, dark green leaves. These are especially fine in the cultivar 'Wachtposten' ('Sentry Tower').

**C.nigra** (L.) Reichard. BLACK-FLOWERING SEDGE.
Clump- or tussock-forming; rhizome slender, occasionally absent. Stems 10–60cm, trigonous, usually scabrid. Leaves shorter than or equal to stems, green to glaucous blue, 3mm wide. Male spike 5–30mm; female spikes 1–3, 10–40mm; female glumes slightly shorter than utricles, oblong-ovate to lanceolate, obtuse or acute, dark black-brown or red-brown; utricle 2–3.5mm, ovate to obovate-elliptic, green. Europe. Z5.

**C.nudata** W. Boott.
Stems densely tufted 30–80cm long, slender, erect or curving outward, longer than leaves; blades 1.75–3.5mm wide, flat, channelled below, leaf sheaths breaking and becoming fibrillose; spikelets 3 or 4, the terminal male, linear, 1.5–3.5cm long, female spikelets approximate, linear, 1–4cm long, lowest bract only occasionally leaflike, much shorter than the stems; female scales oblong to oblong-ovate, narrower than perygynia, perigynia abruptly beaked. W. US.

**C.ornithopoda** Willd. BIRD'S FOOT SEDGE.
5–25cm. Rhizome short, much-branched. Stems closely tufted, terete to 3-angled. Leaves 5–20 × 0.1–0.3cm, blunt-pointed, keeled, soft, margin rough; sheaths red-brown; ligule 0.5–1mm, blunt. Male spike few-flowered, 5–8mm, female spikes 2–3, few-flowered, 5–10mm; female glumes 2–2.5mm, orange-brown, blunt, utricle 3–4mm, pear-shaped, yellow-brown, very short-beaked; stigmas 3. NW Europe. 'Variegata': leaves narrowly striped white. Z7.

**C.oshimensis** Nak. JAPANESE SEDGE GRASS.
Densely tufted perennial, lacking stolons and runners and forming a thick, spilling, hair-like tussock. Leaves to 30 × 0.6cm, slender, smooth and finely tapering, deep, rather glossy green, typically narrowly edged white. Spikes 2–3, slender, rusty brown. Japan. Z6. See note under *C.hachijoensis*. 'Aureavariegata': leaves with a very broad central golden stripe. 'Aureavariegata Nana': as above but far smaller with curving and twisted leaves. 'Evergold': leaves with a creamy-

yellow central stripe. 'Variegata': leaves with a central white stripe.

**C.paniculata** L.
Tussock-forming, 30–150cm. Roots densely matted. Stems rough above. Leaves 20–120×0.5–1cm, dark green, channelled or plicate, tapering to a 3-angled tip, finely serrate; ligule 2–5mm, rounded. Panicle 5–15cm, spikelets 5–8mm, male above, female below; female glumes 3–4mm, triangular, orange-brown, utricle 3–4mm, green-dark brown; stigmas 3. Fruit ovoid, biconvex. Europe (except Mediterranean), Russia. Z7.

**C.pendula** Huds. (*C.maxima* Scop.). WEEPING SEDGE; PENDULOUS SEDGE; DROOPING SEDGE.
Tuft-forming, 50–180cm. Rhizome short. Leaves 20–100 × 1.5–2cm, keeled, flat, yellow-green above, somewhat glaucous beneath, margin rough; lower sheaths red-brown; ligule 30–60mm, pointed. Bracts leaf-like, as long as arching inflorescence; male spikes 1–2, 6–10cm, female spikes 4–5, 50× 5–7mm, becoming pendulous; female glumes red-brown, 2–2.5mm, pointed, utricle 3–3.5mm, glaucous green to brown; stigmas 3. WC & S Europe. Z8.

**C.pennsylvania** Lam. PENNSYLVANIA SEDGE.
Dark or dull green, stoloniferous. Stems 18–45cm, slender, erect, smooth or scaberulous. Leaves 1.2–3.8cm wide, the basal leaf shorter than or sometimes exceeding the culm, old sheaths persistent and fibrillose; lower bract subulate or scale-like, rarely over 1.5cm long; staminate spike sessile or very short-stalked, 1.5–3cm long; pistillate spikes 1–3, short-oblong, few-flowered, sessile, contiguous or the lower somewhat distant; perigynia broadly oval, pubescent, 1-ribbed on each side, narrowed at the base, tipped with a 2-toothed beak; scales ovate, purple-tinted, equalling or a little longer than the perigynia. Early summer. Eastern N America.

**C.pensa** Bail. CALIFORNIA MEADOW SEDGE; WESTERN MEADOW SEDGE.
Stems and leaves arising in scattered tufts from long-creeping rootstocks; stems 15–30cm tall, stiff, suberect or curving, longer than the leaves; blade 1–3mm wide, flat or canaliculate; spikelets few to several, androgynous, clustered to form a rather dense ovate head 1.5–2.5cm long; female scales larger than the perigynia and concealing them; perigynia ovate-lanceolate or elliptic, 3.5–4.5mm long, shining and dark brown at maturity, margin serrulate above the middle, stipitate, tapering into a serrulate beak. Coastal W US.

**C.petriei** Cheesem.
Resembles a smaller *C.buchananii*, of a somewhat weaker and pinker hue with stouter female spikes. Loosely tufted, 10–35cm. Stems terete or compressed, glabrous. Leaves to 40 × 0.05–0.2cm, tinged bronze-red, grooved, margin rough; lower sheaths light brown-red. Male spikes terminal, female spikes 2–5, 10–30 × 3–6mm; female glumes bristle-tipped, more or less equal to utricle, utricle 2.5×1.5mm, usually dark brown, beak light brown; stigmas 3. New Zealand.

**C.phyllocephala** T.Koyama.
Stems tufted, cane-like, to 45cm long, 2.5mm thick, erect, obtusely angled, mainly hidden beneath the leaf sheaths. Lower sheaths without blades, darkest blood-red, true leaves collected together at the top of the culm, the blades spreading like a *Dracaena*, to 20 × 1.3cm, sheaths loosely clasping stem, purple-tinted, minutely hairy. Spikes 8–10 making a capitate head, the terminal spike to 2cm long, tawny, the others female, cylindric with rather small dark brown scales. Japan. 'Sparkler' ('Fuiri Tenjiku-suge'): leaves lime green to dark green with white margins, often with 1–4 strong white stripes or streaks on blade; leaf sheaths tinted purple. Z8.

**C.pilulifera** L. PILL SEDGE.
Dwarf, tufted, to 30cm. Leaves to 20×0.2cm, scabrid above and tapering finely to a point; sheaths red-brown to purple-red. Inflorescence a head, 2–4cm long, male spike 1, female spikes 2–4. Eurasia. Z6. 'Tinney's Princess': very dwarf, each leaf with a creamy yellow central stripe. Lime-hating.

**C.plantaginea** Lam.
Tufted, 25–55cm. Stems covered in bladeless purple sheaths, tinted red at base. Leaves on sterile shoots, evergreen, broad and ribbed, resembling *Plantago*, to 40×1–3cm. Male spike long-stalked, purple, stamens yellow, female spikes 2–4, short-stalked; female glume pointed, utricle 3–4.5mm, 3-angled, beaked, just exceeding glume. Spring. N America. Z7.

*C.polyrrhiza* Wallr. See *C.umbrosa*.
*C.praecox* Jacq. See *C.caryophyllea*.

**C.pseudocyperus** L. CYPERUS SEDGE.
Tufted, 30–90cm. Rhizome short. Stems 3-angled, rough. Leaves 40 × 1–2cm, flat to keeled, longer than stems, yellow-green; lower sheaths pale brown; ligule 10–15mm, blunt. Inflorescence arching; bracts leaf-like, 3–4× longer than inflorescence; female spikes 3–5 on long pedicel, cylindrical, pendulous, 20–60×8–12mm; female glume 5–10mm, utricle 4–5.5mm, beaked, pale brown-green, falling when ripe; stigmas 3. Summer. Temperate regions, nearly cosmopolitan. Z7.

*C.pulchella* Berggr. See *C.comans*.

**C.riparia** Curtis.
To 150cm. Rhizome long, stout, spreading fast. Stem sharply 3-angled. Leaves flat, to 1.5cm wide. Male spikes 3–6, female spikes 2–5, sessile, broadly cylindrical, 3–10 × 0.8–1cm; female glume lanceolate-oblong, pointed, purple-brown, exceeding utricle, utricle more or less beaked, 5–6.5mm, inflated, green-brown; stigmas usually 3. Spring. Widespread, N hemisphere. Z6. 'Variegata': leaves striped or almost wholly pure white.

**C.scaposa** Clarke.
Rhizome stout, short, creeping. Stem obscurely 3-angled. Leaves 30–2 × 2–4cm, lanceolate-elliptic, flat, bright green, margin smooth. Bracts narrow, short-sheathed, shorter than pedicel; spikes 4–6mm, each with a terminal male flowers and lateral female flowers; female glumes lanceolate, somewhat pointed, smooth, utricle 3-angled, beaked; stigmas 3. S China. Z7.

**C.secta** Hook.
Sometimes forming a tussock to 1m high. Stems 3-angled, edges rough. Leaves to 30 × 0.7cm, arching, bright green, grooved, margin and keel rough. Panicle to 45cm; spikes clustered, light brown; female glumes pale brown, just exceeding utricle, utricle 2.5–3 × 1–2mm, smooth, dark brown, beak 2-lobed; stigmas 2. New Zealand. 'Tenuiculmis': leaves long, arching, dark brown. Z7.

**C.siderosticha** Hance.
Deciduous, to 10cm; new shoots tinged red. Leaves broad (i.e. to 2.5cm wide), similar in outline to *C.plantaginea* but rather smoother and more strap-like with stalks more erect. 'Variegata': young growth tinged pink; leaves edged and streaked white. Z7.

**C.spissa** Bail. SAN DIEGO SEDGE.
Plants large, loosely tufted with stout rootstocks; stems 1–2m tall. Leaf blade silver-grey, flat above, 7–14mm wide. Male spikelets 3 or 4, linear, 4–10cm long, female spikelets 3–7, androgynous, approximate or the lower somewhat separate, linear, 6–14cm long, the bracts leaf-like, the lower longer than the inflorescence; female scales lanceolate to ovate-lanceolate, narrower than the perigynia but longer; perigynia broadly obovate, 3–4.5mm long, smooth and only obscurely nerved, abruptly beaked. SW US.

**C.sylvatica** Huds. WOOD SEDGE.
Tuft-forming, 10–80cm. Stem nodding, 3-angled. Leaves plicate or keeled, 5–60 × 0.3–0.6cm, sharp-pointed, yellow-green; sheaths pale brown, not fibrous; ligule around 2mm, blunt. Bracts leaf-like, longer than spikes, shorter than inflorescence; male spike 1–4cm, female spikes 3–4, 2–7cm; female glumes long, thin-pointed, utricle 4–6mm, pale green, beaked; stigmas 3. Europe inc. Scandinavia. Z6.

*C.tenax* Berggr. See *C.buchananii*.

*testacea* Sol. ex Boot.
esembles *C.dipsacea*, but with narrower leaves in a pale
ive, orange-tinted in full sun. Loosely tufted. Stems to
5cm, very slender or slightly scabrid above, occasionally
aching 1.5m in fruit. Leaves 2mm wide, flat, striate, mar-
ins scabrid. Spikes 3–5, to 2.5cm, pale-brown, sessile, termi-
al male, the remainder female, occasionally with a few male
owers below; utricles shorter than or equal to glumes,
roadly ovate, apex tinged purple; beak short with 2 widely
vergent teeth. New Zealand. Z7.

*trifida* Cav.
obust, densely tussock-forming, to 1m. Stems stout, erect,
nooth. Leaves to 150×1.25cm, rigid, striate with scabrid
argins, shiny green above, rather silvery beneath. Spikes
–12, to 12.5×1.5cm, erect or nodding, cylindric, upper 2–4
ale, lower 4–6 female; bracts long, leafy; glumes 8mm, lin-
ar-oblong or lanceolate, deeply bifid, chestnut brown, short-
eaked. New Zealand. Z7.

*uncifolia* Cheesem.
preading, loosely or densely tufted, to 10cm, usually dwarf.
ems short, usually sheathed to the top by the leaves. Leaves
25×0.5cm, pink to red-bronze, concave above, convex
neath, margins finely scabrid above, recurved. Spikes 3–5,
6mm, dark brown, terminal male, slender, erect, the
mainder female, sessile, ovoid or oblong; glumes ovate,

obtuse; utricles exceeding glumes, elliptic-oblong, narrowed
at the base, red-brown to black-brown, very shortly beaked.
New Zealand. Z7.

**C.umbrosa** Host (*C.polyrrhiza* Wallr.).
Densely tufted, to 45cm. Rhizome creeping. Stems bluntly 3-
angled. Leaves 5–45×0.2–0.3cm, pale green; lower sheaths
fibrous, dark or pale brown. Bracts usually leaf-like; male
spike 5–15mm, female spikes 1–3, 5–12mm; female glumes
red-orange brown, obovate-lanceolate, pointed or blunt, utricle 2.5–4mm, white to yellow-green, somewhat hairy, short-
beaked. C Europe to N Spain. 'The Beatles': forms a low,
neat mop of dark green leaves. Z8.

*C.uncinata* L. f. See *Uncinia uncinata*.
*C.variegatus* hort. See *Acorus gramineus* 'Variegatus'.
*C.verna* Chaix. See *C.caryophyllea*.

**C.vulpina** L. FOX SEDGE.
30–100cm. Rhizome short, thick. Stems narrowly winged,
smooth below, rough above. Leaves 60 × 0.5–1cm, dark
green, keeled, margin rough; ligule broadly triangular,
2–5mm. Panicle 3–10cm, irregular; spikes ovoid, red-brown,
8–14mm, upper male, lower female; female glumes
3.5–4mm, elliptic-oblanceolate, brown-orange, pointed, utricle 5–6mm, green, turning dark brown, beaked; stigmas 2. N
Europe, Russia. Z7.

**Cenchrus** L. (From Greek *kenchros*, a type of millet.) Gramineae. Some 22 species of annual or perennial
rasses. Culms slender, erect to procumbent. Leaves narrow, scabrous; sheaths loose, keeled; ligules ciliate.
pikes cylindric; rachis rhombic in section; spikelets solitary or clustered, lanceolate to ovate, to 8 per cluster,
nclosed by a sessile involucre of spines or soft bristles; spikelets 2-flowered, unawned, the lower flower male
r sterile, the upper flower hermaphrodite, abscising together with the spiny involucre at maturity; glumes
horter than spikelet. Summer–autumn. Africa, N America, India. Z9.

ULTIVATION    Native to open grassland in the tropics and noted for its resistance to drought and hard grazing,
e characteristic prickly burrs in *C.ciliaris* are also of ornamental value. They do not persist on drying.
enchrus is reliably perennial only in warm, dry and frost-free climates; in cooler zones, treat as a frost-tender
nual or grow in the cool glasshouse. Grow in very well drained soil in full sun; plants are prone to root rots in
amp soils. Propagate by division or from seed sown under glass in late winter.

*ciliaris* L. (*Pennisetum cenchroides* hort. non Rich.;
*ennisetum ciliare* (L.) Link). BUFFEL GRASS.
erennial. Culms to 90cm, base bent. Leaves to 30×0.4cm,
abrous or pubescent; ligule a dense row of hairs.

Inflorescence dense, to 15×1.6cm, pale green or tinged purple; spikelets solitary or in clusters to 3, surrounded by bristles, inner bristles densely ciliate, basally thickened, united.
Africa to India.

**Chasmanthium** Link. Gramineae. 6 species of perennial grasses. Leaf blades linear to narrow-lanceolate. Inflorescence a loose, racemose panicle; spikelets compressed with overlapping florets, breaking up at maturity, 2–10-flowered plus rachilla extension bearing a rudimentary floret, increasing in size upwards, the lowest 1–2 florets being sterile; lemmas 5–15-nerved, acute. E US, E Mexico. Z6.

CULTIVATION   *C.latifolium* occurs in moist fertile woodlands. It thrives in sun or dappled shade on rich, moisture-retentive soils, sheltered from strong winds. In these conditions it will bear attractive inflorescences similar to those of *Briza* species and suitable for cutting and drying. In overall habit, it is far superior to *Briza*, however, making a fine clump of reedy growth overtopped by nodding, 'spangle' inflorescences which progress from olive to bronze with the onset of autumn. Few grasses are so decorative. Increase by division in spring.

*C.latifolium*   (Michx.) Yates (*Uniola latifolia* Michx.). NORTH AMERICAN WILD OATS; SPANGLE GRASS. Culms 30–100cm, tufted to loosely clumped, slender and rather flimsy to erect, smooth and somewhat bamboo-like. Leaves 10–30×1–2.5cm, carried on culms, linear to narrow-lanceolate, pale glossy olive-green to deep green tinted bronze. Panicle to 30cm, terminal, 1-sided, with slender, arching to drooping branches; spikelets 2–5cm, oblong-lanceolate to broadly ovate, compressed, green to bronze, like a large, flat, angular *Briza*. E US, N Mexico.

*Chasmanthium latifolium*

**Chimonobambusa** Mak. (From Greek *chimon*, winter, and Latin *bambusa*, bamboo, referring to the late appearance of new shoots in native areas.) Gramineae. Some 20 species of small to medium-sized bamboos with running rhizomes. Culms thick-walled; new shoots typically arising in late summer when they may fail to ripen enough to overwinter, but this character is not marked in temperate zones; sheaths thin, ciliate, without auricle or bristles; blades very small or absent; nodes prominent; branches usually 3. Leaves medium-sized, tessellate scaberulous; sheaths hairless, with smooth, or no bristles. S & E Asia. Z6.

CULTIVATION  Handsome bamboos, usually standing at 2m/6ft high and of value in hedging or group plant ing. *C.marmorea* produces lime-green young shoots marbled brown and white and lined pink at the tips; the culms, flushing purple in full sunlight, will bear 3–5 short, tufted branches per node, weighted with slender grass-green leaves. Hardy to –20°C/–4°F, *C.marmorea* can be planted in most soils that are unlikely to dry out in hot spells; it may be invasive. *C.quadrangularis* is slightly less hardy than *C.marmorea*. It slowly forms stands of wide-spaced, erect culms with conspicuously prominent nodes, the upper two-thirds fur nished with gracefully arching tiers of foliage. It is often planted in oriental courtyard gardens, where the spacing of the culms and outline of habit are emphasized by a topdressing of pale gravel or small stones. Propagate by division.

*C.falcata* (Nees) Nak. See *Drepanostachyum falcatum*.
*C.hookeriana* (Munro) Nak. See *Himalayacalamus hookeri-anus*.
*C.jaunsarensis* (Gamble) Bahadur & Naithani. See *Yushania anceps*.

***C.marmorea*** (Mitford) Mak. (*Arundinaria marmorea* (Mitford) Mak.). KAN-CHIKU.
Culms 2–3m × 1–2cm, thin, rounded, smooth, often purple-lined; sheaths fairly persistent, exceeding internodes, mottled purple or pink-brown with white spots when fresh, short-bristly below; branches tufted, 3–5 per node. Leaves 7.5–16×

0.5–1.5cm, sheaths bristly. Long grown in Japan, native coun try uncertain.

***C.quadrangularis*** (Fenzi) Mak. (*Arundinaria quadrangulari* (Fenzi) Mak.; *Tetragonocalamus angulatus* (Munro) Nak.)
SHIKAKUDAKE; SHIHO-CHIKU; SQUARE BAMBOO.
A distinctive species, rampant except in colder areas. Culm 2–10m × 1–3cm, often spinulose, with 4 rounded corners nodes very prominent, somewhat purple below, the lowest with thorny protuberances (aborted rootlets); sheaths hairless soon falling, lacking auricles and bristles. Leaves 10–29 2.7cm, dark green, sheaths with bristles. SE China, Taiwan long naturalized Japan.

**Chionochloa** Zotov. (From Greek *chion*, snow, and *chloe*, grass, referring to the appearance of the leaves. Gramineae. SNOW GRASS. Some 19 species of coarse perennial grasses. Culms erect, tufted, robust or slender Leaves narrow-linear, deeply ridged, acute; old leaf sheaths persistent; ligule a line of hairs. Inflorescence loos or contracted; spikelets several-flowered; awns present; glumes shorter than spikelets; lemmas 7–9-ribbed papery, lobed, longer than glume, the central lobe extended to an awn; awn straight or sharply bent or slightl twisted below. New Zealand, 1 species SE Australia. Z7.

CULTIVATION  An important component of the alpine and subalpine tussock grasslands of New Zealand *Chionochloa* includes a number of highly desirable ornamental grasses for the border, and as specimens in tem perate gardens. They are valued for their strong form and for the graceful flowerheads which in *C.flavescens* ar carried in early summer; the flowers of most species are persistent on the plant and are exceptionally elegant an long lasting when dried. Grow in a light, gritty, well drained but moisture-retentive and fertile soil in a sunn position with shelter from cold drying winds. Most are hardy to at least –10°C/14°F, *C.rigida* to between –1 and –20°C/5 to –4°F. *C.flavicans* is slow growing and requires a hot, dry sunny position to perform well Propagate by seed or division.

***C.conspicua*** (Forst. f.) Zotov (*Cortaderia conspicua* Forst. f.; *Danthonia cunninghamia* Hook. f.). HUNANGAMOLIO GRASS; PLUMED TUSSOCK GRASS.
Culms densely tufted, to 1.5m, robust. Leaves to 120×0.8cm, mid to pale green, midrib orange, rigid, flat to concave, ciliate below. Panicle graceful, erect to pendent, much-branched, compact or loose, to 45cm; spikelets 3–7-flowered, to 1.3cm; lemma pilose at base; awn straight. Summer. New Zealand.

***C.flavescens*** (Hook. f.) Zotov. BROAD-LEAVED SNOW TUSSOCK.
Densely tufted, 60cm–2m. Leaves strongly tinted brown to bronze-red. Panicle to 60cm, much-branched, lax, pale green to white. New Zealand. *C.flavicans* differs in leaf sheaths green, not orange-brown or yellow, leaf blades green, not or scarcely tinted, and panicle branches with spikelets along their entire length.

***C.rigida*** (Raoul) Zotov. SNOW GRASS; SNOW TUSSOCK.
Erect and spreading, to 1.5 tall forming a loose tussock. Lea sheaths persistent, forming a dried, dead mass in centre o clump (thus differing from other species described here); lea blades held more or less erect, V-shaped in section. Panicle t 25cm, much-branched, lax, pale green-white to straw coloured. New Zealand. Resembles a smaller, finer Pampa Grass.

***C.rubra*** Zotov. RED TUSSOCK GRASS.
Dense, tussock-forming, to 1.5m, usually far shorter. Leave narrow, tapering finely, tinged dull red-brown. Panicles smal New Zealand.

**Chimonobambusa**   (a) *C.marmorea*   (b) *C.quadrangularis*

**Chloris** Sw. (From Chloris, Greek goddess of flowers.) FINGER GRASS; WINDMILL GRASS. Gramineae. Some 55 species of tufted or stoloniferous, perennial or annual grasses. Culms slender. Leaves linear, flat or folded, scabrous, midrib present, leaf sheaths compressed, keeled; ligule a ciliate rim. Inflorescence spicate, terminal, digitately compound, rarely racemose; spikelets laterally flattened, in 2 rows on one side of the spike axis, sessile, bearing 1 or more sterile florets; fertile florets 1; rachilla terminating in 1 or more reduced lemmas; glumes membranous, unequal, narrow, acute, 1-ribbed; fertile lemmas keeled, broad, 5-veined, entire or bilobed, sterile lemmas variable, awn slender or rudimentary. Tropics, Subtropics. Z9.

CULTIVATION Occurring in short grassland, especially on poor soils and a ruderal species on disturbed ground, *Chloris* species are usually treated as annuals in cool temperate zones, where their 'fingered' inflorescences are moderately attractive in late summer and autumn; sow *in situ* after danger of frost is past or earlier under glass.

**C.barbata** Sw. (*C.inflata* Link).
Annual, to 45cm. Culms loosely clumped, ascending, base prostrate. Leaves linear, flat, to 25×0.6cm; leaf sheaths compressed. Inflorescence digitate; spikes to 15, to 7.5cm, tinged purple or brown, white-pubescent; lower lemma obtuse, to 3mm, awn to 4mm. Autumn. Tropics.

**C.berroi** Arech. GIANT FINGER GRASS; URUGUAY GRASS.
Perennial, to 70cm. Culms clumped. Inflorescence cylindric, spicate, digitate; spikes to 5, to 7cm; spikelets densely arranged, to 3mm. S America (Uruguay).

*C.crinita* Lagasca. See *Trichloris crinita*.
*C.elegans* HBK. See *C.virgata*.
*C.inflata* Link. See *C.barbata*.

**C.radiata** (L.) Sw.
Annual, to 45cm. Culms rooting at nodes, ascending. Leaves flat to folded, to 0.6cm wide, apex blunt; leaf sheaths flattened. Inflorescence digitate; spikes to 10, slender, to 7.5cm, green, occasionally tinged purple; lowest lemmas to 3mm, awn to 1cm. Summer–autumn. Tropical America.

**C.truncata** R. Br. STAR GRASS; CREEPING WINDMILL GRASS.
Stoloniferous perennial, to 30cm. Culms erect, slender. Leaves flat or folded, 0.2–0.3cm diam., apex obtuse. Inflorescence digitate; spikes to 10, to 8cm, horizontal or reflexed; spikelets glabrous, lowest to 3mm, becoming black when mature, to 3mm, awns to 13mm. Autumn. Australia.

**C.ventricosa** R. Br. AUSTRALIAN WEEPING GRASS.
Perennial, to 90cm. Culms procumbent, prostrate or spreading to erect. Inflorescence digitate; spikes to 6, to 10cm, flexible; spikelets as for *C.truncata* but upper lemma blunt, longer than glumes, awns shorter, to 6mm, glabrous. Australia.

**C.virgata** Sw. (*C.elegans* HBK).
Annual to 60cm. Culms clumped, ascending to erect. Leaves flat, to 0.6cm diam.; leaf sheaths inflated. Inflorescence digitate; spikes to 12, to 7.5cm, soft-pubescent to feathery, green to purple-tinged; spikelets pubescent; lowest lemma to 3mm, awn to 1cm. Summer–autumn. Tropics.

**Chusquea** Kunth. (From the vernacular name used in NE South America.) Gramineae. Some 120 species of bamboos. Culms clumped, terete, pith-filled; sheaths persistent with auricles and bristles; branches basically 3 but these soon branching densely and near to the base thus giving the appearance of many primary divisions. Mexico to Chile. Z7.

CULTIVATION *Chusquea* species may be recognized by their closely sheathed, slender culms and densely tufted, alternate-branched whorls. *C.culeou* is most often encountered in cultivation and forms slow-growing clumps to 5m/16ft, bowed with the weight of a profusion of small, pale green leaves on short branches or distinguished by the spear-like new culms clothed with pale papery sheaths. They require a minimum temperature of −15°C/5°F (frost-free for *C.coronalis*) in sites rich in damp humus and sheltered from harsh winds. Most can be raised easily from seed sown thinly in a heated case or covered pots. They can also be propagated with some difficulty by division in spring or layer young culms by weighting and notching below buried nodes.

*C.breviglumis* hort. See *C.culeou* 'Tenuis'.

**C.coronalis** Söderstr. & Cald.
Delicate, clump-forming, with slender somewhat arching or sprawling culms; branches completely encircling the culms and pushing away the sheaths as they develop. Leaves 2.3–7.0×0.3–0.8cm, not tessellate. C America.

**C.culeou** E. Desv.
A variable yet distinctive stately and graceful species with yellow-green to olive, erect, stout culms, 4–6m × 1.5–3cm; sheaths persistent until the branches break through in the sec-

ond year, long-tapering, longer than the internodes, papery white and decorative when fresh, pubescent above; nodes prominent with white wax below at first; branches 10–80cm, from alternate sides of the culm, appearing unbranched and densely tufted. Leaves 6–13×0.5–1.5cm, tessellate, scaberulous, hairless; sheaths lacking auricles or bristles. Chile. 'Tenuis' (*C.culeou* var. *tenuis* D. McClintock; *C.breviglumis* hort.): culms thinner, shorter, to 2m × 0.5cm, growing out at an angle of 45°.

*C.culeou* var. *tenuis* D. McClintock. See *C.culeou* 'Tenuis'.

**Chusquea**   (a) *C.culeou*  (b) *C.coronalis*

**Cladium** P. Browne. Cyperaceae. 2 species of perennial, rhizomatous, grass-like herbs. Stems terete. Leaves grass-like, linear, tip triangular in section. Inflorescence a terminal panicle of clustered spikelets; spikelets with 1–3 flowers each in the axil of the spirally arranged floral bracts (glumes); lower glumes sterile, middle glume female, upper glume hermaphrodite; flowers very small, sepals and petals absent; stamens 2, occasionally 3; style very thin, stigmas 3, occasionally 2. Fruit a 3-angled nut. Cosmopolitan.

CULTIVATION    As for *Carex*. *C.mariscus* (Elk Sedge) historically used for thatching in the British Isles.

*C.mariscus* (L.) Pohl. ELK SEDGE.
125–250cm. Rhizome stout, creeping great distances. Stems hollow, leafy. Leaves to 200 × 1–1.5cm, tapering to the tip, keeled, margin finely sharp-toothed. Panicle 30–70×5–12cm; spikelets 3–4mm; glumes bristle-like, lanceolate, yellow-brown; stigmas 3. Fruit around 3mm, glossy brown. Summer. N Africa to Scandinavia and Siberia. Z3.

**Coix** L. (From Greek *koix*, a reed-leaved plant.) Gramineae. Some 6 species of monoecious annual or perennial grasses, often rhizomatous. Leaves flat; ligules membranous. Inflorescence grouped in upper leaf axils, compound, consisting of many racemes; peduncles long; spikelets unisexual; awns absent; female raceme of 3 spikelets, 2 sterile, 1 fertile; lodicule absent; female fertile spikelet sessile, solitary, protected by a hollow, globose to pyriform sheath or utricle; glumes membranous; palea and lemmas papery; male spikelets in fascicles of 2–3, 1–2-flowered, exserted from the bract sheath; lodicules absent; glumes membranous; sterile spikelets with a tubular glume as long as the fertile spikelet. Tropical Asia. Z9.

CULTIVATION    Grown for its curious arching inflorescences which bear pendulous, tear-shaped utricles that give rise to the common name, Job's Tears, in cool temperate climates *C.lacryma-jobi* needs a long hot summer to flower well. Sow seed in late winter under glass and plant out in a warm sunny position in a well drained soil, but ensure plentiful water when in growth. In warmer climates sow *in situ* in spring.

*C.lacryma-jobi* L. JOB'S TEARS; CHRIST'S TEARS; ADLAY.
Annual to 1.5m. Culms robust, erect. Leaves narrow-lanceolate, to 60 × 5cm. Inflorescence arching gracefully; male racemes to 5cm; spikelets to 1cm; utricles ovoid-globose, hard, to 13mm, white to grey tinged blue to brown. Autumn. SE Asia.

**Cortaderia** Stapf. (From *Cortadera*, the Argentine name for these plants.) PAMPAS GRASS. Gramineae. Some 24 species of dense, tussock-forming perennial grasses, to 3m. Culms robust, coarse. Leaves crowded at base, long, stiff, narrow-linear, glaucous, flat, margins rough to very sharp; ligule a silky fringe. Inflorescence a dense, showy, plume-like panicle, silver-white to pale rose or golden; spikelets unisexual, occasionally hermaphrodite, 3–7-flowered; spikelet axis pubescent, glumes longer than lowest lemma, narrow, membranous, entire to acuminate or bidentate, veins 3, glabrous in male spikelets, basally pubescent in female spikelets, long-awned; palea glabrous or softly hairy; lemmas white, papery. Summer–autumn. S America, New Zealand, New Guinea (1 species).

CULTIVATION    Grown as lawn specimens, in the border and by lake and streamside, where the roots are well above water level, *Cortaderia* species are valued for their graceful tussock-forming habit and for the elegant plume-like inflorescence (more showy in female forms) which often persists into winter and dries well if cut when the panicles first emerge. The dwarf form *C.selloana* 'Bertinii' is ideally suited to smaller gardens. Most species tolerate short-lived frosts to about −7°C/20°F, *C.selloana* occasional lows to −15°C/5°F, where winters are not excessively wet. In addition to those described, the 'dwarf' *C.dioica* Spegazz. is sometimes available.

Generally undemanding in cultivation, requiring a sunny position in any fertile well drained soil, although since specimens can be long lived thorough pre-planting soil improvement is recommended. In essentially frost-free zones, *C.richardii* can be grown as a water marginal. These grasses accumulate a great deal of dead material at the centre of the tussock, and this should be cleared annually to show the habit and form to best advantage. The commonly recommended practice of burning out old growth in spring, demands a guaranteed fast dry burn and impeccable timing, since there is risk of killing new growth, and, in uncontrolled burning, of killing off the rootstock. With regular spring maintenance, cutting back and removing the dead material with gloved hands, burning is usually unnecessary. Propagate by division in late spring. Seed-raised stock is variable, and seed has short viability; where specimens of both genders are grown, the female form *C.selloana* may give rise to small numbers of variegated seedlings.

*C.argentea* (Nees) Stapf. See *C.selloana* 'Elegans'.
*C.conspicua* F. Forst. See *Chionochloa conspicua*.

*C.fulvida* (Buch.) Zotov.
To 1.8m. Leaves long, curved, narrow, nerved, forming large tussocks. Panicles pale, tawny, 30–60cm; spikelets 1–3-flowered; outer glumes 1–1.6cm. New Zealand. Z8.

*C.jubata* (Lemoine) Stapf (*C.quila* (Nees & Mey.) Stapf; *Gynerium jubatum* Lemoine).
Resembles *C.selloana* but inflorescence looser, branches nodding; spikelets smaller. To 3m. Leaves to 120×1.2cm, arching, scabrous, tough. Panicles to 60cm; branches nodding or drooping, yellow tinged red to purple. Autumn. W Tropical America. Sometimes misnamed *C.rudiuscula*. Z9.

*C.quila* (Nees & Mey.) Stapf. See *C.jubata*.

*C.richardii* (Endl.) Zotov (*Arundo richardii* Endl.). TOE TOE; PLUMED TUSSOCK.
To 3m. Leaves to 120cm, coriaceous, recurved. Panicle to 60cm, very plume-like, white, tinged yellow or silver. Summer–autumn. New Zealand. Z8.

*C.rosea* hort. See *C.selloana* 'Rosea'.
*C.rubra* hort. See *C.selloana* 'Violacea'.

*C.selloana* (Schult. & Schult.f.) Asch. & Gräbn. (*C.argentea* (Nees) Stapf; *C.rosea* hort.; *C.rubra* hort.; *Arundo selloana* Schult.).
To 3m. Leaves narrow, to 270cm, somewhat glaucous, arching, margins very scabrous. Panicle oblong to pyramidal, to 120cm, silver-white, tinged red or purple. Autumn. Temperate S America. 'Albolineata' ('Silver Stripe'): compact, slow-growing; leaves edged white. 'Aureolineata' ('Gold Band'): habit small; leaves broadly edged rich yellow, later deep gold; hardy. 'Bertinii': dwarf to 1m; all parts miniature. 'Carminea Rendatleri': culms very tall, weak; panicles pink flushed purple. 'Carnea': panicles soft pink. 'Elegans' ('Argenteum'): panicles silvery white.

'Marabout': culms erect; panicles large, dense, pure white. 'Monstrosa': large, mound-forming, to 3m; panicles to 70cm. 'Pumila': dwarf, to 1.5m, compact; floriferous; winter-hardy. 'Roi des Roses': panicles flushed pink. 'Rosea' ('Rosa Feder'): panicles lightly tinted pink. 'Silver Beacon': leaves rigid, more erect, edged white; culms tinted purple; panicles very fine and silvery. 'Silver Comet': leaves striped cream; stems to 1.2m; panicles very silvery. 'Sunningdale Silver': habit large, culms to 3.5m; plumes dense, white; wind-resistant, sturdy. 'Sun Stripe': leaves yellow-striped. 'Variegata': leaves striped cream. 'Violacea': panicles tinted violet. 'White Feather': panicles to 2m high, white. Z5.

**Ctenium** Panzer (From Greek *ktenos*, a comb, referring to the shape of the inflorescence.) Gramineae. 20 species, perennial grasses from the Americas and Africa with distinctive comb-like, 1-sided flower spikes. One species is grown in native plant collections in the US – *C.aromaticum* Panzer (TOOTHACHE GRASS), an erect, shortly rhizomatous perennial. Culms to 1.5m. Leaves basal and on base of stem; blades to 40 × 0.1–0.5cm, scaberulous above, glabrous beneath. Spike solitary, erect, recurved, rachis flattened, hairy, margins ciliate; spikelets in 2 rows on one side of rachis, 5–6mm long, sessile, awned and toothed. Grain glossy dark red. SE US. Z9.

**Cymbopogon** Spreng. (From Greek *kymbe*, boat, and *pogon*, beard.) Gramineae. Some 56 species of aromatic, usually perennial grasses. Culms tufted. Leaves linear to lanceolate; ligules membranous. Inflorescence a much-branched panicle; branches racemose, paired, enclosed by spathes; spikelets paired, one sessile, fertile, or stipitate, male; sessile spikelets awned; lower glume flat or concave, laterally bi-keeled; upper lemma bilobed with a short, glabrous awn, awn occasionally absent; pedicelled spikelet well-developed, awnless. Old World Tropics, warm temperate. Z9.

CULTIVATION Occurring in open habitats in dry soils and frequently in savannah, *Cymbopogon* species are grown as ornamentals for foliage and for the light elegant flowering panicles; *C.citratus* also for culinary use, the strongly lemon-scented foliage widely used as a flavouring in southeast Asian cuisine. The flavour will taint other foods in storage unless well-wrapped. Grow in a moisture-retentive potting mix in full sun and maintain a moderately humid but buoyant atmosphere with a minimum temperature of 10–13°C/50–55°F. Water plentifully when in growth. Propagate by division or seed when available.

*C.citratus* (DC. ex Nees) Stapf (*Andropogon citratus* DC. ex Nees; *Andropogon schoenanthus* L.). LEMON GRASS.
Perennial, to 1.5m+, all parts strongly scented of lemon. Culms clumped, slender to robust, often cane-like and sheathed. Leaves blue-green, to 90×1.3cm, sheaths coarsely pubescent, apex and base attenuate, midrib stout, margins rough. Inflorescence lax; rachis pubescent; spathes long, linear-lanceolate, to 1.6cm; spikelets awnless. S India, Ceylon.

*C.confertiflorus* Stapf. See *C.nardus*.

*C.nardus* (L.) Rendle (*Andropogon nardus* L.). CITRONELLA; MANA GRASS.
Lemon-scented perennial, to 2m+. Culms robust, erect, smooth. Leaves to 90 × 1.8cm; ligule blunt, to 0.3cm.

Inflorescence compact, to 60cm; rachis ciliate; spathes elliptic, to 2cm; spikelets awnless. Tropical Asia.

*C.schoenanthus* (L.) Spreng. CAMEL GRASS
Perennial, densely tufted, to 30cm. Culms erect, slender, glabrous. Leaves to 25cm, arching, filiform, semi-terete, scabrous, glaucous. Inflorescence narrow, to 30cm, composed of clustered racemes; joints long-pubescent; spikelets to 7.5cm, linear-lanceolate; glumes equal, chartaceous; lower floret reduced to a linear-oblong valve, slightly shorter than glume, upper floret to 3.75cm, cuneate-linear, shortly bifid; awn long, slightly twisted below the middle. N Africa to N India.

**Cynodon** Rich. (From Greek *kuon*, dog, and *odous*, tooth, referring to the sharp scales of the rhizome.) Gramineae. 8 species of creeping perennial grasses with rhizomes or stolons, rooting at nodes. Culms erect or ascending. Leaves flat, linear to filiform, ligules hyaline or a pubescent fringe. Inflorescence terminal, with 2–6 spikes arranged in an umbel; spikelets 1-flowered, laterally compressed, sessile, spikelet axis breaking at nodes above glumes; glumes subequal, narrow, acute, papery, keeled, shorter than lemma; palea 2-ribbed, lemma keeled, entire, awns absent; stamens 3. Tropical and Southern Africa, *C.dactylon* in many tropical and warm temperate regions.

CULTIVATION   A genus most frequently used as warm-season turf grasses, although *C.dactylon* is occasionally used as an ornamental in cool temperate zones, valued for its interesting inflorescence and as a cover for warm sunny banks. Tolerant to about –10°C/14°F. Grow in a warm sunny position in well drained soil. Propagate by division, or by plugs.

**C.dactylon** (L.) Pers. BERMUDA GRASS; BAHAMA GRASS; KWEEK; DOOB; COUCH; STAR GRASS; CREEPING DOG'S TOOTH GRASS; CREEPING FINGER GRASS; ST LUCIE'S GRASS. Rhizomatous, stoloniferous perennial, to 30cm. Culms slender. Leaves to 15cm × 1.8cm, grey to grey-green; ligule a dense row of hairs. Inflorescence spikes 3–6, slender, to 5cm; spikelets 1-sided, overlapping in 2 rows, to 0.3cm. Summer–autumn. Cosmopolitan. Z7.

**C.transvaalensis** Burtt-Davy. Close to *C.dactylon*, but differing in its finer leaves, spikes to 2cm, narrower spikelets and shorter glumes. S Africa, introduced US. Z8.

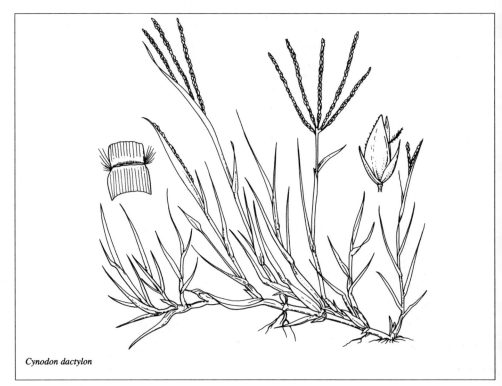

*Cynodon dactylon*

**Cynosurus** L. (From Greek, *kuon*, dog, and *oura*, tail.) Gramineae. DOG'S TAIL GRASS. Some 7 species of annual or perennial grasses to 60cm or more. Stems clumped, slender. Leaves glabrous; ligules membranous. Panicle spicate, secund; spikelets paired, sessile, inner spikelet fertile, to 5-flowered, outer spikelets sterile; fertile spikelets shorter than and obscured by sterile spikelets, lemmas leathery, 5 ribbed, rough above, smoother beneath, apex acute to awned; sterile spikelets a cluster of sterile lemmas. Summer–autumn. Mediterranean, Europe, N Africa. Z7.

CULTIVATION  *C.cristatus*, the Crested Dog's Tail Grass, is occasionally used in wildflower meadow mixtures and in amenity turf grass mixtures, especially for sports turf. *C.echinatus* is cultivated for its moderately decorative inflorescence. Both are tolerant of a wide range of soils, damp or dry. Sow seed *in situ* in spring, in any light well-drained soil in sun.

**C.echinatus** L. ROUGH DOG'S TAIL GRASS.
Annual, to 110cm. Leaves glabrous, to 20×1cm; ligule blunt, to 1cm. Panicle ovoid to globose, to 8 × 2cm, sometimes tinted purple; spikelets to 1.4cm, to 5-flowered; glumes narrow-lanceolate, to 1.3cm; lemmas convex, 5-ribbed, tipped with long, fine awns.

**C.elegans** Desf.
Resembles *C.echinatus*, but to 45cm. Leaves narrower. Inflorescence narrower, to 2.5cm long, more lax; spikelets smaller; awns silky, longer than lemmas.

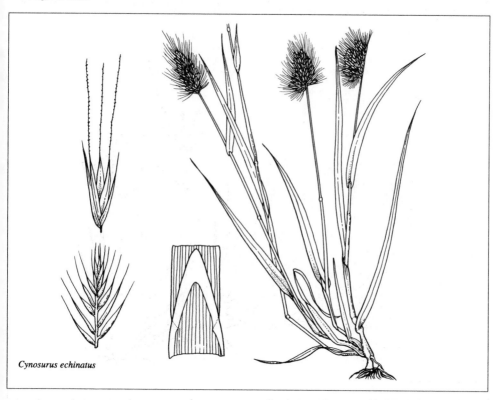

*Cynosurus echinatus*

**Cyperus** L. (From the Greek name for this plant.) Cyperaceae. Around 600 species of annual to perennial, rhizomatous herbs, to 3m. Stems terete, 3-angled or winged, usually leafy below. Leaves narrow, grass-like. Inflorescence a terminal umbel with spikelets clustered at the tips of its rays, rays usually subtended by a collar of leaf-like bracts; spikelets linear, terete or compressed, few- to many-flowered; flowers very small, hermaphrodite, arranged in 2 rows in the axils of scale-like bracts (glumes); sepals and petals absent; stamens 1–3; stigmas 2–3. Fruit a nut, compressed in species with 2 stigmas, 3-angled in species with 3 stigmas, usually chestnut-brown. Cosmopolitan, except very cold regions.

CULTIVATION Grown for their graceful foliage and umbrella-like inflorescence, *Cyperus* species generally enjoy conditions similar to *Carex*; many, such as *C.alternifolius, C.papyrus* and *C.longus*, are also happy with up to 15–30cm/6–12in. of water above the root zone. *The* papyrus of Egypt, *C.papyrus* is a large and attractive ornamental for tropical and subtropical gadens, large pools under glass or, in cooler regions, for standing outdoors in summer in water-filled tubs; its young inflorescences are also increasingly used as cut flowers. *C.longus, C.esculentus* and *C.haspan* should tolerate temperatures as low as –15°C/5°F, but most others require to be overwintered at minimums of 7–10°C/45–50°F and are therefore more easily cultivated as marginals for conservatory ponds in cool temperate areas. Grow tropical species at a minimum of 18°C/65°F. Smaller species such as *C.alternifolius, C.albostriatus* and *C.prolifer* make attractive houseplants that prefer to stand permanently with the pot base in a shallow tray of water. Propagate in spring from seed, by division and by stem cuttings of the umbels rooted in water, *C.esculentus* by tuber.

**C.albostriatus** Schräd. (*C.diffusus* hort. non Vahl; *C.elegans* hort. non L.).
Perennial, 20–50cm. Rhizome woody. Stems thin, firm. Leaves numerous, to same length as stem, 8–16mm wide, 3-nerved, nerves pale, prominent; sheaths purple. Bracts 6–9, leaf-like, exceeding inflorescence, 16mm broad, spreading. Umbel compound, lax; rays 8–24, 2.5–10cm; spikelets 8–24-flowered; glume brown; glume blunt, tip short-bristled, brown, midrib green; stamens 3; style short, stigmas 3. S Africa. Z9. 'Variegatus': leaves and bracts striped white.

**C.alopecuroides** Rottb.
Perennial to 120cm. Rhizome woody, short, thick. Stem robust, 3-angled, smooth at base. Leaves 6–12mm wide, often exceeding stems in length, nerves 3, prominent above. Bracts 4–7, lower bracts exceeding inflorescence. Umbel compound; rays 5–10, 18cm; spikes oblong or oblong-cylindrical, 15–40 ×7–15mm, spikelets 4–8 × 2–3mm, 10–30-flowered; glumes membranous, 2–2.5mm, 7-nerved; stamens 2, rarely 3; style long, thin, stigmas 2. Tropical Africa. Z10.

**C.alternifolius** L. UMBRELLA PLANT.
Tufted, leafless perennial, 45–90cm. Rhizome woody, short. Stems numerous, erect, dark green, crowded. Bracts 11–25, leaf-like, 6–10mm wide, flat, longer than inflorescence. Rays 20–25, 6cm; spikelets brown, 6–8×1–2mm, 8–20-flowered; glumes pale brown-yellow, 1–2 × 1mm, pointed, bristle-tipped; stamens 3; stigmas 3. Madagascar. Z10. 'Variegatus': stems, leaves and bracts striped white or wholly white.

**C.compressus** L.
Glabrous, tufted, annual herb, 10–40cm. Leaves 3–5mm across, two-thirds stem length. Bracts leaf-like, lowest exceeding inflorescence. Umbels simple or nearly so; spikelets in clusters of 3–10, 12–20-flowered, 12–25 × 4–6mm, compressed, green becoming yellow; glumes ovate, many-veined, keel sharp, bristle-tipped; stamens 3; style long, thin, stigmas 3. Tropical Africa. Z10.

**C.congestus** (Vahl) C.B. Clarke (*Pycreus congestus* (Vahl) Hayek).
Tufted annual or perennial, to around 30cm. Rhizome short. Stems sparsely leafy. Leaves equal to or exceeding stems, 2–12mm wide. Inflorescence lax, compound or simple, rays 2–12; 8cm; spikelets 3–25-flowered, 8–30mm, stalked, brown, falling whole from plant; bracts 4–6, exceeding inflorescence; glumes 3–5mm, red-purple, pointed, bristle-tipped, keel green; stamens 8. S Africa, Australia. Z9.

**C.cyperoides** (L.) Kuntze (*Mariscus sieberianus* Nees; *Mariscus cyperoides* (L.) Urban; *Scirpus cyperoides* L.; *C.sumula* hort.).
30–75cm. Rhizome small. Stem slender, 3-angled. Leaves two-thirds stem length, bright green; lower sheaths red.

Umbel simple, 2–12cm across; rays 5–12, 2–10cm; spikelets numerous, linear-lanceolate; glumes straw-yellow, lowest 2 and uppermost glumes sterile; stamens 2–3; style 3-branched. Africa (widespread). Z9.

**C.diffusus** hort. non Vahl. See *C.albostriatus*.
**C.elegans** hort. non L. See *C.albostriatus*.

**C.eragrostis** Lam. (*C.vegetus* Willd.).
Perennial, 60–90cm. Rhizome short. Stems stout, roughly 3-angled above, leafy. Leaves as long as stem, 4–8mm wide, net-veined, margin rough. Bracts 5–8, exceeding inflorescence Umbel compound, spreading; rays 8–10, to 12mm; spikelets in clusters of 20, pale green, 8–13mm; glumes 2–3 × 1–1.5mm, straw yellow to red, midrib tinged green; stamens 1; stigmas 3. Summer. SW US and warm temperate S America. Z8.

**C.erythrorrhizos** Muhlenb.
Annual, 15–120cm. Stem bluntly 3-angled. Leaves linear, pale beneath, 4–8mm wide, margin rough. Bracts 3–10, leaf-like, exceeding inflorescence. Umbel simple, sometimes compound; rays 3–12, 8cm; spikelets numerous, crowded, 12–15-flowered, 3–10mm, narrow-linear, spreading, yellow-brown; glume lanceolate, bristle-tipped, 3-nerved; stamens 3; style long, stigmas 3. Summer. N America. Z7.

**C.esculentus** L. (*C.melanorhizus* Del.). CHUFA.
Annual or perennial, 5–90cm. Rhizome long, thin, creeping, occasionally producing tubers. Stems solitary, rigid, 3-angled, often compressed. Leaves 3–10mm wide, somewhat rigid, bright green, shorter than or exceeding stem length. Bracts 4–8, some exceeding inflorescence. Umbel simple or compound; rays 4–10, to 10cm; glumes 2–3 × 2mm, blunt, pale yellow-brown; stamens 3; stigmas 3. S Europe to India. var. *sativus* Boeck. Tubers present, banded, rarely producing flowers. Z8.

**C.fertilis** Boeck.
4–10cm. Rhizome short. Stem 3-angled, compressed. Leaves lanceolate or linear-lanceolate, 8–14mm wide, prominently 3-nerved, very finely toothed. Bracts 5–7, far longer than rays. Umbel simple, lax; rays 5–8, 40mm; spikelets 2–3, 5–10× 3.5mm, 8–12-flowered; glumes blunt, oblong-ovate, many-nerved, margin white; stamens 3; stigmas 3. W Africa. Z10.

**C.filicinus** Vahl. See *Pycreus filicinus*.
**C.flabelliformis** Rottb. See *C.involucratus*.

**C.haspan** L.
Perennial, 10–50cm. Rhizome creeping. Stem 3-angled above. Leaves 3–5mm wide, linear, flat, smooth, much shorter than stem. Bracts 2–3, shorter than rays. Umbel compound; rays 4–12, 4–12cm; spikelets 10–15 × 1–2mm, strongly compressed, 10–40 flowered; glumes 1–1.5mm, blunt, light green-brown; stamens 3, sometimes 2; style around 1.5mm, stigmas 3. C Asia. Z8.

**C.involucratus** Rottb. (*C.flabelliformis* Rottb.).
45–200cm. Stem 3-angled to subterete, pale green. Bracts 12–28, 6–40×1–18mm, finely grooved, tapering from base, apex pointed; rays 14–32, tapering to tip, slender, 5–130mm; spikelets ovate, rarely obovate, 4–10mm; glume red-brown, 1.5–2mm. Fruit brown-white, less than 1mm. Africa, naturalized elsewhere. Z9.

*C.isocladus* Kunth. See *C.profiler*.

**C.longus** L. (*Pycreus longus* (L.) Hayek). GALINGALE.
Perennial, 60–150cm. Rhizome long, knotted. Stem solitary, 3-angled, stiff. Leaves 2–3, on lower part of stem, 2–10mm wide, bright glossy green and grooved above, pale and sharply keeled beneath, margin rough; sheath red-brown at base. Bracts 2–6, some exceeding rays. Umbel lax; rays 2–10 to 35mm; spikelets linear, 4–50mm, tapering to both ends; glumes 2–3×1.5–2mm, chestnut red, midrib green, 3–5-nerved; stamens 3. Europe, N America. Z7.

*C.melanorhizus* Delile. See *C.esculentus*.
*C.nuttallii* Eddy. See *Pycreus filicinus*.

**C.papyrus** L. PAPYRUS; EGYPTIAN PAPER REED.
Robust, leafless perennial, 2–5m. Rhizome short, woody, thick. Stem stout, dark green, bluntly 3-angled, not jointed.

Bracts 4–10, papery and broad, spathe-like, shorter than rays. Umbel compound; rays 100–200, slender, erect to weeping, 12–30cm; spikelets brown 8–20mm, thin, in clusters of 20–30; glumes blunt, 1.5–2×1mm, red to pale brown; stamens 3; stigmas 3. Africa. Z9.

**C.profiler** Lam. (*C.isocladus* Kunth). MINIATURE PAPYRUS.
30–75cm. Rhizome creeping, 3–4mm across, brown, scaly. Stem triangular or terete. Leaves absent. Bracts very much shorter than inflorescence. Umbel simple; rays 50–100, 2.5–10mm, slender; spikelets to 17.5×2mm, 6–12-flowered, compressed, brown; glumes ovate, 3-nerved, slightly bristle-tipped; stamens 3; style slender, as long as fruit, stigmas 3. S Africa. Z9.

*C.sumula* hort. See *C.cyperoides*.

**C.umbilensis** C.B. Clarke ex Will. Wats.
Robust perennial 60–90cm. Stems 3-angled, rough above. Leaves 6–8mm wide, two-thirds stem length, midrib often rough below. Bracts 4–8, 30–60cm. Umbel large, compound; spikes 6–25mm; spikelets 3–4-flowered, as long as broad, falling whole from plant; glume around 2mm; stamens 3; style 2-branched. S Africa. Z9.

*C.vegetus* Willd. See *C.eragrostis*

**Dactylis** L. (From Greek *daktylos*, a finger, referring to the stiff branches of the panicle.) Gramineae. 1 species, a perennial grass to 1m in flower. Culms tufted. Leaves to 45 × 1.5cm, linear, flat or inrolled, green to grey-green; ligules papery, to 12mm; leaf sheaths keeled, compressed. Inflorescence an open, 1-sided panicle, spicate, to 20cm, green-purple to yellow; spikelets short-stipitate, laterally compressed, in dense fascicles at branch apices, to 1cm; flowers to 5 per spikelet; glumes unequal, persistent, keels ciliate; lemmas entire or 2-toothed, to 7mm, keels ciliate, mucronate, 5-ribbed; palea equalling lemma, awns absent or to 2mm. Summer–autumn. Europe, N Africa, temperate Asia. Z5.

CULTIVATION Scarcely ornamental, it is best suited to its common use as a fodder and pasture grass; only the variegated forms can be considered decorative, used as ground cover or as specimens on the larger rockery. Grow in light, well drained soil in sun or part shade; plants fail to thrive on heavy and poorly drained soils. Propagate by division.

**D.glomerata** L. COCKSFOOT; ORCHARD GRASS.
'Aurea': a yellow-leaved form probably now lost to horticulture. 'Elegantissima': dwarf, hummock-forming, to 15cm, graceful; leaves striped white; a useful bedding plant. 'Variegata': dense, to 25cm; leaves striped green and white, somewhat coarser than 'Elegantissima'.

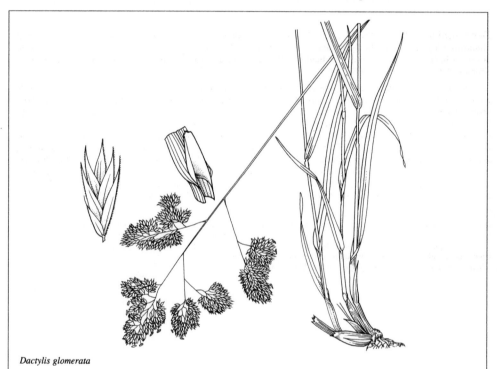

*Dactylis glomerata*

**Danthonia** DC. Gramineae. 10 species of perennial, rarely annual tufted grasses. Inflorescence a crowded, compact, solitary panicle; spikelets dimorphic: normal spikelets several-flowered, cleistogamous spikelets 1–2-flowered; glumes equal, broad, papery, acute; lemma boat-shaped, bifid at apex, lobes acute with a geniculate awn between them. Subtropics &Tropics. Z9.

CULTIVATION   As for the closely related *Chionochloa*.

*D.cunninghamia* Hook. f. See *Chionochloa conspicua*.

**D.purpurea** (Thunb.) P. Beauv. ex Roem. & Schult.
Culms tufted, forming tussocks, to 22cm. Leaves to 3cm, wide, curved, pubescent. Panicle with few to several spikelets on slender pedicels; spikelets to 0.5cm; glumes purple, florets to 3mm; lemma pilose in lower half, sparsely tomentose; awn to 3mm. S Africa.

**D.semiannularis** (Labill.) R. Br.
Culms to 1m, clustered in tussock. Leaves narrow, often inrolled. Outer glumes acute, to 2.5cm; lemmas to 8, pilose in lower part; apical lobes acute or mucronate, awn to 2.5cm. Australia, Tasmania.

**D.setacea** R. Br. WALLABY GRASS.
As for *D.semiannularis*, except to 30cm. Leaves setaceous. Panicles always crowded; glumes surrounding spikelet; lemmas with narrow lobes, long-mucronate. Australia.

**Dendrocalamus** Nees. (From Greek *dendron*, tree, and *kalamos*, reed.) Gramineae. 30 species of giant, clump-forming bamboos, chiefly separated from *Bambusa* by floral characters and size: *Bambusa* species are never so huge, and mostly have the blade of the culm sheath broader. SE Asia. Cultivated throughout the Tropics. Z10.

CULTIVATION   *D.giganteus* has become a popular interior landscape plant for very large glasshouses and the atria of public buildings, although care should be taken to accommodate both its height and fast-spreading, powerful rhizomes. Other species sometimes cultivated include *D.asper* (Schult.) Backer & Heyne with downy-bristly culms and *D.strictus* (Roxb.) Nees with downy leaves. These giant bamboos require a minimum temperature of 10°C/50°F, high humidity, and bright filtered sunlight. Pot in large tubs or beds where the spreading rhizomes can be carefully controlled. They are gross feeders and succeed best in a medium that is rich in leafmould and well rotted farmyard manure. Syringe foliage regularly during hot dry weather to prevent leaf-curling and red spider mite. Propagate by division or cuttings of young culms set sideways in sphagnum moss in a heated case. Seed will germinate readily in a heated greenhouse, but young plants may remain in their juvenile state – one of weedy, low-growing culms and small leaves – for some years: this may be broken by cutting the juvenile culms to ground level.

*D.giganteus* Munro (*Sinocalamus giganteus* (Munro) Keng f.). KYO-CHIKU.
Culms 15–40m × 15–35cm, usually branchless below, bristly below the sheath scar; sheaths very large, caducous, rough, ciliate, with auricles and bristles, ligule ciliate, blade small.

Leaves 20–55×3–12cm, denticulate, glabrous above, glabrescent beneath, not tessellate; sheaths glabrescent, with auricles, ligule long. S Tropical Asia, China, widely cultivated elsewhere.

**Deschampsia** P.Beauv. HAIR GRASS. (For Louis Deschamps (1765–1842), French naturalist.) Gramineae. Some 50 species of tufted, perennial or occasionally annual grasses. Culms slender, erect, unbranched. Leaves threadlike to linear, flat, folded or rolled, scabrous; ligules membranous, firm. Inflorescence paniculate, loose to compact, axis pubescent; spikelets laterally compressed, 2-flowered, lower floret sessile, light green, occasionally tinged purple; glumes equal, keeled, glossy, persistent, as long as floret, 1–3-veined; lemmas shorter than glumes, membranous to tough, flexible, apex 2–4-dentate; awns straight to slightly bent. Temperate zones. Z4.

CULTIVATION  A genus of distinctive tussock-forming grasses valued for the light, elegant and glistening panicles of flower, *Deschampsia* shows good cold-tolerance at temperatures of –25°C/–13°F and below. Suitable for the flower border and especially for the woodland garden in approximation of habitat, they tolerate a range of soil and light conditions but perform best in moist, fertile soils in semi-shade; *D.flexuosa* prefers a humus-rich acidic soil; it will, however, tolerate dry shade. Propagate by seed in autumn or by division in spring.

*D.caespitosa* auctt. See *D.cespitosa.*

**D.cespitosa** (L.) P. Beauv. (*Aira cespitosa* L.) HASSOCKS; TUFTED HAIR GRASS; TUSSOCK GRASS.
Perennial to 1.5m. Culms densely tufted, slender to robust, erect, smooth. Leaves to 60 × 1cm, rigid, ridged above, margins and midrib rough; ligules acute, to 1.5cm. Panicle loose, open, to 45 × 20cm; spikelets to 0.6cm, green suffused silver, purple or variegated; spikelet axis prolonged above upper flower as a bearded appendage; lemma smooth, apex toothed; awns to 0.4cm, straight. Summer. Temperate Eurasia, tropical Africa, Asia (mts). 'Bronzeschleier' ('Bronze Veil'): panicles large, tinted bronze. 'Goldhänge' ('Gold Shower'): panicles golden yellow, late-flowering. 'Goldschleier' ('Gold Veil'): spikelets bright yellow tinted silver. 'Goldstaub' ('Gold Dust'): panicles yellow. 'Goldtau' ('Golden Dew'): compact; flower heads small, becoming a rich golden yellow. 'Schottland' ('Scotland'): tufted mound, to 40cm high; leaves dark green. 'Tardiflora': small; panicles tufted, silver, late-flowering. 'Tautraeger' ('Dew Carrier'): small and slender; inflorescence blue-tinted. var. **parviflora** (Thuin) Coss. & Germain. Leaf blades narrow; spikelets smaller. var. *vivipara* ('Fairy's Joke'): panicles with slender, wiry branches weighed down with plantlets instead of seed.

**D.flexuosa** (L.) Trin. (*Aira flexuosa* L.) CRINKLED HAIR GRASS; WAVY HAIR GRASS; COMMON HAIR GRASS.
To 90cm. Culms smooth, slender, wiry. Leaves to 20cm, glabrous, wavy, terete, threadlike; ligules blunt, to 0.3cm. Inflorescence open, loose, to 12 × 7.5cm, shiny; spikelets to 6mm, tinged purple, brown or silver; lemma scabrous; awns to 0.8cm, twisted, bent. Summer. Europe, Asia, NE & S America. 'Mückenschwarm' ('Fly Swarm'): spikelets small, dark and very abundant. 'Aurea' ('Tatra Gold'): tuft-forming, to 50cm; leaves needle-like, remaining bright yellow-green all season long, arching; flowerheads soft bronze; usually comes true from seed.

**Deschampsia**   (a) *D.cespitosa* (b) *D.flexuosa*

**Dichanthium** Willem. Gramineae. Some 20–30 species of annual or perennial herbs. Leaves flat or inrolled; ligules membranous. Inflorescence of stalked, solitary or somewhat compound racemes; spikelets sessile or on pedicels, sessile spikelet bisexual, dorsally compressed; lower glume boat-shaped; lower florets reduced to a lemma; upper lemma entire to minutely emarginate, awned. Tropics & Subtropics. Z9.

CULTIVATION  As for *Cymbopogon*.

*D.annulatum* (Forssk.) Stapf (*Andropogon annulatus* Forssk.). DIAZ BLUE STEM; RINGED BEARD GRASS; BRAHMAM GRASS; KLEBERG GRASS.
Perennial to 1m, culms twisting, ascending. Leaves to 30× 0.7cm. Inflorescence compound, consisting of 1–15 racemes, each to 7cm; stalks glabrous; spikelets subimbricate, sessile spikelet narrowly oblong, to 6mm; lower glume pubescent to villous; awn to 2.5cm. India, China, introduced subtropics. var. *papillosum* (A. Rich.) De Wet & Harlan. Lower glume of sessile spikelet pilose to villous, with a distinct fringe of bulbous-based hairs below apex.

*D.aristatum* (Poir.) C. Hubb. (*Andropogon nodosus* (Willem.) Nash). ANGLETON GRASS; ANGLETON BLUESTEM.
Perennial to 1m. Leaves to 25×0.5cm. Inflorescence of somewhat compound racemes, each to 8cm; stalks pubescent; spikelets imbricate, sessile spikelets elliptic to obovate, to 5mm; lower glume with a distinct midrib, pilose below middle, glabrous or slightly ciliate toward apex; awn to 2cm. India to SE Asia.

*D.caricosum* (L.) A. Camus (*Andropogon caricosus* L.).
As for *D.annulatum*, sheaths flattened; ligule membranous, ciliate. Lower glumes of bisexual spikelets obovate or oblong-truncate, midrib absent. India.

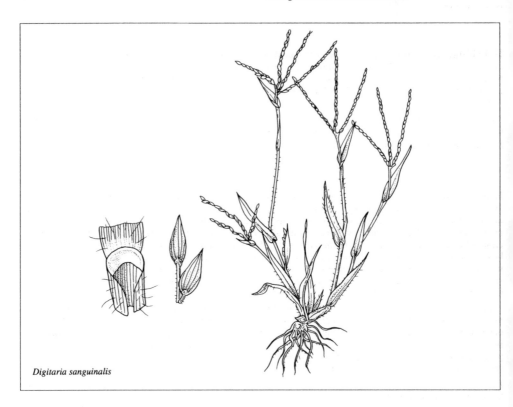

*Digitaria sanguinalis*

**Digitaria** Haller. (From Latin *digitus*, finger, referring to the shape of the inflorescence.) FINGER GRASS; CRAB GRASS. Gramineae. Some 230 species of annual or perennial grasses. Leaves linear to lanceolate; ligules papery. Inflorescence borne at stem apex or along main axis, secund, slender, consisting of digitately arranged spicate racemes; spikelets overlapping to diffuse, lanceolate or elliptic, to 3-fascicled, rarely solitary, rounded on one side, nearly flat on the other, subsessile or short-stipitate, alternate, borne in 2 rows on one side of a triangular, occasionally winged rachis; upper floret fertile; lower glume tiny or absent, sterile lemmas equal to or longer than upper glume; fertile lemma tough, membranous, margin thin, pale, awnless. Tropical, warm temperate regions. Z7.

CULTIVATION  A native of sandy soils, often on cultivated ground, *Digitaria sanguinalis* is grown for its distinctive slender-fingered ornamental inflorescence. Sow seed *in situ* in spring in a warm sunny position on well drained light fertile soil.

*D.sanguinalis* (L.) Scop. (*Panicum sanguinale* L.). HAIRY FINGER GRASS; CRAB GRASS.
Annual to 30cm, often tinged purple. Culms slender, ascending from a decumbent base. Leaves narrow-lanceolate, to 10 × 1cm, pubescent. Racemes more or less terminal , to 15cm, slender, spreading, to 10 per inflorescence; spikelets paired, to 0.3cm, elliptic; lower glume minute. Autumn.

**Drepanostachyum** Keng f. (From Greek *drepanon*, sickle, and *stachys*, spike, in token of the sickle-shaped inflorescence.) Gramineae. Clumping bamboos, several of the species formerly included here now combined in *Himalayacalamus*. Culms tall, hollow, terete, glabrous; sheaths thin, glabrous or tomentose, lacking auricles or bristles, blades very small; branches bushy, produced at the nodal line. Leaves almost hairless, scaberulous on one margin, tessellation obscure; sheaths glabrous, ligule long. Himalaya. Z8. Differs from *Himalayacalamus* in having 2 or more florets per spikelet.

CULTIVATION   A slow-growing, tall bamboo forming dense clumps of (ultimately) glossy yellow canes. It is intolerant of winter wet and of temperatures below –6°C/21°F and is best grown in large tubs of friable, sandy loam with added leafmould, kept evenly moist in a cool conservatory.

*D.falcatum* (Nees) Keng f. (*Arundinaria falcata* Nees; *Chimonobambusa falcata* (Nees) Nak.; *Sinarundinaria falcata* (Nees) Chao & Renvoize; *Thamnocalamus falcatus* (Nees) Camus).
Culms 2–6m × 1.5–2.5cm, slender, often white-powdery at first; sheaths fairly persistent, long-tapering, ciliate, glabrescent, ligule long, blade narrow. Leaves 11–35 × 1–2.5cm. E Himalaya. *D.khasianum* differs in having culms of a rich, polished, coppery red.

*D.falconeri* (Hook. ex Munro) J.J.N. Campbell ex D. McClintock. See *Himalayacalamus falconeri*.
*D.hookerianum* (Munro) Keng f.   See *Himalayacalamus hookerianus*.
*D.hookerianum* auctt.   See *Himalayacalamus falconeri* 'Damarapa'.

**Dulichium** Pers. Cyperaceae. 1 species, a perennial herb 30–100cm tall. Stems hollow, terete, jointed. Leaves numerous, with short blades, lowermost bladeless, uppermost with sheaths overlapping or reduced to sheaths; blades 5–10cm. Inflorescence terminal, short-stalked; spikes 1–3cm, appearing as alternating branches, forming a loosely oblong, spiky head; spikelets distichous, 4–10-flowered, linear, 1–2.5cm, scales narrow, acuminate, 5–7mm. E US. Z6.

CULTIVATION   A tall perennial, reed-like sedge becoming widely available in North America, where it is used in damp gardens, on water margins and in wetland restoration.

*D.arundinaceum* (L.) Britt. THREE-WAY SEDGE.

**Echinochloa** P.Beauv. (From Greek *echinos*, a hedgehog, and *chloe*, grass, due to the prickly overall appearance of the spikelets of some members of the genus.) Gramineae. Some 30 species of annual or perennial, tufted or stoloniferous grasses. Culms stout, sometimes succulent. Leaves linear to narrow-lanceolate, flat, glabrous, midrib prominent, ligules ciliate or absent. Inflorescence racemose or paniculate, spicate, compact or loose, arranged on a central axis; spikelets usually in 4 rows, short-stipitate, often hispid, cuspidate or awned, rounded beneath, flat above, 2-flowered, lower flower sterile, upper flower hermaphrodite; glumes acute to acuminate, membranous, often stiffly hairy, lower glume to half length of spikelet; upper palea reflexed, upper lemma hard, smooth, shiny, enclosing seed, lower lemma membranous, often stiffly hairy, often awned, as long as glumes. Cosmopolitan, temperate and tropics.

CULTIVATION   *Echinochloa* species occur in a wide range of damp and ruderal habitats, and are valued as ornamentals for their dense racemes. These are well suited to drying, gathered when the flower heads are fully developed. Sow seed in late winter/early spring under glass and plant out after danger of frost is passed. *E.crus-galli* is grown for its long-awned inflorescences, often purple-tinted. Although usually an annual and treated as such, plants may be potted at the end of summer and overwintered under glass. *E.polystachya* is the less ornamental species, and requires glasshouse protection if it is to be grown as a perennial in cool temperate zones.

*E.crus-galli* (L.) P.Beauv. (*Panicum crus-galli* L.). COCKSPUR; BARNYARD GRASS; BARNYARD MILLET.
Annual, to 120cm. Culms clumped, erect to spreading or decumbent. Leaves elongate, to 30 × 2cm, glabrous; sheaths glabrous; ligules absent. Inflorescence erect or pendent, lanceolate to ovate, to 25 × 8cm; racemes spreading, tinged purple, to 10cm; spikelets to 0.8cm, awn to 5cm or absent. Summer–autumn. Distribution as for the genus. Z6.

*E.polystachya* (HBK) A. Hitchc. (*Panicum polystachyum* HBK).
Creeping perennial to 180cm. Culms coarse, pubescent, hairs adpressed. Leaves to 2.5cm diam., sheaths smooth; ligules stiff yellow-pubescent. Inflorescence a slender, dense panicle, pendent, to 30cm; lower racemes to 7cm; spikelets to 0.5cm, awn to 2cm. Summer. Americas. Z8.

**Ehrharta** Thunb. (For Jakob Friedrich Ehrhart (1742–1795), pupil of Linnaeus and government botanist at Herrenhausen, 1780–1795.) Gramineae. Some 25 species of annual and perennial grasses. Culms erect to spreading. Leaves flat, linear to rolled; ligules membranous, occasionally ciliate. Inflorescence usually paniculate, narrow; spikelets laterally flattened, 3-flowered, the lower flowers sterile lemmas; lodicules 2; glumes ovate, obscurely keeled, minute to conspicuous, persistent; spikelet axis fracturing at nodes; fertile flower with sterile florets reduced to 2 lemmas, coriaceous, abscising together; lemmas of sterile flowers membranous, unequal, often wrinkled. Africa, Mascarene Is., New Zealand.

CULTIVATION   *E.erecta* is reliably perennial only in regions where temperatures seldom fall much below –5°C/23°F but is easily grown from seed in spring in any moderately fertile soil in sun and treated as a half-hardy annual.

**E.erecta** Lam. VELD GRASS
Perennial to 90cm. Culms erect, tufted. Leaves to 15cm × 0.9cm; ligules membranous. Inflorescence to 20cm, pale green; spikelets to 4mm, awns absent; second lemma rugose. Autumn. S Africa. Z8.

**Eleocharis** R. Br. (From Greek *helodes*, marshy, and *charis*, grace or loveliness.) SPIKE RUSH. Cyperaceae. Around 150 species of annual or perennial grass-like herbs. Stems 3–4-angled, compressed or grooved. Leaves reduced to bladeless sheaths or lower leaves with very short blades. Inflorescence a single, terminal spikelet, lacking large bracts; spikelets few- to many-flowered; flowers very small, hermaphrodite, each subtended by a scale-like bract (glume), spirally arranged; sepals and petals represented by 1–12 barbed bristles or absent; stamens 2–3; stigmas 2–3. More or less cosmopolitan.

CULTIVATION   Cultivated as for *Carex* in bog gardens and at pond margins: small species may be cultivated in indoor aquaria as submersed aquatics. Most prefer a slightly acid medium; *E.parvula* grows in saline conditions. Propagate *E.vivipara* by rooting the upper stem portions, *E.dulcis* by tuber. *E.dulcis* is grown as a crop given the same conditions as rice; the tubers are the Chinese waterchestnut, cultivated in flooded fields and harvested to provide the crunchy white staple of much Chinese cuisine.

**E.acicularis** (L.) Roem. and Schult. NEEDLE SPIKE RUSH; SLENDER SPIKE RUSH; HAIR GRASS.
Perennial, 5–30cm. Stems and stolons very thin, mat-forming. Stems tufted, slender. Leaf sheaths blunt. Spikelets compressed, pointed, 3–10-flowered, 3–6 × 1mm; glumes oblong, blunt, becoming pointed higher up the spikelet, pale green with brown bands on either side of midrib, soon falling; bristles 3–4, shorter than fruit, fragile; stamens 3; stigmas 3. Fruit pale, 3-angled. Summer. Widespread N America, Europe, Asia. Z7.

**E.dulcis** (Burm.) Trin. ex Hens. CHINESE WATER CHESTNUT.
Perennial, 40–120cm. Rootstock red-brown, elongating, stolons bearing edible white-fleshed tubers. Stems round, 3–5mm thick, jointed, nodes 2–5cm long, numerous false septa also present. Leaf sheaths red-brown, pointed, membranous, soon decaying. Spikelets 50-flowered, cylindrical, pointed, 15–60 × 3–5mm; glumes 4–5mm, straw-yellow to grey, pointed or blunt; bristles light brown, 6–8, exceeding fruit stamens 3; stigmas 2–3. Fruit around 2mm. Widespread in Asia. Z9.

**E.palustris** (L.) Roem. and Schult. (*Scirpus palustris* L.). CREEPING SPIKE RUSH.
Perennial, 30–160cm. Roots horizontal. Stems stout but flexible, round or compressed, grooved. Lower leaf sheaths brown, sometimes bearing a short blade, upper sheaths never bearing a blade. Spikelets 6–25 × 3–4mm, cylindrical, many-flowered; glumes ovate-lanceolate, pale green to brown with a green midrib; bristles 4, slender, barbed, exceeding fruit; stamens 2–3; stigmas 2–3. Fruit yellow, smooth. Late summer. N America, Europe, Asia. Very variable, often split into a number of subspecies.

**E.parvula** (Roem. & Schult.) Link ex Bluff, Nees & Schauer (*E.pygmaea* Torr. *Scirpus parvulus* Roem. & Schult.).
Tuft-forming perennial, to 8cm. Stolons very thin, producing white tubers. Stems thin. Leaf sheaths very thin, pale brown. Spikelets 2–5mm, 3–8-flowered, green; lowest glume large, blunt, nearly as long as spikelet; bristles exceeding fruits; stamens 3; stigmas 3. Fruit 3-angled, 1mm, pale yellow, smooth. Europe, SE Russia. Z8.

**E.pygmaea** Torr. See *E.parvula.*

**E.vivipara** Link.
Perennial, 10–30cm. Rootstock stout, vertical, deep brown. Stems slender, 0.5mm wide, light green. Leaf sheaths yellow, purple at base, tip pointed, often tinged purple. Spikelet 3–8mm, cylindrical, pointed, many-flowered; glumes tightly packed, blunt, 2mm, dark brown, margin white; bristles barbed, red-brown, 1mm; stigmas 3. Fruit 3-angled, 1mm, dark grey; fruit rarely seen – offsets usually produced within inflorescence. Florida, SE US. Z9.

**Eleusine** Gaertn. (From Eleusis, site of the temple of Ceres, goddess of the harvest – this genus includes several cereal crops.) Gramineae. Some 9 species of tufted annual and perennial stoloniferous grasses. Leaves linear, flat or folded; sheaths compressed; ligule membranous, ciliate. Inflorescence terminal, composed of digitately arranged spikes; spikelets 2 to many, arranged in 2 closely overlapping rows, racemes terminating in a fertile spikelet; spikelets 1–10-flowered, sessile, laterally flattened, usually fracturing at nodes; glumes unequal, persistent, upper glume to 3-ribbed, without awn; lemmas narrow, acute or obtuse, membranous, smooth, conspicuously keeled, longer than glumes, 3–5-ribbed, without awn. E, NE Africa, S America. Z9.

CULTIVATION  Occurring in savanna and upland grassland, *Eleusine* is grown for its unusual digitate panicles which may be cut and dried before the inflorescence is fully developed. Sow seed *in situ* in spring or earlier under glass, and grow in any ordinary soil in full sun.

*E.barcinonensis* Costa. See *E.tristachya*.

*E.coracana* (L.) Gaertn. FINGER MILLET; CORACAN; KURAKKAN; RAGI; AFRICAN MILLET.
Densely tufted annual to 1.5m. Leaves flat or folded, tough, to 50×1.5cm, rounded to tip. Inflorescence 1–11-spicate, to 8 × 1cm; spikes to 1.6×1.5cm, curving inwards with maturity; spikelets 6–9-flowered, to 9mm; lemmas broad-ovate. Old World Tropics.

*E.indica* (L.) Gaertn. YARD GRASS; WIRE GRASS; GOOSE GRASS.
Annual to 1.5m. Leaves to 30cm × 0.8cm, glabrous; ligules very short. Inflorescence 2–5-spicate; spikes to 17 × 0.5cm. Summer. Old World Tropics, a naturalized weed elsewhere.

*E.oligostachya* Link. See *E.tristachya*.

*E.tristachya* (Lam.) Lem.
Similar to *E.indica* but smaller with narrower leaf blades. Spikes usually 3, to 3cm. Old World Tropics.

**Elymus** L. (From Greek *elymos*, classical name for millet.) WILD RYE; LYME GRASS. Gramineae. Some 150 species of tufted or rhizomatous perennial grasses. Culms slender to stout, usually erect. Leaves linear, flat or rarely rolled; ligules membranous. Inflorescence linear, stout to slender, spicate, bristled; spikelets sessile, laterally flattened, to 9-flowered, situated at rachilla nodes, alternate, fracturing at nodes above glumes; glumes asymmetric, lanceolate to subulate, equal, membranous to rigid, acute or awned, to 9-veined, apex keeled; lemmas rounded below, acute, obtuse or awned, coriaceous, persistent, veins to 5. N temperate Asia. *Elymus* is contiguous with *Leymus*, and splitting of the two genera is often contentious.

CULTIVATION  From diverse habitats including steppes, woodland and dunes, *Elymus* species are valued for the often bright blue-greens of the foliage clumps, topped by flowering spikes in shades of buff and parchment. *E.magellanicus* is one of the most intensely blue of all grasses. Most are tolerant of temperatures to at least –10°C/14°F, with *E.canadensis* and *E.sibiricus* hardy where temperatures fall below –25°C/–13°F. Treatment in cultivation varies slightly according to habitat but most thrive in any moderately fertile, moisture-retentive soil in sun. Propagate by division.

*E.angustus* Trin. See *Leymus angustus*.
*E.aralensis* Reg. See *Leymus multicaulis*.
*E.arenarius* L. See *Leymus arenarius*.

*E.canadensis* L. (*E.glaucifolius* Muhlenb.). CANADA WILD RYE.
Perennial. Culms clumped, to 180cm. Leaves to 45×2cm, flat or rolled, scabrous or slightly hispid, mid-green to glaucous grey-blue, apex long-acute. Spikes dense, usually nodding, bristled, red-tinted, to 25cm; spikelets in groups to 4, 2–5-flowered, to 16mm; glumes long, linear, stiff, scabrous; awn to 16mm; lemmas lanceolate, to 16mm, glabrous or hairy, rough, awn curved, to 15mm. Summer–autumn. N America. Z3. A variable species with several colour forms and small-leaved, low-growing variants.

*E.chinensis* (Trin.) Keng. See *Leymus chinensis*.
*E.condensatus* Presl & Presl. See *Leymus condensatus*.

*E.elongatus* (Host) Runem. (*Agropyron elongatum* (Host) P. Beauv.)
Culms to 90cm. Leaves scabrous above, flat at first, becoming involute, rigid. Spikelets borne in 2 rows, with 4–8 flowers; glumes ovate-oblong, obtuse, 5–7-nerved; lemma truncate or apically notched. Eurasia. Z5.

*E.farctus* (Viv.) Runem. ex Melderis.
Rhizomatous perennial. Culms 30–60cm, rigid, glabrous. Leaves 2–5mm diam., convolute or flat with inrolled margins, rigid, glaucous-green, pubescent on ribs of upper surface, glabrous beneath. Spike to 25cm; glumes to 1.8cm, narrow-lanceolate or oblong, 5–12-veined, keeled, glabrous, obtuse; lemma to 1.8cm, glabrous, obtuse; anthers to 1cm. Turkey, Mediterranean. Z5.

*E.giganteus* Vahl. See *Leymus racemosus*.
*E.glaucifolius* Muhlenb. See *E.canadensis*.
*E.glaucus* Buckley. See *Leymus secalinus*.
*E.glaucus* hort. non Buckley. See *Leymus racemosus*.

*E.hispidus* (Opiz) Melderis (*Agropyron glaucum* hort; *Agropyron intermedium* (hort.) P. Beauv.). INTERMEDIATE WHEATGRASS.
Culms to 1.2m. Leaves held upright, margin hardened, scabrous, sparsely hispid above, glabrous beneath, intense grey-blue. Spikes to 20cm, dense, fading to papery, pale straw-yellow; spikelets 3–8-flowered; glumes to 8 × 3mm, about 5-nerved, lemma to 11mm, often ending in a short point. Eurasia. Z5. a variable plant, often grown in its smaller, low-lying forms.

*E.junceus* Fisch. See *E.farctus*.

*E.magellanicus* (Desv.) A. Löve (*Agropyron magellanicum* Desv.; *Agropyron pubiflorum* (Steud.) Parodi).
Laxly tufted, rhizomes present or not; culms to 130cm, often far shorter. Leaves to 35×0.7cm, erect to somewhat flattened and spreading, flat or folded, glabrous or puberulous, vivid blue-grey; ligule to 0.2cm, truncate; sheath glabrous or puberulous. Spikes to 20cm, dense; spikelets to 3cm, 2–7-flowered; glumes to 1.6cm, equal or the upper slightly longer. Often grown under the name *Agropyron glaucum*. *A.glaucum* of gardens is *Elymus hispidus*. *Elymus glaucus* of Buckley is the Blue Wildrye, referrable to *Leymus secalinus*. The name *E.glaucus* in gardens has also been applied to *Leymus racemosus*.

*E.mollis* Trin. See *Leymus mollis*.
*E.multicaulis* Karel. & Kir. See *Leymus multicaulis*.
*E.racemosus* Lam. See *Leymus racemosus*.

*E.repens* (L.) Gould. See *Agropyron repens*.

*E.sibiricus* L. (*Clinelymus sibiricus* (L.) Nevski).
Perennial, to 90cm. Culms slender, erect, clumped, nodes black. Leaves to 23 × 1.3cm, scabrous, thin, apex finely pointed. Spikes pendent, dense; spikelets in pairs, 3–7-flowered, to 2.5cm, green or green tinged purple; glumes linear to subulate-linear, to 5mm, awn to 5mm or more; lemmas oblong-lanceolate to lanceolate, to 1cm, very rough, tapering, awn to 2.5cm, curved. Summer. E Europe, temperate Asia. Z5.

*E.spicatus* (Pursh) Gould (*Agropyron spicatum* (Pursh) Scribn. & J.G. Sm.; *Agropyron inerme* (Scribn. & J.G. Sm.) Rydb.).
Culms to 90cm, clustered, erect. Leaf sheaths glabrous; leaves less than 2mm diam., pubescent above, flat to loosely involute. Spikes to 15cm, slender; spikelets mostly 6–8-flowered; glumes narrow; lemma about 1cm, with strongly divergent awn. US. Z5.

*E.trachycaulos* Link) Gould ex Shinn. (*Agropyron pauciflorum* (Schweinf.) A. Hitchc.; *Agropyron tenerum* Vasey). SLENDER WHEATGRASS.
Culms over 90cm, erect, clustered. Leaf sheaths usually glabrous, rarely pubescent; leaves to 2mm diam. Spikes to 25cm, slender, erect to nodding; spikelets slightly overlapping, few-flowered; glumes broad, awnless; lemma glabrous, unawned. US. Z5.

*E.virginicus* L. VIRGINIA WILD-RYE.
Culms erect, tufted to 1.2cm. Leaf sheaths glabrous; leaves to 15cm, flat, scabrous, dark green becoming brown-tinted if grown in full sun on lean soils. Spike to 15cm, compact, erect to drooping; glumes to 2mm diam., strongly veined, firm, tinged yellow, veins absent at base; lemmas glabrous and unveined beneath, scabrous and veined above. N America (Newfoundland to Alberta, S to Florida and Arizona). Z3.

**Eragrostis** Wolf. (From Greek *eros*, love, and, *agrostis*, grass, a reference to the plant's beauty.) LOVE GRASS. Gramineae. Some 250 species of annual or perennial grasses. Leaves narrow, flat or rolled, glabrous, scabrous to sparsely pubescent; sheaths often glandular; ligules ciliate, occasionally papery. Inflorescence paniculate, similar to those of *Poa* or *Briza*, open or dense, sparingly to profusely branched; spikelets closely overlapping, 2- to many-flowered, laterally compressed, stipitate, awned, main axis of spikelets breaking at nodes between lemmas, or persistent, exceeding glumes; glumes subequal, abscising, convex, keeled; palea persistent; lemma entire, contiguous or overlapping, acute to acuminate, longer than glume, usually glabrous but pubescent in some Old World species, to 3-ribbed. Tropics, subtropics, N & S America, S Africa, Australia.

CULTIVATION   *Eragrostis* species are ubiquitous in the tropics and subtropics, especially on poor soils. *E.tef* is cultivated for its edible seeds in Ethiopia, *E.chloromelas* is a drought-tolerant soil-binder; *E.curvula* is cultivated for pasture, hay, and its soil-binding properties. Their chief ornamental value resides in the subtle leaden or bronzed colours of the showy branched inflorescence: *E.tef* commonly shows a range of colours; *E.spectabilis* is tinted purple-red. The panicles of *E.trichodes* are so tall that they flop over surrounding plants toward the end of the season, a characteristic particularly evident in the aptly named cultivar 'Bend'. All are attractive late summer companion plants, used where their inflorescences may overtop darker, stronger forms, and are excellent for naturalizing in meadow gardens. Most species will tolerate frosts, to between –5 and –10°C/23–14°F and are easily grown in any well drained soil in sun; *E.curvula* needs a hot, dry position and is slightly less cold-tolerant. Propagate annuals by seed sown *in situ* in spring or earlier under glass, perennials by division.

*E.abyssinica* (Jacq.) Link. See *E.tef*.
*E.amabilis* (L.) Wight & Arn. See *E.tenella*.

*E.aspera* (Jacq.) Nees.
Annual, tufted. Culms erect, to 1m. Leaves to 30cm, flat, tapering finely. Panicles to 40cm, very loose and open; spikelets narrow, pale or purple-tinted, long-stalked, 4–16-flowered. Late summer. Z8.

*E.capillaris* (L.) Nees. LACE GRASS.
Annual. Culms to 50cm, tufted, erect. Leaf sheaths pilose, bearded at ligule. Leaves to 3mm diam., erect, flat, pilose above. Panicle oblong or elliptic, open, very finely diffuse, with capillary branches; spikelets to 3mm, long-stalked, 3–5-flowered; glumes to 1mm, acute; lemmas about 1.5mm, acute, obscurely nerved, rounded on the back, minutely scabious towards apex. Mid to late summer. US. Z7.

*E.chloromelas* Steud. BOER LOVE GRASS.
Perennial. Habit densely clumping. Culms erect, to 90cm. Leaves to 50cm, elongate, tinted blue, curling. Panicle to 20cm; branches very fine, purple-tinted, falling in one direction; spikelets dark olive. Late summer. S Africa. Introduced SW US. Z9.

*E.cilianensis* (All.) Lutati. STRONG SCENTED LOVE GRASS.
Annual. Culms to 50cm, loosely tufted, ascending or spreading. Leaf sheaths pilose at the throat; leaves to 7mm wide, tapering finely, margins glandular, flat. Panicle to 20cm, open, erect, dark leaden grey-green, with ascending branches; spikelets to 15 × 3mm oblong, compressed, with 10–40 flowers; lemmas about 2.5mm, ovate, subacute. US, Mexico, W Indies, south to Argentina. Weed-like, with an unpleasant odour when fresh. Z9.

*E.curvula* (Schräd.) Nees (*Poa curvula* Schräd.). WEEPING LOVE GRASS; AFRICAN LOVE GRASS.
Perennial, to 120cm. Culms flimsy to stout, densely clumped. Leaves to 30 × 0.3cm+, dark green, scabrous above; leaf sheaths pubescent at base. Panicle erect to nodding, to 30cm, pyramidal, very lax with weeping branches; spikelets short-stipitate, narrow-oblong to oblong, 3–18-flowered, a striking olive-grey, to 13 × 2mm; glume to 2mm; lemma lanceolate-oblong, 1–2mm, apex obtuse. Summer–autumn. S Africa. Z7.

*E.elegans* Nees. See *E.tenella*.
*E.interrupta* P.Beauv. See *E.tenella*.

*E.japonica* (Thunb.) Trin. (*E.tenuissima* Schräd.; *E.namaquensis* Nees). JAPANESE LOVE GRASS.
Annual to 60cm. Culms flimsy, erect, clumped. Leaves to 20 × 0.4cm, glabrous. Panicles open, loose or dense, ovoid-cylindric to ovoid-oblong, to 25cm, stiff, branches whorled, spreading; spikelets short-stipitate, scattered, linear, 4–10-flowered, tinged purple, to 4 × 1mm; glumes minute, to 1mm; lemma lanceolate-oblong, very small, abscising. Summer–autumn. Tropical Asia, Australia. Z8.

*E.mexicana* (Hornem.) Link (*Poa mexicana* Hornem.). MEXICAN LOVE GRASS.
Annual, loosely tufted, to 90cm. Culms erect or ascending, flimsy, branching or simple. Leaves to 23 × 0.6cm, flat, scabrous above. Panicle oblong to ovoid, to 38 × 20cm; spikelets stipitate, narrow-ovate to narrow-oblong, 4–12-flowered, tinged purple or grey, to 8mm; glumes lanceolate, to 2mm; lemma ovate to ovate-oblong, to 3mm. Summer–autumn. US, Mexico. Z8.

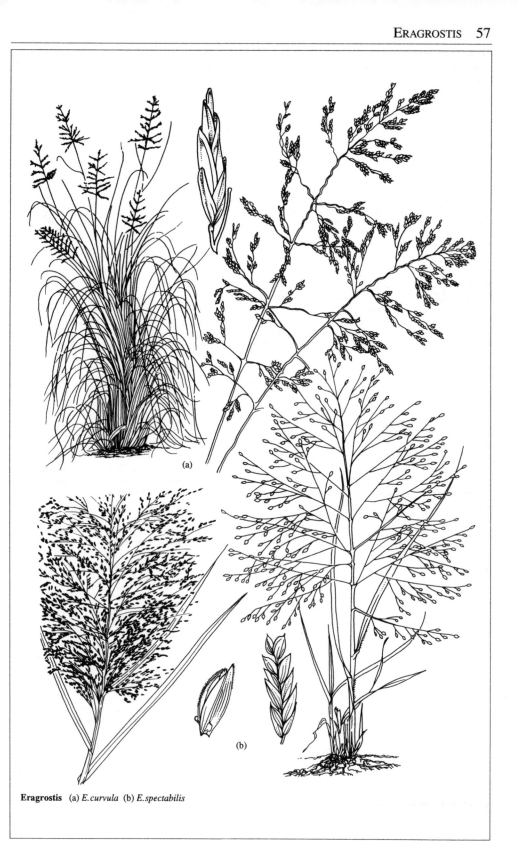

**Eragrostis** (a) *E. curvula* (b) *E. spectabilis*

*E.namaquensis* Nees. See *E.japonica.*

**E.obtusa** Munro ex Stapf (*Briza geniculata* Thunb.).
Perennial, to 45cm. Culms densely clumped. Leaves to 15 ×
0.3cm, usually involute, finely pointed. Panicle ovoid, open to
compressed, to 18 × 8cm; spikelets resemble those of *Briza
media,* slender-stalked, ovate, flattened, 8–20-flowered, to
1cm × 3mm; glume, palea and lemma navicular, keeled.
Autumn. S Africa. Z7.

*E.pellucida* Steud. See *E.pilosa.*

**E.pilosa** (L.) P.Beauv. (*Poa pilosa* L.; *E.pellucida* Steud.).
SOFT LOVE GRASS.
Annual, to 45cm. Culms very flimsy, spreading to erect.
Leaves to 20 × 0.3cm. Panicle ovoid to pyramidal, to 30 ×
10cm; spikelets to 14-flowered, stipitate, to 8 × 1mm, grey or
tinged purple; glumes unequal; lemma narrow-ovate to lanceolate-oblong, obtuse to acute. Summer–autumn. Warm temperate regions. Z7.

*E.plumosa* (Retz.) Link. See *E.tenella.*

**E.spectabilis** (Pursh) Steud. PURPLE LOVE GRASS; TUMBLE
GRASS.
Perennial, densely tufted. Culms to 60cm, erect to spreading.
Leaf sheaths glabrous or pilose, bearded at top; leaves 30 ×
0.8cm, flat or folded, rigid, ascending, finely tapering,
glabrous or occasionally pilose. Panicle to 40×30cm, lax, diffuse, much-branched; branches dark green to purple-red,
spreading with age; spikelets to 8mm, green, olive to leaden
grey, white pink or purple, oblong to linear, with 6–12 flowers; glumes over 1mm, acute; lemmas about 1.5mm, acute,
slightly scabrous at apex. US. Z6.

**E.superba** Peyr. in Sitzb.
Densely tufted. Culms erect or ascending, to 3m, rather stout,
smooth, 2-noded. Sheaths smooth, bearded at opening; ligule
a fringe of short hairs; leaf blades 5–25 × 0.3–1cm, firm,
rather rigid, spreading, usually involute or convolute, smooth
beneath, scaberulous above. Panicle 10–35cm, narrow, linear-
oblong, erect, branches erect, solitary, filiform, usually simply racemose, to 8cm; spikelets distant or clustered toward
tips of branches, strongly compressed, suborbicular to
oblong-ovate, to 6mm, straw-coloured or purple-tinted;
glumes subequal, lanceolate. S & Tropical Africa. Z8.

**E.tef** (Zucc.) Trotter (*E.abyssinica* (Jacq.) Link). TEFF.
Annual, to 1.5m. Culms solitary to tufted, upright to spreading.
Leaves glabrous, to 45×1cm, flat. Panicle pendent, loose, open
or contracted, to 60cm, branches flimsy; spikelets lanceolate,
long-stipitate, 3–18-flowered, variously green, white or violet,
to 15 × 3mm; lemma lanceolate-oblong, to 3mm, persistent.
Summer–autumn. NE Africa, Ethiopia, SW Arabia. Z9.

**E.tenella** (L.) P.Beauv. (*E.elegans* Nees; *Poa plumosa* Retz.;
*Eragrostis plumosa* (Retz.) Link; *E.amabilis* (L.) Wight &
Arn.; *E.interrupta* P.Beauv.).
Loose, tufted annual. Culms flimsy, erect or spreading, to
60cm. Leaves to 10 × 0.3cm, flat, glabrous, narrow-acuminate. Panicle ovoid to oblong, to 13cm; spikelets short-stipitate, oblong, 3–10-flowered, to 4×2mm, tinged green or purple; glumes to 1mm; lemma oblong, blunt, to 1.5mm.
Summer–autumn. Old World Tropics. Z9.

*E.tenuissima* Schräd. See *E.japonica.*

**E.trichodes** (Nutt.) Alph. Wood (*Poa trichodes* Nutt.). SAND
LOVE GRASS.
Tufted perennial to 120cm. Culms flimsy, erect, stiff. Leaves
to 60×0.6cm, flat, long-acuminate, glossy, dark green, arching; leaf sheaths stiffly pubescent near apex. Panicle loose,
open, oblong to elliptic, to 60×35cm; branches fine, numerous, erect then outspread; spikelets long-stipitate, lanceolate
to lanceolate-oblong, 3–9-flowered, green or purple, to 1cm;
glumes narrow lanceolate, to 3mm; lemma ovate-oblong to
narrow-ovate, to 4mm. Summer–autumn. US. 'Bend': inflorescence red-bronze; panicle arching to weeping with age. Z7.

**E.unioloides** (Retz.) Nees ex Steud.
To 45cm. Panicles loose or dense, lanceolate to ovoid, to 15×
6cm; spikelets ovoid to cylindric, compressed, many-flowered,
to 13×3mm, tinged green to purple; glumes to 2mm; lemma
ovate, overlapping, to 2mm. Autumn. Tropical Asia. Z9.

**Eriochloa** HBK. (From Greek *erion*, wool, and *chloe*, grass.) CUP GRASS. Gramineae. Some 30 species of
clump-forming annual or perennial grasses. Culms clumped. Leaf ligules a row of hairs. Inflorescence a racemose
panicle; spikelets 2-flowered, solitary to clustered, elliptic, pubescent, bearing a beadlike structure at the base; terminal flower hermaphrodite, fertile, lower flower sterile, occasionally male; glumes usually unequal, upper glume
as long as spikelet, usually awned, lower glume reduced; lemma coriaceous, mucronate. S US, Mexico.

CULTIVATION   Grown for the large distinctive inflorescence, with the characteristic small bead-like structure at
the base of each spikelet, especially useful for dried arrangements, *E.villosa* is easily grown from seed sown *in
situ* in spring in a fertile and moisture-retentive soil with a warm position in full sun.

**E.villosa** (Thunb.) Kunth. HAIRY CUP GRASS.
Annual, to 60cm. Culms loosely clumped. Panicle of 3–15
racemes, each to 2cm; spikelets elliptic, pubescent, to 6mm;
inflorescence axis and pedicels with long, conspicuous hairs;
lower floret sterile; upper lemma mucronate. Asia. Z9.

**Eriophorum** L. (From Greek *erion*, wool, and *phoreo*, to bear, alluding to the hairy fruits.) COTTON GRASS.
Cyperaceae. 20 species of perennial, grass-like herbs. Tufted or with slender, scaly and creeping rhizomes.
Stems tufted, leafy, 3-angled. Leaves grasslike, slender, flat, remaining green in winter, usually dead at flowering. Inflorescence subtended by leaflike bracts, an umbel of many-flowered spikelets or an erect, single, terminal spikelet; flowers hermaphrodite, minute, spirally arranged, each subtended by a glume; sepals and petals
represented by numerous soft, pale, cotton-like hairs, elongating in fruit to aid dispersal; stamens 3; style 3-
branched. Fruit a nut, 3-angled, remains of style persistent. Summer. Europe, N America, S Africa.

CULTIVATION   Attractive sedges hailing from bogs, moors and marshes where they form large colonies of rather
dull, grassy foliage brought alive in summer by massed flowerheads thick with silvery white, cotton-like hairs.
They grow easily in full sun on damp or saturated neutral to acid soils, especially in bog gardens and at pond
margins. They spread quickly and may require containment. Propagate by division.

**E.alpinum** L. (*Scirpus alpinus* (L.) dalla Torre).
Densely tufted, 15–40cm. Rhizome short. Stems triangular in
section, scabrid above. Leaf blades 1–3cm, basal sheaths awl-
tipped. Spikelets 8–12-flowered, 5–7×3mm, ellipsoid-lanceo-
late; glumes brown-yellow, oblong-lanceolate, blunt; bristles
4–6, white, to 2.5cm. Nut shiny, dark brown. Summer.
Northern N America, N & C Europe. Z7.

*E.angustifolium* Honck. COMMON COTTON GRASS.
Creeping far with long, scaly rhizomes. To 75cm. Stem subterete below, distinctly 3-angled above. Leaves grooved, to 30
× 0.7cm, tapering to a 3-sided tip; sheaths of stem leaves
inflated. Inflorescence an umbel; spikelets 3–7, drooping,
rays unequal; glumes brown-red, dark at tip, lanceolate-ovate,
around 7mm, margin broad, transparent, hairs showy, white,
cotton- or down-like, densely tufted, to 5cm when in fruit. N
Europe, N America, including Siberia and Arctic regions.
'Heidelicht': showy; glumes white. Z4.

*E.chamissonis* C.A. Mey.
10–80cm. Stem soft, subterete. Leaves slender, sulcate, flowering stem leaves bladeless, sheaths inflated. Inflorescence a
solitary spikelet, 1.5–2cm; glumes blunt, brown to dark grey,
margins broad, white; hairs red-tinged to tawny, to 5cm.
Western N America. var. *albidum* (Nyl.) Fern. Inflorescence
hairs white. Z7.

*E.latifolium* Hoppe (*E.paniculatum* Druce). BROAD-LEAVED
COTTON GRASS.
Similar to *E.angustifolium* except rhizome tufted, not spreading, stems 3-angled throughout, leaves broader, 3–8mm wide,
stem leaf sheaths close-fitting; inflorescence larger, showier,
with 2–12 pure white spikelets. Europe, Turkey, Siberia, N
America. Z8.

*E.paniculatum* Druce. See *E.latifolium*.

*E.×polystachion* L. (*E.angustifolium* × *E.latifolium*.)
Similar to *E.angustifolium* except to 30cm, stem sheaths
somewhat inflated, inflorescence hairs 3–4cm in fruit.
Europe, Siberia, N America. Z6.

*E.scheuchzeri* Hoppe.
Similar to *E.vaginatum* except to 40cm, stems terete, leaves
to 2mm wide, sheaths yellow-brown, hairs to 3cm, often
deflexed. N Europe to the Urals, C Europe (mts). Z6.

*E.vaginatum* L. COTTON GRASS; HARE'S TALE.
Tussock-forming, scarcely running, 30–80cm. Stem terete at
base, 3-angled above. Leaves shorter than stem, around 1mm
wide, bristly, sheaths fibrous, brown; flowering stem leaves
bladeless or with very short dark blades, sheaths inflated.
Inflorescence a solitary terminal spikelet, 2cm at flowering,
bracts absent; glumes around 7mm, lanceolate-ovate, 1-
nerved, silver below, dark above, those at base of spikelet
sterile; hairs around 2cm, white. N Temperate regions. Z7.

*E.viridicarinatum* (Engelm.) Fern.
20–90cm. Leaves flat, 2–6mm across, folded from the midpoint to the tip. Bracts ridged, 15–150 × 15–40mm. Spikelets
2–10, 10–20mm at anthesis, 25–45mm in fruit; glumes
4–10mm; hairs bright white. Summer. Northern N America,
Eurasia. Z7.

**Fargesia** Franch. (For Paul Guillame Farges (1844–1912), French missionary and naturalist in Central China).
Gramineae. Some 40 species of clumping bamboos. Culms to medium height, slender and graceful, rather thin-
walled, often with white powder below the nodes; culm sheaths persistent or falling late, leaving prominent scars;
primary branches 3 to many per node, clustered. Leaves comparatively small, tessellate. China, Himalaya. Z6.

CULTIVATION Beautiful bamboos native to damp, semi-wooded regions. Their characteristic growth habit
is a dense, erect thicket of canes adorned in their first season with pale, persisting sheaths, later bowed with
the weight of cascading branches. The leaves are delicate, weeping and produced in great profusion. The
two best known species are *F.nitida*, the Fountain Bamboo, and *F.murielae*, the Umbrella Bamboo, both
common – albeit distinguished – features of western gardens. Other, more recent introductions include
*F.dracocephala* with weeping branches of fine leaves in a frond-like arrangement and the tall, purple-
stemmed and sea-green-leaved *F.utilis*. They will tolerate temperatures to −25°C/−13°F (even lower for
*F.utilis*) and a wide range of soils and sites. Ideally, however, they should be planted in sheltered, light
shade, in a rich, well drained, turfy loam and never allowed to dry out. *F.nitida* will make a fine specimen
or may be informally group-planted to create dense, emerald green groves. All will succeed in tubs, provided they are never allowed to dry out. If grown in this way, they should be kept far from hot, dry terraces
where bright and arid conditions will cause the foliage to curl and wilt. In prolonged *cold* weather, similar
symptoms usually betoken physiological drought, best counteracted by light hosing of the leaves and regular
soaking of the containers.

*F.crassinoda* Yi. See *Thamnocalamus crassinodus*

*F.dracocephala* Yi.
Culms 3–5m × 0.4–0.8cm, rather thin and lax in appearance,
forming a neat, fountain-like clump with arching tops;
branches numerous, growing vertically and close to the culm
at first, later spreading to drooping. Leaves 5–17 × 0.5–2cm,
matt green, held horizontally in an outspread, frond-like
arrangement. China (Sichuan). Introduced to European horticulture under the name 'Daba-Shan 2', after its place of origin.

*F.maling* (Gamble) Simon ex D. McClintock. See *Yushania
maling.*

*F.murielae* (Gamble) Yi (*Arundinaria murielae* Gamble;
*Arundinaria spathacea* (Franch.). D. McClintock; *Fargesia
spathacea* Franch; *Sinarundinaria murielae* (Gamble) Nak.;
*Thamnocalamus spathaceus* (Franch.) Söderstr.) UMBRELLA
BAMBOO.
Close in general appearance to *F.nitida*, differing in its white-
powdery, soon yellow-green, later yellow culms, the pale
brown sheaths, sometimes ciliate, gradually standing away
from the culms. Culms usually branching in the first year.
Leaves 6–15 × 0.5–1.8cm, apple green with a very long,
drawn-out apex, slightly scaberulous on our margin; sheaths

often with bristles, stalks rarely tinted purple. C China.

*F.nitida* (Mitford) Keng f. & Yi (*Arundinaria nitida* Mitford;
*Sinarundinaria nitida* (Mitford) Nak.). FOUNTAIN BAMBOO.
Culms rather slender, 3–4m × 0.4–0.8cm, with greyish powder at
first, finally lined purple-brown, not branching in the first year,
branchlets usually tinted purple, finally cascading and capable of
weighing over the culms; sheaths persistent for a year, purple-
green, pubescent at first, lacking auricles or bristles; blades narrow. Leaves rather small, 4–11 × 0.5–1.1cm, dark green alternate,
apex finely tapered, not or hardly scaberulous, tending to flicker
in the breeze; sheaths purple, usually without bristles, stalks purple. C China. Z5. 'Eisenach': to 2m; culms very upright, black-
tinted with prominent nodes. 'Ems River': to 4m, hardy; culms
dark-tinted, appearing almost black. 'Maclureana': to 6m; culms
arching widely, pearly grey, pruinose. 'Nymphenburg': culms
arching to weeping, dull grey-green.

*F.spathacea* Franch. see *F.murielae*.

*F. utilis* Yi.
Fast-growing to 5m. Culms tall, arching, pale green becoming
deep purple; sheaths straw-coloured, pubescent; branches
numerous, purple. Leaves narrow, matt sea-green with
strongly glaucous undersides. China (Yunnan). Z6.

*Fargesia murielae*

*Fargesia nitida*

**Festuca** L. (Latin *festuca*, stalk or stem; in Pliny the wild oat growing among barley.) FESCUE. Gramineae. Some 300 species of perennial grasses to 2m, usually very much shorter, rhizomatous or tufted. Culms flimsy or rigid. Leaves flat, folded or involute, often becoming rolled in high humidity; ligules translucent, papery. Flowers in a dense or loose panicle; spikelets on pedicels, 2- to many-flowered, flattened laterally; upper floret largely vestigial; glumes shorter than lemmas, 1–3-ribbed, acute to short-awned at apex; lemmas lanceolate, convex, stiff, 5- to 7-nerved, acute, obtuse or awned at apex, rarely faintly emarginate; palea with 2 keels, papery, brittle. Summer. Cosmopolitan (mainly cold and temperate regions). Z5.

CULTIVATION The fescues are a large, useful and sometimes beautiful genus; *F.longifolia*, *F.rubra*, *F.rubra* var. *commutata*, *F.ovina* and *F.tenuifolia* are used in cool-season turf grass mixtures, other species are used as pasture grasses. Those described here are ornamental, valued for the attractive, often brilliant blue-greens of the low, hair-like foliage tussocks and for the fine soft and light inflorescences. With the possible exception of *F.punctoria*, most are tolerant of cold to at least $-15°C/5°F$, although some species such as *F.cinerea* do not tolerate heavy snow cover.

*Festuca* species occur in a diversity of habitats and suit a number of situations in the garden. *F.gigantea* is more tolerant than most grasses of shade and is well suited to the woodland and wild garden, *F.vivipara* is well placed at the waterside of lakes and streams. *F.eskia*, a calcifuge species of scree and rocky pasture, with soft brilliant green foliage and a dense carpeting habit, is useful on the rock garden. *F.glauca* and *F.g.* 'Caesia' are among the best blues, the latter especially brilliant; good colour can be maintained by clipping over in spring and summer, and by regular removal of dead leaves. Both give good textural and colour contrasts in the border, and are occasionally grown as lawns, cut with the lawn mower at high settings, although they do not withstand heavy wear and have a tendency to die out at the centre of the clump when overcrowded; lift, divide and replant every third of fourth year. Most species are easily grown in light, well drained soil in sun; *F.vivipara* prefers an acid soil rich in organic matter. Propagate by division or seed, although species hybridize readily and seed may not come true.

*F.alpina* Suter.
To 28cm. Leaves glabrous, to 0.5mm diam., veins 3–5, rib 1; leaf sheaths closed at the mouth, somewhat brown; ligules short-auriculate, glabrous. Panicle to 3.5cm; spikelets 6mm, 2–4-flowered, yellow-green, rarely somewhat violet-tinged; upper glume linear to oblong-lanceolate, to $3.5 \times 1$mm, long-acuminate; lemma oblong-lanceolate, to $4 \times 1.5$mm; awn to 4mm. Alps, Pyrenees.

*F.altissima* All. FEED FESCUE; WOOD FESCUE.
Densely tufted perennial. Culms to 1.2m, stout or slender. Leaves to 60cm, flat, tapering finely, glabrous, scabrous throughout or on margins only. Panicles to $20 \times 15$cm, lax, open-branched; spikelets 2–5-flowered, oblong-cuneate; lemmas finely pointed. Late spring to summer. Z7.

*F.amethystina* L. TUFTED FESCUE; LARGE BLUE FESCUE.
To 50cm. Stems flimsy, forming dense tussocks, flushed mauve. Leaves with very narrow, soft, involute, blue-grey blades to 25cm; sheaths rolled and furrowed below. Panicle narrow, flexuous, to 10cm; branches paired, hispidulous, tinged violet; spikelets 3–7-flowered, to 3cm; glumes not equal; lemma translucent in upper part, acute to shortly awned. C Europe. 'Aprilgrün' ('April Green'): leaves olive green. 'Bronzeglanz' ('Bronze-Glazed'): leaves tinted bronze. 'Klose': leaves olive. 'Superba': leaves blue; stems amethyst when flowering.

*F.arundinacea* Schreb. See *F.elatior*.
*F.arvensis* Augier, Kerguelen & Markgr. See *F.glauca*.
*F.capillata* Lam. See *F.tenuifolia*.

*F.cinerea* Vill.
To 35cm, densely tufted. Culms glabrous or scabridulous above. Leaves somewhat hard, to 1mm diam., blue-grey, glaucous to pruinose, usually scabrous, veins 7–9, ribs 3–5; sheaths closed below, glabrous. Panicle rather dense, to 6cm; branches glabrous or short-pubescent; spikelets to 7.5mm, 4–7-flowered, glaucous, pruinose, violet-tinged; upper glume lanceolate, to $4 \times 1.5$mm, acuminate, scabridulous at ciliate above; lemma lanceolate, to $5 \times 2$mm, acuminate, scabrous, dense-pubescent or ciliate above, awn to 1.5mm. S France, NW Italy. Commonly confused with *F.glauca*, or held to be synonymous with it. Many 'blue' cultivars are assigned to this species. Pending a more thorough investigation, these are listed under *F. glauca*.

*F.crinum-ursi* hort. non Rum. See *F.gautieri*.

*F.elatior* L. (*F.pratensis* Huds.). MEADOW FESCUE; TALL FESCUE.
Perennial with stout culms to 120cm, short-rhizomatous, rarely stoloniferous. Leaves flat, narrow-linear, elongate, to 8mm diam.; ligules very short. Panicles erect or nodding at apex, scarcely to much-branched, narrow-ovate, to 20cm; branches ascending, paired, scabrous, bearing spikelets almost to the base; spikelets pale green to purple-tinted, 5–8-flowered, to 12mm; glumes lanceolate, to 5mm, rather acute to obtuse; lemmas coriaceous of chartaceous, oblong-lanceolate, to 7mm, subacute, glabrous, obscurely 5-nerved, scarious at apex, rarely short-awned. Summer. Europe, Siberia, introduced to California from Europe.

*F.eskia* Ramond ex DC.
To 45cm. Culms short, erect, densely tufted, very compact, rounded cushions, ultimately carpet-forming. Leaves very deep green, filiform; blades to 20cm, stiff, acute to pungent, involute, thus terete, glabrous. Panicles loose, pendent, ovoid, to 10cm; spikelets narrowly ovate, to 1cm, 6–8-flowered, green, marked yellow and orange; lemmas acute, to 0.5cm, smooth or slightly rough, mucronate. Europe (Pyrenees). *F.flavescens* Bell differs in its yellow-tinted panicles.

*F.gautieri* (Hackel) K. Richt.
Culms to 50cm × 1.5mm. Leaves glabrous, pungent, often curved, to 0.5mm diam., 5–7-veined, rib 1, towards apex only; sheaths mostly closed; ligule rounded, to 1mm, dense short-ciliate. Panicles dense, to 7cm; branches short-pubescent; spikelets few, to 12mm, yellow-green to stramineous; upper glume ovate-lanceolate, to 6mm, short-acuminate; lemma lanceolate, to 7.5mm, acuminate, mucronate or unawned. SW France, NE Spain. 'Pic Carlit': habit dwarf.

*F.gigantea* (L.) Vill. GIANT FESCUE; GIANT BROME.
To 1.5m. Culms tufted, smooth, nodes purple. Leaves bright green, glabrous, rough above, shiny beneath; sheaths glabrous; blades flat to U-shaped, linear, acute, to $60 \times 2$cm, arching, occasionally scabrous above; ligules to 3mm, translucent, papery. Panicles loose, nodding, lanceolate to ovate, to 50cm; branches drooping; spikelets to 2cm, 3–10-flowered; glumes to 1cm; lemmas awned, awn to 2cm, bent. Europe, temperate Asia, N Africa. 'Striata': leaves faintly striped white.

*F.glacialis* Miégev.
To 15cm. Culms flimsy, forming dense tussocks, clothed at base with old leaf sheaths. Leaf blades filiform, obtuse, glabrous, 5-ribbed, blue-grey. Panicles branched, to 2.5cm; spikelets usually solitary, rarely 2 or 3 on each branch, 3–5-flowered, to 0.5cm; lemmas acute, to 0.5cm, not awned or mucronate. Europe (Pyrenees and Alps).

**Festuca**   (a) *F. elatior*  (b) *F. gigantea*

**F.glauca** Vill. non Lam. (*F.ovina* var. *glauca* (Lam.) Hackel). BLUE FESCUE; GREY FESCUE.
To 40cm, usually far shorter. Culms upright, flimsy, densely tufted, blue-green. Leaves glabrous, glaucous, blue-green, blades filiform, terete, obtuse, rigid, straight to arching, 9-ribbed. Panicles dense, obovate, shortly branched, erect, to 10cm, blue-green; rachis glabrous; spikelets 4–7-flowered, elliptic to oblong, to 0.5cm; lemmas to 5mm, with a short spine at apex, glabrous. Often confused with *F.cinerea* and *F.rubra* and sold as *F.ovina* 'Glauca'. The name is retained here but should perhaps be considered a group epithet. Europe. 'Azurit': tall, to 30cm, brilliant blue. 'Blaufink' ('Blue Finch'): tall, to 15cm; leaves dull blue. 'Blauglut' ('Blue Ember'): leaves intense silver-blue. 'Blausilber' ('Blue Silver'): leaves strong blue silver. 'Caesia' (var. *caesia* Sm.): leaves more slender, vivid blue. 'Daeumling' ('Tom Thumb'): dwarf, to 10cm; leaves blue. 'Elijah Blue': dwarf, tufted, of an intense blue – one of the most striking small blue grasses. 'Frühlingsblau' ('Spring Blue'): leaves strong blue. 'Harz': leaves dark olive, sometimes tinted plum at tips in summer. 'Meerblau' ('Sea Blue'): leaves blue-green; strong-growing. 'Palatinat': leaves blue tinted green. 'Seeigel' ('Sea Urchin'): leaves very fine, green. 'Silberreiher' ('Silvery Heron'): leaves silvery blue. 'Soehrenwald': tall, to 20cm; leaves olive green. 'Bergsilber', 'Glaucantha' and 'Kentucky Blue' are also listed. Differs from *F.rubra* in stouter inflorescences produced in mid, not early summer, its split, not entire, leaf sheaths, and its densely tufted, not running habit. Its foliage is also of a more intense blue-grey.

**F.mairei** St Yves. ATLAS FESCUE.
To 1.2m, clump-forming. Leaves to 60 × 0.5cm, flat, semirigid, scabrous-serrate, grey-green. Panicles sparsely branched, slender. Morocco. Z7.

**F.nigrescens** Lam.
To 90cm, densely tufted, stolons absent or very short. Leaves soft, usually glabrous, dark green, to 1mm diam., 5–7-veined, sheaths closed to the mouth, somewhat pink, decaying into fibres; ligule a minute rim. Panicle to 10cm, secund; branches scabrous; spikelets to 9.5mm; upper glume ovate-lanceolate to lanceolate, to 5 × 1.5mm; lemma lanceolate, to 6×2.5mm, long-acuminate, uually glabrous; awn to 3mm. S, W, & C Europe.

**F.ovina** L. SHEEP'S FESCUE.
To 60cm. Culms forming dense tussocks, flimsy, upright or somewhat spreading. Leaves glabrous to very scabrous, green or slightly glaucous blue-green; blades finely filiform, closely infolded, stiff, 5–7-ribbed, to 25cm, scabrous; sheaths split to almost half their length. Panicles erect, lanceolate to narrowly oblong, open to dense, to 12cm, tinged-purple; spikelets borne mainly on one side of rachis, elliptic to oblong, to 1cm, 3–9-flowered; glumes almost equal, acute, stiff; lemmas acute, to 5mm, glabrous or very finely hairy, awn to 2mm at apex. N temperate regions. Of many subspecies, ssp. *coxii* is perhaps the most attractive variant, with bright blue-grey leaves. It does not flower. The blue fescues, *F.glauca* and *F.rubra*, are often confused with this species and their cultivars assigned to it. It differs from the former in its leaf sheaths, which are rounded behind, smooth and have rounded auricles, and in its leaf blades, which have blunt, not pointed tips that may curl. It differs from *F.rubra* in its split leaf sheaths on sterile shoots, its lack of runners and its larger awns.

*F.ovina* var. *glauca* (Lam.) Hackel. See *F.glauca*.

**F.paniculata** (L.) Schinz & Thell.
To 1m. Stems erect, densely tufted; some stems infertile. Leaves stiff, long, to 0.5cm wide; ligules to 3mm. Panicles ovate to obovate, dense or loose, to 15cm; spikelets 3–5-flowered, broadly ovate, to 0.5cm; lemmas acute, scabrous, without awns, red-brown or chestnut brown, edged purple-brown. S Europe (mts).

*F.pumila* Vill. See *F.quadriflora*.

**F.punctoria** Sibth. & Sm. HEDGEHOG FESCUE.
To 30cm. Culms rigid, tufted. Leaf blades tightly inrolle thus terete, filiform, acute, pungent, with 7–9 ribs, to 7.5cr very glaucous, blue-grey – the whole appearing rounded ar spiky. Panicles lanceolate to narrowly obovate, dense, to 5cr spikelets 4–7-flowered, obovate, to 1cm; lemmas to 0.5cr glaucous, awn to 2mm. N Turkey (mts).

**F.pyrenaica** Reut.
To 32cm, loosely tufted. Rhizome long; culms ascendin Basal leaves to 0.5mm diam., soft, slightly pubescent, obtus veins 5–7, ribs prominent, long-pubescent; sheaths somewh pink, hirtellous, closed to the mouth, decaying into fibre Panicles more or less unbranched, to 3.5cm; spikelets 6.5mm, grey-violet variegated; upper glume oblong-lance late, to 3.5 × 1.5mm, long-acuminate; lemma lanceolate, to 4 1.5mm, long-acuminate, awn to 1.5mm. C & E Pyrenees.

**F.quadriflora** Honck. (*F.pumila* Vill.).
To 30cm. Culms scabrous below panicle. Leaves slight pungent, to 1mm diam., 5–7-veined, 1-ribbed, sometim with low lateral ribs; sheaths somewhat scabrous; ligu truncate, to 1mm. Panicles lax, erect to 4cm; branche scabrous; spikelets few, to 1cm, usually glaucous, marke with intense violet; upper glume ovate-lanceolate, to 5 2mm, long-acuminate; lemma broad-lanceolate, acuminat with a scarious margin above; awn to 1.5mm. Europe.

**F.rubra** L. RED FESCUE; STRONG CREEPING RED FESCUE.
To 110cm. Loosely tufted, usually long-rhizomatous. Leave to 1mm diam. when inrolled from margins, typically gree obtuse or abruptly acute, pubescent; veins 5–7, ribs 3–sheaths closed to the mouth, somewhat purple-pink, decayin into fibres, glabrous to sparse-pubescent; ligule a minute rir Panicles lax, to 14cm, branches glabrous or pubescen spikelets to 1cm, bright green or glaucous, rarely pruinos upper glume lanceolate, to 1.5mm diam., acuminate; lemm to 7 × 2.5mm, glabrous or subglabrous, awn variable, 4.5mm. Europe. Differs from *F.glauca* in its running, sprea ing, not densely tufted habit, entire leaf sheaths, narrow infl rescences produced in late spring and early summer, an foliage which is either green or of a weaker grey-green. c also *F.ovina*. 'Silver Needles': often referred to this species, fact a cultivar of *Agrostis canina* q.v. The true variegate *F.rubra* is 'Variegata'.

**F.rupicola** Heuff. (*F.sulcata* (Hackel) Nyman).
Densely tufted. Culms to 65cm, scabrous above. Leaves 32.5cm × 1mm, scabrous, 5–7-veined, rarely pruinose, ribbed; sheaths open to base, scabrous-pubescent or glabrou Panicles lax, to 9.5cm; branches scabrous; spikelets green, 8mm; florets 3–5 per spikelet; upper glume ovate-lanceolat to 5 × 2mm, acuminate; lemma lanceolate to ovate-lanceolat to 5 × 2.5mm, glabrous or pubescent, awn to 2.5mm. Europe, Balkans.

*F.scoparia* A. Kerner ex Nyman non Hook. f. See *F.gautieri F.sulcata* (Hackel) Nyman. See *F.rupicola*.

**F.tenuifolia** Sibth. (*F.capillata* Lam.).
To 55cm, densely caespitose. Culms usually scabrou above. Leaves to 0.5mm diam., veins 5–7, rib 1; sheath open to base. Panicles lax, to 8cm; branches scabrou spikelets to 6.5mm, somewhat green, occasionally prolife ating; upper glumes oblong-lanceolate, to 4 × 1mm, acum nate, scabrous above; lemma ovate-lanceolate, to 4.5 1.5mm, scabrous above or occasionally pubescent dorsall sometimes mucronate, mucro to 0.5mm; awn absent. W & Europe.

**F.valesiaca** Schleich. ex Gaudin.
To 50cm. Culms scabridulous above. Leaves to 0.5m diam., scabrous, pruinose, 5-veined, ribs 3; sheaths glabrou open to base; lamina sometimes deciduous. Panicle som what interrupted, to 7cm; branches scabridulous; spikele pruinose, to 6.5mm; florets 3–5 per spikelet; upper glum oblong-lanceolate, to 4 × 1.5mm, abrupt-acuminate; lemm subulate, to 5 × 1.5mm, glabrous or ciliate. Europ

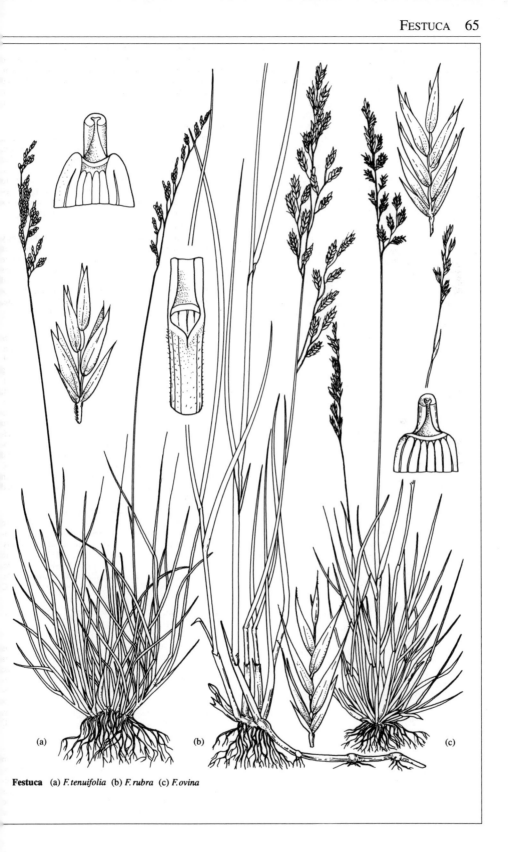

**Festuca**    (a) *F. tenuifolia*   (b) *F. rubra*   (c) *F. ovina*

'Silbersee' ('Silver Sea', 'Seven Seas'): dwarf and compact; leaves silvery blue.

**F.varia** Haenke.
To 55cm. Culms scabrous above. Leaves pungent, glabrous, to 1mm diam., 5–9-veined, 3–7-ribbed, ribs inconspicuous; sheaths partly closed; ligule to 2.5mm. Panicle to 9cm lax; branches very densely short-pubescent; spikelets few, to 11mm, glaucous, violet-tinted; florets 3–8 per spikelet; upper glume broadly lanceolate to broadly ovate-lanceolate, to 6 ×

3mm, short-acuminate, with a wide scarious margin; lemma ovate-lanceolate, to 7 × 3mm, scabrous above, ciliolate, with narrow scarious margin towards apex, awn to 1mm, palea to 7mm, scabrous. E Alps.

**F.vivipara** (L.) Sm.
Very similar to *F.ovina* except spikelets producing small plantlets instead of florets. N Europe (moors and mts). 'In': very dwarf, to 5cm.

**Glyceria** R. Br. (From Greek *glykys*, sweet, referring to the vegetation which attracts grazing livestock.) SWEET GRASS; MANNA GRASS. Gramineae. 16 species of perennial, glabrous, aquatic or marsh grasses to 2.5m. Rhizomes creeping, culms pithy, reed-like, some sterile. Leaves with folded blades, becoming flat; sheaths tubular; ligules translucent, papery. Flowers in open or contracted, deltoid panicles; spikelets many, stalked, linear to ovate, often laterally flattened, 3 to many-flowered with a reduced, terminal floret; glumes unequal, transparent, 1-ribbed, shorter than lemmas; lemmas exserted, ovate to oblong-lanceolate, convex dorsally, 5–11-ribbed with hyaline or acute apices, not awned; palea as long as lemma, keeled. Wet places, mainly N Temperate, some from Australia, New Zealand, S America.

CULTIVATION  Hailing from lakesides, ponds and slow-moving rivers, sometimes submerged to depths of 75cm/30in., *Glyceria* is used in the bog garden, and other garden situations approximating to those in habitat and is especially useful in the wider landscape in controlling erosion of riverbanks. Both the species and its variegated form (amongst the most brightly coloured of the variegated grasses) thrive in fertile, moisture-retentive soil in the herbaceous and mixed border. Grow in full sun. *Glyceria* spreads vigorously in water and should be confined in a tub or basket in smaller pools. Propagate by division.

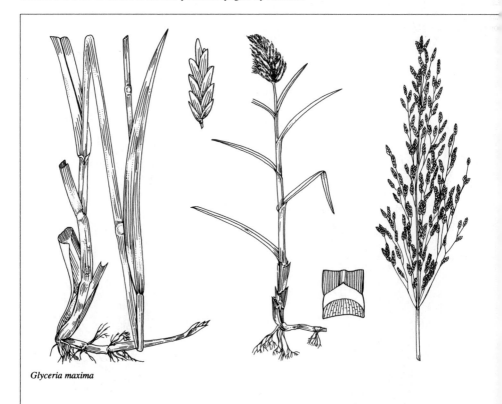

*Glyceria maxima*

*,aquatica* L. See *G.maxima*.

*,maxima* (Hartm.) Holmb. (*G.aquatica* L.). GREAT WATER ₨ASS; REED MANNA GRASS; REED SWEET GRASS; REED ℇADOW GRASS; SWEET HAY.
ₒ 2.5m. Rhizome stout, creeping; culms erect, forming large ₐnds, smooth, and glabrous toward apex. Leaves with tessel-ₜed venation; blades with a brown mark at point where ₑath and blade diverge, acute, 30–60×2cm, with a central, ₙgitudinal keel; sheaths splitting; ligule abruptly acuminate.

Panicles loose to dense, broadly ovate to oblong, much-branched, to 45cm, often tinged purple; spikelets 4–10-flow-ered, oblong, to 10×3mm; lemmas obtuse, to 5mm. Summer. Temperate Eurasia. 'Pallida': leaves boldly variegated off-white. 'Variegata' (var. *variegata* Boom & Ruys): new growth strongly flushed coral-pink on emergence, leaf blades striped green, cream, white, broadly so, and often retaining some pink/red tint or lining. Usually smaller than the type (to about 70cm), with slightly broader, shorter leaf blades. Z5.

**₣ynerium** Willd. (From Greek *gyne*, woman, and *erion*, wool, referring to the hairy female lemmas.) ₗamineae. 1 species, a dioecious, perennial, aquatic reed to 10m. Rhizomes large, creeping; culms coarse, ₑct, tufted, to 2.5cm diam., clothed with old leaf bases. Leaves distributed along culm, forming a fan-like ₐrangement at apex, flat, scabrous; blades to 200 × 10cm; ligule short. Panicles plumose, to 1m, grey-green to ₑy, resembling those of a Pampas Grass; branches pendent; spikelets several-flowered; female spikelet 2-flow-ₑd; upper glume longer than lemma, spreading, curved, 3–5-ribbed, translucent, papery; lemmas narrow, ₙtire, silky; male glumes and lemmas hyaline; lemmas smaller than female, glabrous. Early autumn. Tropical ₘerica, W Indies. Z9.

ₗLTIVATION    Occurring in wet tropical habitats, *Gynerium* is a handsome species for the warm glasshouse in ₒol temperate zones, grown for foliage effects and for the large, graceful plume-like inflorescence. Grow in full ₙn in a fertile and moisture-retentive potting mix in large tubs, submerged at the base in water. Maintain a win-ₑr minimum temperature of 13–16°C/55–60°F. Propagate by division in late winter/early spring, or by seed.

*jubatum* Lemoine. See *Cortaderia jubata*.
*saccharoides* Humbert & Bonpl. See *G.sagittatum*.

*,sagittatum* (Aubl.) P. Beauv. (*G.saccharoides* Humbert & ₒnpl.). WILD CANE; UVA GRASS.

**Hakonechloa** Mak. ex Honda (From Hakone in Honshu, Japan.) Gramineae. 1 species, a perennial grass to 75cm. Rhizomes spreading, scaly; culms ascending or spreading, flimsy arching to form close, tousled clumps. Leaves smooth; blades linear-lanceolate, acute, tapering at base, to 25 × 1.5cm, rich green, becoming orange bronze in autumn in mild climates; ligules finely hairy. Panicles loose, lanceolate to ovoid, nodding, to 18cm; rachis with hairy joints; spikelets stalked, oblong, somewhat flattened, 3–5-flowered, to 2cm, pale green; glumes acute, 3–5-ribbed, papery, translucent, lower shorter; lemmas exserted, on a short basal stalk, imbricate, convex, 3-ribbed, to 3mm, papery, translucent, ciliate, with straight awn, to 5mm. Autumn. Japan (mts). Z5.

CULTIVATION  The variegated cultivars of *Hakonechloa* provide dramatic colour and textural contrasts in the border, at their best where space allows drift plantings; they often retain their beautiful red flushed autumn foliage into winter. Given shelter from cold drying winds, they are hardy to between −15 and −20°C/5 to −4°F. Easily grown in any moderately fertile, moisture-retentive soil with additional leafmould. Light shade will enhance leaf colour; plants spread slowly once established to form substantial clumps. Also excellent for container cultivation. Propagate by division in spring.

*H.macra* (Munro) Mak. HAKONE GRASS.
'Albo-aurea': leaves striped off-white and gold, with very little green – this and the next cultivar of poor vigour and often short-lived. 'Albovariegata': leaves striped white. 'Aureola' ('Urahajusa Zuku'): leaves yellow, striped with narrow green lines, tinted pink-red in autumn – the most commonly grown cultivar.

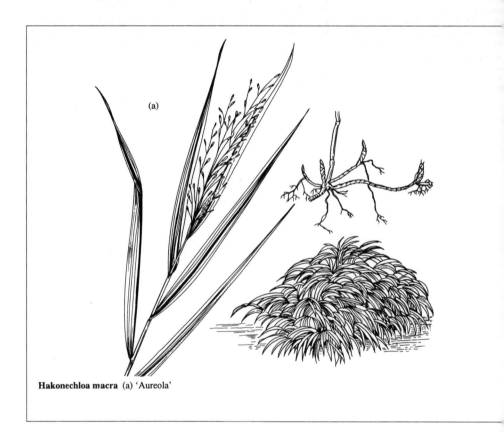

**Hakonechloa macra** (a) 'Aureola'

**Helictotrichon** Besser ex Roem. & Schult. (From Greek *heliktos*, twisted, and *thrix*, *trichos*, hair or bristle, referring to the shape of the awn.) OATGRASS. Gramineae. 60 species of perennial grasses to 1.2m. Culms slender, forming tussocks. Leaf blades flat, folded or revolute, conspicuously ribbed; ligules truncate or papery, dentate, translucent. Panicles erect or nodding, narrow; rachis flimsy, hairy; spikelets 2–6-flowered, oblong, laterally flattened, glistening; 2–4 florets fertile, 1–2 florets sterile; glumes lanceolate, acute, unequal, 1–5-ribbed, thin; lemmas exserted, narrowly lanceolate, acute, margins and apex hyaline, 5–11-ribbed, glabrous, awned from centre of dorsal surface; awn twisted, abruptly bent; palea enclosed by lemma, emarginate, hairy-keeled. Summer. Temperate N Hemisphere, S Africa. Z4.

CULTIVATION   Occurring in meadow, on dry hillsides and at woodland edge, with *H.sempervirens* in rocky and stony pasture on calcareous soils. Grown for their strong and attractive form, the arching inflorescence held clear above the rounded basal hummocks of foliage, *Helictotrichon* species are sometimes grown in the border, but are best given space as a specimen to achieve their full architectural potential. The vivid blue *H.sempervirens* is especially suited to plantings in gravel or in pebble gardens. Grow in well drained soil in sun, give mulch protection where winter temperatures fall much below –15°C/5°F. Remove dead foliage regularly to retain form. Propagate by division or seed in spring.

*H.planiculme* (Schrad.) Pilger.
To 1m. Culms stout, forming tussocks. Leaf blades flat or folded, acute, 7.5–18cm, smooth; lower sheaths finely scabrous; ligules to 1cm. Panicles contracted, dense, to 27.5 × 2cm; spikelets 4–7-flowered, linear to oblong, to 2.5cm, tinged purple; glumes to 1.5cm; lemmas to 1.5cm, hairy at base; awns to 2.5cm. C & SE Europe, SW Asia.

*H.sempervirens* (Vill.) Pilger (*Avena sempervirens* Vill.; *Avena candida* hort.).
To 1.2m. Culms erect, rigid, tufted. Leaf blades tightly rolled or flat, to 22.5 × 0.1–0.3cm, tapering finely, rigid, vividly grey-blue, glaucous; sheaths glabrous; ligules minute. Panicle loose, open, arching, one-sided, to 17.5cm; spikelets to 3-flowered, oblong, to 1.5cm, straw-coloured, marked with purple at first; fertile florets 2; glumes to 1cm, 1–3-ribbed; lemmas to 1cm, hairy at base, jagged-toothed; awns to 2.5cm. SW Europe. 'Pendulum': inflorescence more nodding. 'Saphirsprudel': leaves strongly tinted steel blue; wind-resistant.

× **Hibanobambusa** Maruyama & H. Okamura. (Bamboo of Mount Hiba.) Gramineae. 1 species, a bamboo thought by some to be of fairly recent hybrid origin, the parents being species in *Phyllostachys* and *Sasa*. Culm 2–5m × 1–3cm, curved at the base; sheaths deciduous with scattered hairs, auricles and bristles; blade narrow branches 1 per node, rarely 2. Leaves 15–25 × 3.5–5cm, scaberulous, glabrous; sheaths hairless, with auricle and bristles. Japan. Z8.

CULTIVATION   As for *Semiarundinaria*.

× **H.tranquillans** (Koidz.) Maruyama & H. Okamura (*Phyllostachys tranquillans* (Koidz.) Muroi; *Semiarundinaria tranquillans* Koidz.). INYOU-CHIKUZOKU.

'Shiroshima': culms, sheaths and leaves striped white and yellow.

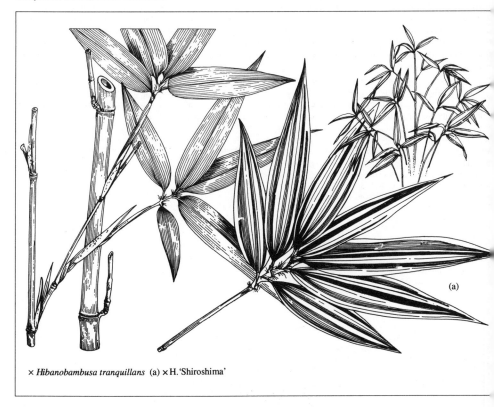

× *Hibanobambusa tranquillans* (a) × H. 'Shiroshima'

**Hilaria** Kunth. (For A. St. Hilaire (1779–1853), French botanist.) Gramineae. Some 7 species of perenni. grasses to 1m. Culms erect to spreading, rigid, solid, some infertile. Leaf blades flat to inrolled, narrow, occa sionally pubescent. Panicles usually obcuneate in outline; spikelets in clusters of 3, 2 lateral spikelets usuall staminate, 2-flowered, central spikelet hermaphrodite, 1-flowered; glumes in flexuous arrangement, awned Summer. N America. Z7.

CULTIVATION   Occurring on dry plains, and other arid and semi-arid habitats, *Hilaria* may be useful in dry garden and in collections of native plants. Grow in perfectly drained sandy soils in full sun. Propagate by seed or division.

**H.berlangeri** (Steud.) Nash. CURLY MESQUITE.
To 30cm. Culms flexuous, tufted producing stolons to 20cm with tough, slender internodes. Leaf blades to 0.3cm wide, rough, with soft, straight hairs. Panicle open, to 3cm; rachis flexuous; spikelets in 4–8 clusters of 3, to 0.5cm; glumes convergent at base, stiff, rough, 2–3-ribbed, awned or mucronate. SW US.

**H.jamesii** (Torr.) Benth. GALLETA.
To 30cm. Rhizome rough, bumpy; culms erect, tufted. Leaf blades to 5 × 0.3cm, not arching, stiff, becoming inrolled. Panicle open; spikelets to 1.5cm, hairy in lower part; glumes of staminate spikelets acuminate, awned. W US.

**H.mutica** (Buckley) Benth. TOBOSA GRASS.
To 60cm. Culms tufted, glabrous except at nodes. Leaf blades flat to inrolled, to 0.3cm wide. Panicle spreading, to 6cm;

spikelets to 5mm, densely hairy in lower part; glumes of sta minate spikelets distinctly unequal, obtuse, downy at ape glumes of hermaphrodite spikelet keeled, with numerou awns and narrow lobes at apex. SW US.

**H.rigida** (Thunb.) Benth. ex Scribn. BIG GALLETA.
To 1m. Culms numerous, erect, stiff, softly downy wi upright branches. Leaf blades to 5 × 0.3cm, more or le inrolled, downy to glabrous; sheath villous at junction wi blade. Spikelets in groups of 3 to 0.5cm, villous at bas glumes unequal, 7-ribbed, 1–3 ribs extending at awns; lemm membranous, bifid, with a short awn. SW US.

**Himalayacalamus** Keng.f. (From Himalaya, their place of origin, and Greek *kalamos*, a reed.) Gramineae. PARANG. Some 5 species of clumping bamboos with tall, hollow culms. They differ from *Drepanostachyum* in having only 1 (not 2 or more) florets per spikelet, from *Arundinaria* and *Chimonobambusa* in their tightly clump-forming (not running) rhizomes, and from *Fargesia* in their open inflorescences, which lack spathes. Himalaya. Z8.

CULTIVATION  Introduced to European gardens in the 1840s, two species of *Himalayacalamus* have long, if rather confusing, histories in cultivation. *H.falconeri* produces a close, V-shaped clump of elegant, slender culms in olive green with a purple ring beneath each node (ultimately yellow). Its foliage is fine and feathery, produced in great abundance. *H.falconeri* 'Damarapa' is none other than the celebrated striped bamboo known for so long to horticulturalists as *Arundinaria hookeriana*. This is even nobler in form than *H.falconeri sensu stricto*, with sprays of delicate blue-green leaves and culms that progress from pink striped lime green to a warm amber-gold irregularly and finely striped dark olive. The true *Arundinaria hookeriana* (now *Himalayacalamus hookerianus*) is the Blue-Stemmed Bamboo, another Himalayan native found in Nepal, India and Bhutan. A tall, elegant, V-shaped clump-former, this bamboo produces blue-green culms densely covered in glaucous wax. At a distance and seen in a mature specimen, this coloration gives the impression of a misty blue column topped with exceptionally graceful foliage.

Each of these species requires a friable, sandy soil rich in leafmould and never dry. They resent exposed sites, harsh weather and temperatures below −10°C. For these reasons, they should be attempted outdoors only in mild regions of zones 7 and over, ideally given the protection of surrounding woodland. Elsewhere, they make excellent tub plants for cold glasshouses and conservatories.

**H.falconeri** (Hook. ex Munro) Keng f. (*Arundinaria falconeri* (Hook. ex Munro) Benth. & Hook. f.; *Drepanostachyum falconeri* (Hook. ex Munro), J.J.N. Campbell ex D. McClintock; *Sinarundinaria falconeri* (Hook. ex Munro) Chao & Renvoize; *Thamnocalamus falconeri* Hook. ex Munro).
Culms 5–10m × 2–3cm, thin-walled, glabrous, yellow-green with a distinctive purple-flushed ring below the nodes; sheaths almost glabrous, fairly persistent, ciliate. Leaves 7–16 × 0.8–2.7cm, ligule short, dark-fringed. C Himalaya. 'Damarapa' (*Arundinaria hookeriana* hort. non Munro; *Drepanostachyum hookerianum* hort. non (Munro) Keng f.; *Himalayacalamus hookerianus* hort. non (Munro) C. A. Stapleton; *Sinarundinaria hookeriana* hort. non (Munro) Chao & Renvoize): culms emerging pink with lime green stripes,

later a deep amber or red-brown with irregular dark olive stripes.

*H. hookerianus* hort. non (Munro) C.A. Stapleton. see *H.falconeri* 'Damarapa'.

**H.hookerianus** (Munro) C.A. Stapleton (*Arundinaria hookeriana* Munro; *Chimonobambusa hookeriana* (Munro) Nak.; *Drepanostachyum hookerianum* (Munro) Keng f.; *Sinarundinaria hookeriana* (Munro) Chao & Renvoize). BLUE-STEMMED BAMBOO.
Culms 6–9m × 1.5–3cm, blue-green, thickly white powdered at first, nodes with a blue-grey ring below, later yellow-green or purple-red, especially having endured hard winters; sheaths to 35cm, tough, smooth, glabrous or scaberulous at apex, tapering to a long, narrow, involute tip. Leaves to 18 × 2cm, glabrous, lacking tessellation, ligule hairy. E Himalaya.

*Himalayacalamus hookerianus*

**Holcus** L. (From Greek *holkos*, a kind of cereal.) Gramineae. 8 species of annual or perennial grasses to 1m. Culms erect, flimsy, tufted. Leaf blades linear, flat or folded; ligules papery, translucent. Panicles spicate, dense or open; spikelets stalked, laterally flattened, falling at maturity, 2-flowered; lower floret bisexual; upper floret staminate; glumes equal, longer than lemmas, keeled, lower 1-ribbed, upper 3-ribbed; lemmas obtuse, to 2-toothed at apex, shiny, lemma of staminate floret awned. Summer. Europe, temperate Asia, N & S Africa. Z5.

CULTIVATION *H.lanatus* is sometimes found as a week of fine turf and may be spot-treated with translocated herbicide. Although very attractive in flower in grassland managed as ornamental meadow, it may out-compete less vigorous fine grasses. *H.mollis*, found on predominantly acid soils in open woodland and shady hedge-banks, is more suitable for wild flower meadow mixes; in more controlled gardens, it is usually deemed a noxious weed. *H.mollis* 'Variegatus', considerably less invasive than the type, is grown in the larger rock garden, as border edging and as groundcover; when established it will withstand mowing and can be used in multi-coloured lawns. The tactile nature of the soft velvety foliage is of great value in gardens designed for people with visual handicaps. Grow in sun in well drained but moisture-retentive soil. Clip over *H.m.* 'Variegatus' after flowering. Propagate by division, species also by seed.

*H.bicolor* L. See *Sorghum bicolor*.

*H.halapensis* (L.) Brot. See *Sorghum halapense*.

**H.lanatus** L. VELVET GRASS; YORKSHIRE FOG.
Perennial to 1m. Rhizome creeping; culms erect, clustered, downy, joints glabrous. Leaves acuminate, to 5mm wide, grey-green, downy; ligule obtuse, to 5mm. Panicle soft, contracted, dense, to 15cm; spikelets to 3mm, light green to pink-purple; glumes shaggy-haired, except on veins; lower glume narrower than upper, lemma of bisexual floret with a hooked awn. Europe.

**H.mollis** L. CREEPING SOFTGRASS.
Perennial to 45cm. Rhizome creeping; culms erect, joints hairy. Leaves acute, to 1cm wide, grey-green, glabrous to slightly hairy, scabrous to smooth. Panicles narrowly oblong to ovate, dense to loose, to 12cm; spikelets to 5mm; glumes hairy, papery; lemma of bisexual floret with a bent awn. Europe. 'Variegatus' ('Albovariegatus'): leaves broadly edged white with a narrow green central stripe.

*H.sorbus* L. See *Sorghum bicolor*.

**Holcus** (a) *H. mollis* (b) *H. lanatus*

**Hordeum** L. (Latin *hordeum*, barley, a corruption of *horridus*, bearded with bristles, referring to the bristle-like, long-awned glumes and vestigial spikelets.) BARLEY. Gramineae. Some 20 species of annual or perennial grasses to 120cm. Culms flimsy to stout. Leaves linear, flat or rolled; sheaths auricled at apex; ligules short, membranous. Flowers in dense, narrow, usually cylindric, occasionally laterally flattened, spike-like panicle; rachis articulate or unjointed; spikelets in alternate groups of 3 at each joint of rachis in 2 ranks, 1- (rarely 2-) flowered; lateral pair of spikelets usually staminate or infertile, rarely bisexual, subsessile, occasionally vestigial as 3 awns; central spikelet usually hermaphrodite, rarely aborted, sessile; glumes narrow, often bristle-like, stiff, usually long-awned; lemma of central spikelet convex on dorsal surface, 5-ribbed, rigid, awned or unarmed; palea narrower than lemma, keeled. Summer. Temperate N hemisphere and S America.

CULTIVATION   The species described are grown for the graceful, feathery flowerheads which dry and dye well. *H.jubatum* is a highly ornamental barley with delicate, long, straight and silky awns faintly tinged crimson. Pick immature heads and air dry. Sow seed *in situ* in autumn or spring in light, well drained, dryish soils in full sun.

*H.hystrix* Roth. MEDITERRANEAN BARLEY.
Annual to 37.5cm. Culms solitary or tufted, geniculate. Leaf blades flat, acuminate, to 7.5 × 0.5cm, downy; ligules obtuse, papery, translucent. Inflorescence ovate to oblong-ovate, grey-green to purple-tinged, to 6 × 2cm; lateral spikelets with stiff, bristle-like, glumes to 2cm; lemma to 1cm, unarmed; central spikelet with bristle-like glumes to 2.5cm; lemma lanceolate, to 1cm, awn to 2cm. Mediterranean, C Asia. Z7.

*H.jubatum* L. SQUIRRELTAIL BARLEY; FOX-TAIL BARLEY.
Annual or perennial to 60cm. Culms numerous or solitary, erect or spreading, smooth. Leaf blades erect or arching to 15 × 0.5cm, scabrous. Inflorescence dense, bristly, soft, nodding, pale green to purple-tinged, to 12.5 × 8cm; lateral spikelets asexual, reduced to awns; middle spikelet with fine, bristle-like, spreading glumes to 7.5cm; lemma lanceolate, to 0.5cm, awned; awn hair-like, to 8cm, pale green. N America, NE Asia. Z5.

**Hyparrhenia** Anderss. ex Fourn. (From Greek *hypo*, under, and *arren*, male, referring to the pair of staminate spikelets at raceme base.) Gramineae. 53 species of annual or perennial grasses, 2–6m. Culms tufted, erect, flimsy to stout, solid. Leaf blades arching or erect, linear-acuminate, flat. Panicles large, elaborate, composed of paired, spike-like racemes; racemes stalked, with a spathe at base; spikelets in pairs, dissimilar, one sessile, the other stalked, sessile spikelet 2–flowered; lower floret reduced to a membranous lemma, upper floret hermaphrodite; glumes equal lemma, 2-toothed at apex, hairy, tapering into a geniculate awn; stalked spikelets staminate or asexual, without awns, occasionally with an apical hair; spikelet pair(s) at base of racemes equal, staminate, without awns. Africa, Madagascar, Tropical America, Tropical Asia. Z9.

CULTIVATION   Found almost exclusively in tropical savannah, *Hyparrhenia* species are best treated as frost-tender perennials and given warm glasshouse protection in cool temperate zones, although they may be grown in sheltered situations in favoured temperate gardens, they will require protection from frost and excessive winter wet. The intricately branched inflorescence develops as days shorten in late summer. Grow in large pots of fertile loamy soil, in sun and water moderately when in grown. Maintain good ventilation and a minimum temperature of 1–13°C/50–55°F. Propagate by seed sown in gentle heat in spring.

*H.cymbaria* (L.) Stapf.
Perennial; culms stout, to 6m. Leaf blades to 45 × 1cm. Panicle much-branched, to 60cm; spathes ovate-navicular, acute, glabrous, red or maroon; racemes to 1cm; awns to 1.5cm. Late summer–autumn. Tropical & S Africa.

*H.rufa* (Nees) Stapf.
Perennial; culms 1–3m. Leaf blades to 75 × 1.5cm, margins scabrous. Panicle loose or contracted, to 60cm; spathes narrowly lanceolate, green; racemes to 3cm, pubescent, straw-yellow to red-brown; spikelets dark red-pubescent; awns to 3cm, twisted, red-brown, bristly. Summer. Tropical Africa.

**Hystrix** Moench. (From Greek *hystrix*, porcupine, referring to the spiky panicles.) BOTTLE-BRUSH GRASS. Gramineae. 6 species of perennial grasses to 120cm. Culms erect, tufted. Leaf blades flat, arching, linear, flimsy, acuminate; ligules papery, translucent. Flowers in a narrow, spike-like, loose, bristly raceme; spikelets sessile, 1–4 clustered per joint, 1–4-flowered; glumes reduced to small hooks or bristles or absent in upper spikelets; lemmas lanceolate, twice length of glumes, convex, 5-ribbed, stiff, awned. N America, N India, China, New Zealand.

CULTIVATION   Native to moist woodland, often found amongst rocks, *Hystrix* is grown for the curious and distinctive flowerheads which given rise to the common name bottlebrush grass: *H.patula* produces slender grey stems and swaying heads of long-awned, horizontally spreading spikelets. The flowerheads should be picked when immature for air drying. It grows best on moisture-retentive but well drained soil in sun or light dappled shade, but also tolerates dry shade – thus ideal for naturalistic woodland gardens. Propagate by division or seed in spring.

*H.patula* Moench.
Loosely tufted herbaceous perennial to 120cm, usually shorter. Culms erect. Leaf blades to 15 × 1cm wide. Panicles arching, to 15cm, resembling a bottle-brush; spikelets usually paired, held horizontally, pale green to pink-tinted, spreading, to 1.5cm; awns to 4cm. Summer. N America. Z4.

**Imperata** Cyr. Gramineae. 8 species, clump-forming to running perennial herbs with slender culms and narrow leaves resembling *Miscanthus*. Inflorescence a spike-like panicle; spikelets paired, small, 2-flowered, enclosed by long silky hairs. Late summer to autumn. Cosmopolitan.

CULTIVATION   In hot, humid climates, *I.cylindrica* is an invasive spreader achieving heights of 1.5m. In cool temperate cultivation, it is far shorter and altogether less aggressive which is, perhaps, unfortunate because it is among the most beautifully coloured of all grasses. It earns the name Blood Grass by its slender, upright leaves tinted bright ruby red and becoming a more solid, deep garnet toward the end of summer. It will grow in most damp but well drained soils, preferably in full sunlight. Hard frosts and winter wet may cause young plants to deteriorate or perish – their crowns may need a dry mulch. Increase by division or seed.

*I.brevifolia* Vasey (*I.hookeri* Rupr.) SATIN TAIL.
Culms 1–1.5m tall from scaly rhizomes; ligule villous; leaf blades 10–50 × 0.8–1.5cm, glabrous; uppermost leaves reduced. Panicle 15–25cm, somewhat tawny or pinkish, soft-silky; spikelets 3mm long, with hairs twice as long. S California to Utah, Mexico. Grown for its beautiful, soft inflorescences. Z8.

*I.cylindrica* (L.) Beauv. BLOOD GRASS; CHIGAYA; FUSHIGE-CHIGAYA.
Rhizomes slowly spreading; culms to 80cm, slender, firm, long-pubescent on nodes. Leaf sheaths smooth or often long-pilose; leaf blades to 60 × 1.5cm, flat, linear, narrowly acuminate; ligules very short, truncate. Panicles to 20cm, nearly erect, with silver-white hairs, cylindric; racemes less than 3cm; spikelets to 4.5cm, lanceolate, glabrous, terete, acute; glumes lanceolate, acuminate, membranous, few-nerved. Late summer. Japan. 'Red Baron' ('Rubra') (BLOOD GRASS; JAPANESE BLOOD GRASS): leaves strongly flushed bright ruby red, later dark burgundy, red-tipped when young. The form most often seen in cultivation, especially in Japan, where it originates and is often used as a bonsai. Seldom flowers. Z7.

*I.hookeri* Rupr. See *I.brevifolia*.

**Indocalamus** Nak. (From *indo*, Indian, and Greek *kalamos*, reed.) Gramineae.  Some 25 species of relatively small, running bamboos, allied to *Sasa*, differing chiefly in floral characters, e.g. 3 not 6 stamens; its nodes are less prominent, with white waxy powder below. Branches, of similar thickness to the culms, usually 1, occasionally 2 or 3. Leaves very large, thick with scaberulous margins, 1–3 per branch. China, Japan, Malaysia.

CULTIVATION  These have the largest leaves among the hardy bamboos, forming dense, spreading mounds of luxuriant foliage. They are resilient and spreading, but rarely exceed 1.5m in heigh in cool temperate gardens. Ideal for creating a low, exotic screen or focal point. General treatment as for *Sasa*.

*I.latifolius* (Keng) McClure (*Arundinaria latifolia* Keng).
Culms 0.5–1m × 0.5cm, pubescent; sheaths ciliate and often hairy, shorter than the internodes. Leaves 10–40 × 1.5–8cm, slightly hairy beneath. E China.

*I.tessellatus* (Munro) Keng (*Sasa tessellata* (Munro) Mak. & Shib.; *Sasamorpha tessellata* (Munro) Nak.; *Arundinaria ragamowskii* (Nichols.) Pfitz.).
Differs from *I.latifolius* in growing in untidy mounds, curved downwards by the weight of the large leaves. Culms 1–2.5m × 0.5–1.5cm, subglabrous; sheaths ciliate, glabrescent, exceeding internodes. Leaves to 60 × 10cm (the largest of any hardy bamboo, especially in shade) ovate-lanceolate, rather dull green, glabrous. C China, Japan. Z8.

**Indocalamus**  (a) *I.tessellatus*  (b) *I.latifolius*

**Isolepis** R. Br. Cyperaceae. Around 40 species of grasslike, annual or perennial herbs, to 30cm. Stems usually leafy, upright, sometimes prostrate, terete. Leaves grasslike, often terete, narrow, sulcate. Inflorescence a terminal spikelet or umbel of spikelets subtended by a leaflike bract; flowers hermaphrodite, minute, spirally arranged, subtended by a scale-like glume; sepals and petals absent; stamens 1–3; style 2–3-branched; ovary single. Fruit a nut, obovate to elliptic, style persistent. Widespread in temperate regions and on tropical mountains. Z8.

CULTIVATION As for *Scirpus*, i.e. damp, rather acidic soils. *I.cernua* and *I.setacea* are grown indoors.

*I.cernua* (Vahl) Roem. & Schult. (*Scirpus cernuus* Vahl; *Scirpus filiformis* Savi non Burm. f.).
A densely tufted bright or dark green rush with very fine hair-like stems and leaves which arch and form a dense mop-like head, usually grown under glass, where it falls like green hair over the sides of small pots or glasshouse borders. It differs from *I.setacea* in that its inflorescence bract is often shorter than the inflorescence, its spikelets are solitary, its glumes green, spotted red and its fruit not shiny, but smooth and brown-red. W & S Europe, N Africa.

*I.prolifera* R. Br. (*Scirpus prolifer* Rottb.).
Rhizome creeping. Stems tufted, bearing bladeless leaf sheath. Bract short, glume-like, concealed by inflorescence; rays 1–2, 12–50mm; spikelets several to many, in dense clusters, pale brown; glumes blunt, ovate, striped brown. Fruit smooth to granular. S Africa, Australia.

*I.setacea* (L.) R. Br. (*Scirpus setaceus* L.). BRISTLE SCIRPUS.
Tufted annual, 3–30cm. Stems very thin. Leaves shorter than stem, around 1–6cm, filiform, dark green; sheaths often purple at base. Bract longer than inflorescence, to 15mm; spikelets 2–3, 5mm; glumes around 1mm, bristle-tipped, brown-purple, midrib green, margin transparent. Fruit around 1mm, 3-angled, dark, shiny brown, ribbed. Europe to S Africa and W Asia.

**Juncus** L. (From Latin *iuncus*, a rush, derived from *iugere*, to join, from the use of the stems in tying.) RUSH. Juncaceae. About 225 species of rhizomatous grassy plants growing in wet places. Stems several, terete, simple, tufted, glabrous. Leaves basal and stem-like or cauline and reduced, often subtending inflorescence, terete or flat, sometimes represented by basal leaf sheaths only. Inflorescence a terminal, or (when the stem continues as an apical bract) apparently lateral cyme; flowers green or brown, small; bracteoles absent or one bifid and two or more simple and forming an involucre; perianth of 6 rigid chaffy parts in 2 whorls of 3; stamens 3 or 6, short. Fruit a 3-locular, or rarely 1-locular, capsule; seeds many, often striated. Cosmopolitan, but rare in the tropics.

CULTIVATION *Juncus* species inhabit wetland and water margins and are sometimes used in plantings by larger, semi-natural ponds where their fast-spreading habits will not prove too competitive and their form will give height and foliage contrast; in wilder gardens they also provide useful bird cover on boggy ground. Of greatest horticultural value are cultivars of the soft rush, *J. effusus*, such as 'Spiralis', with spiralling corkscrew stems, and 'Vittatus', variegated with alternating bands of green and yellow; any reversions to the more vigorous straight or green foliage of the type should be promptly removed.

Give a sunny to partially shaded site and plant in spring on heavy soils, level with or slightly above the water surface; shallow water to a depth of 8cm/3in. suits *J.effusus, J.ensifolius* and *J.inflexus; J.subnodulosus* prefers well drained, gravelly pond margins. For easier management, plant into plunged tubs or other containers. *J.leseurii* is useful for planting on coastal sites and will withstand brackish water. Most will, however, thrive even far from water, provided they never dry out. On damp sandy soils, the annual *J.bufonius* becomes very invasive. Propagate cultivars by division in spring, species from seed or by division.

*J.arcticus* ssp. *balticus* (Willd.) Hylander. See *J.balticus*.

*J.articulatus* L. (*J.lampocarpus* Ehrh. ex Hoffm.). JOINTED RUSH.
Perennial, tufted or with a creeping rhizome; stems to 60cm, or to 100cm when growing in water, erect or ascending, with to 2 basal sheaths. Inflorescence with 5–20 heads each of to 15 flowers, ebracteolate; perianth segments to 3.5mm, ovate or lanceolate; stamens 6, anthers equalling or longer than filaments. Capsule to 4mm, usually ovoid, shiny, black; seeds to 0.6mm. Boreal regions, Australia. Z4.

*J.balticus* Willd. (*J.arcticus* ssp. *balticus* (Willd.) Hylander).
BALTIC RUSH.
Stems to 100cm, erect, stiff, glaucous; lowest bract to one-third the length of the stem. Inflorescence lateral, loose, with to 60 flowers; perianth segments to 5mm, subequal, ovate; anthers to 1.5mm, to twice the length of the filaments. Capsule to 4.5mm, dark brown; seeds to 1mm, ovoid. Boreal regions. Z3.

*J.bufonius* L. TOAD RUSH.
Stems many, to 50cm, often much shorter, weak, spreading to sprawling. Leaves all cauline, to 12 × 0.2cm, flat, pungent. Inflorescence loose, apical, occupying the upper half of the stem; flowers with involucral bracteoles; perianth segments to 8mm, unequal, narrow-ovate; stamens 6, anthers shorter than or equalling the filaments. Capsule to 5mm, ovoid to ellip-soid; seeds ellipsoid, to 0.55mm, smooth or striate. Temperate regions. Z5.

*J.bulbosus* L. (*J.supinus* Moench; *J.kochii* F.W. Schultz).
Tufted perennial; stems to 30cm, or to 100cm when growing in water, often prostrate or floating, with a bulbous basal swelling. Leaves basal, terete, slightly compressed or furrowed. Inflorescence of 3–20 heads, each with 2–15 flowers; flowers sometimes replaced by adventitious shoots; perianth segments to 3mm; stamens 3–6, anthers one-third as long as or equalling the filaments. Capsule to 3.5mm, 3-angled; seeds to 0.6mm. Temperate regions. Z5.

*J.capitatus* Weigel. DWARF RUSH.
Tufted annual to 5cm. Stems erect, branching below. Leaves to 4cm, radially disposed, channelled. Inflorescence a solitary, green to red-brown head subtended by radial bracts.

*J.castaneus* Sm. CHESTNUT RUSH.
Perennial with a creeping rhizome; stems solitary, to 32 × 0.3cm, with basal sheaths. Leaves basal, sometimes also one on the upper stem, to 10 × 0.4cm, the leaves and one long bract narrowing to a blunt end. Inflorescence of 1–3 heads, each with to 10 flowers and leaflike bracts; perianth segments to 5.5mm, dark brown, ovate; stamens to 6, anthers to one-third the length of the filaments. Capsule to 7.5mm, red-brown at apex, lighter at base; seeds to 3mm. Boreal and arctic regions. Z3.

**Juncus**  (a) *J.effusus* (b) *J.ensifolius* (c) *J.bufonius* (d) *J.articulatus* (e) *J.bulbosus*

**J.chamissonis** Kunth (*J.imbricatus* Laharpe).
Densely tufted perennial; stems to 40cm, with several basal sheaths. Leaves to 30 × 0.1cm, canaliculate. Inflorescence compact, with to 25 flowers; perianth segments equal, outer segment ovate, inner segment lanceolate; stamens to two-thirds the length of perianth, anthers equalling filaments. Capsule to 6mm, exceeding perianth; seeds 0.4mm, ovoid. Temperate S America. Z8.

*J.communis* hort. See *J.effusus*.

**J.compressus** Jacq. ROUND-FRUITED RUSH.
Loosely tufted or with a creeping rhiz; stems to 40cm, with to 3 basal sheaths. Leaves basal and on upper stem, to 25 × 0.2cm, glaucous. Inflorescence loose; flowers to 60; perianth segmentsto 3mm; anthers to twice length of filaments. Capsule to 1.5 times length of perianth segments, globose to ovoid; seeds to 0.5mm, with to 12 striations. Temperate regions. Z5.

*J.decipiens* See under *J.effusus*.

**J.effusus** L. (*J.communis* hort.). COMMON RUSH; SOFT RUSH.
Perennial, to 150cm, often shorter; aerial parts smooth or faintly striated; stems soft, erect or arching, densely tufted, with red-brown basal sheaths. Leaves all basal, cylindrical, equalling stems. Inflorescence a diffuse 'lateral' cyme, to 5cm long; perianth segments to 3mm, pale brown or yellow-green; stamens 3–6, anthers to 0.8mm, equalling filaments. Capsule obtuse, shorter than perianth; seeds to 0.5mm, ovoid, reticulate. Eurasia, N America, Australia, New Zealand. 'Aureostriatus': stem banded ivory to lime. 'Spiralis' (CORKSCREW RUSH): spiralling strongly. A smaller, intriguing plant with with more tightly spiralled leaves and stems is *J.decipiens* 'Curly Wurly'. It forms a low, straggling, dark green mass resembling coiled florist's wire. 'Vittatus': narrowly banded ivory. 'Zebrinus': deeply banded green-white. Z4.

**J.ensifolius** Wikstr.
Perennial, to 80cm, often shorter, loosely tufted or with a creeping rhizome; stem compressed and narrowly winged, with a few basal sheaths. Stem leaves 4–6 on flowering stems only, sword-shaped, to 5mm wide, sheathing, equitant, resembling a very small flag iris. Inflorescence of 1–6 globose, many-flowered heads, each to 0.8cm diam., deep glossy chestnut to maroon-black, terminating an erect scape; perianth segments to 4mm; stamens to half length of perianth segments, anthers to 0.8mm, equalling filaments. Capsule slightly

exceeding perianth; seeds to 0.7mm, dark, both ends acute. Western N America. Z3. Another rush with Iris-like foliage is the larger *J.xiphioides* – the inflorescences are red-brown and the leaf margins papery, appearing variegated.

*J.glaucus* Sibth. See *J.inflexus*.
*J.imbricatus* Laharpe. See *J.chamissonis*.

**J.inflexus** L. (*J.glaucus* Sibth.; *J.longicornis* Bast.). HARD RUSH.
Tufted perennial, to 120cm, with red-brown basal sheaths; lowest bract long. Leaves and stems glaucous, with 10–20 striations; pith interrupted. Inflorescence apparently lateral, lax, with straight branches; flowers many; perianth segments to 4mm, unequal, ovate; stamens 6, anthers to half as long again as the filaments. Capsule exceeding perianth, ovoid to ellipsoid; seeds to 0.5mm, reticulate. Temperate regions. 'Afro': to 60cm, compact; leaves blue-green, in close spirals; flowers brown. Z4.

*J.kochii* F.W. Schultz. See *J.bulbosus*.
*J.lampocarpus* Ehrh. ex Hoffm. See *J.articulatus*.

**J.leseurii** Bolander. SALT RUSH.
Rhizome creeping. Stems to 90cm, thick, smooth. Flowers in 3–4-branched cymes to 3.5cm; perianth segments deep brown edged purple-maroon. N America, Pacific Coast. Z6.

*J.longicornis* Bast. See *J.inflexus*.
*J.nodosus* Weber. See *J.subnodulosus*.

**J.pusillus** Buch.
Dwarf tufted perennial. Stems to 1.5 x 0.3cm, creeping and rooting. Leaves filiform, to 0.2cm diam. Inflorescence usually solitary.

*J.obtusifolius* Ehrh. ex Hoffm. See *J.subnodulosus*.
*J.spiralis* hort. See *J.effusus* 'Spiralis'.

**J.subnodulosus** Schrank (*J.obtusifolius* Ehrh. ex Hoffm.; *J.nodosus* Weber). BLUNT-FLOWERED RUSH.
Perennial; stems to 130cm, with 3–4 basal sheaths and 1–2 stem leaves; non-flowering stems short, with one long leaf. Leaves to 10 × 0.4cm, sword-shaped. Inflorescence of 50 hemispherical heads each with to 30 flowers; perianth segments to 2.5mm, elliptical; stamens 6, shorter than perianth segments, anthers to twice length of filaments. Capsule slightly longer than perianth segments, flattened; seeds to 0.6mm, striated. Europe; boreal Asia. Z6.

*J.supinus* Moench. See *J.bulbosus*.
*J.xiphioides* See under *J.ensifolius*.

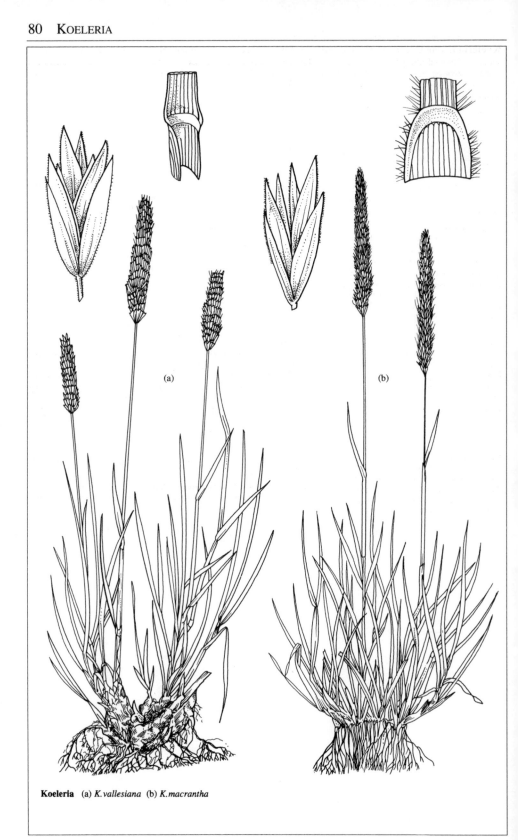

**Koeleria**  (a) *K.vallesiana*  (b) *K.macrantha*

**Koeleria** L. (For G.L. Koeler (1765–1806), botanist specializing in grasses.) HAIR GRASS. Gramineae. Some 25 species of annual and perennial grasses to 60cm. Stems flimsy, erect. Leaf blades narrow; lower sheaths occasionally swollen, appearing bulbous; ligules papery, translucent. Panicle shining, cylindric, lobed or disjointed, very dense, spike-like, to 10cm; spikelets subsessile, laterally compressed, 2–8-flowered; glumes unequal, acute, papery, translucent; lower glume 1-ribbed, upper glume 3–5-ribbed, longer than lower; lemma exserted beyond glumes, narrow, lanceolate, acute to obtuse, 3–5-ribbed, keeled, margins very thin, occasionally short-awned; palea equal to, or shorter than lemma, 2-keeled, with 2 apical lobes. Summer. Temperate regions, Tropical Africa.

CULTIVATION Grown for the shining spike-like inflorescences carried on slender stems above basal clumps of fine, sometimes blue-grey foliage, *Koeleria* species are used in the flower border, particularly useful over early spring bulbs. *K.glauca* and *K.vallesiana* have fine, blue leaves overtopped by slender, parchment-coloured spikes. The flower spikes are used fresh or dried in floral arrangements. Grow in full sun in any not too fertile well drained soil including calcareous soils and shallow soils over chalk. Propagate by seed or division in autumn or spring.

**K.argentea** Griseb.
Lemmas definitely awned on the dorsal surface; lower sheaths scarious, shiny, silvery; glumes subequal; spikelets 5mm; awns to 2mm. Himalaya. Z6.

**K.brevis** Steven. See *K.lobata*.
**K.cristata** (L.) Pers. See *K.macrantha* or *K.pyramidata*.
**K.cristata** var. *glauca*. See *K.glauca*.
**K.degenii** Domin. See *K.lobata*.
**K.generensis** Domin. See *K.pyramidata*.

**K.glauca** Coleman ex Willk. & Lange.
Perennial, to 60cm. Culms thickened at base. Some sterile leaf blades flat or involute, strongly glaucous-blue above, adpressed-pubescent beneath, scabrous; sheaths splitting. Panicles cylindric, to 10cm, spikelets 2–3-flowered, to 5mm; glumes obtuse, glaucous; lemma, glaucous, pubescent. C Europe, Siberia. Z4.

**K.gracilis** Pers. See *K.macrantha*.

**K.lobata** (Bieb.) Roem. & Schult. (*K.brevis* Steven; *K.degenii* Domin).
To 35cm, densely caespitose, forming small, compact tufts. Culms enlarged and bulbous at base. Leaves convolute, to 5cm × 1.5mm, glabrous, glaucous; sheaths persistent, dilated, glabrous, parallel- or corrugated-veined. Panicles dense, to 2.5 x 0.8cm; spikelets to 7mm; pedicels short or absent; glumes unequal, acuminate, light green or green-purple; scabrous or pubescent; lemma glabrous to somewhat pubescent, awn to 1mm. SE Europe. Z5.

**K.macrantha** (Ledeb.) Schult. (*K.cristata* (L.) Pers.; *K.gracilis* Pers.). CRESTED HAIR GRASS.
Perennial, 10–50cm. Culms in compact clumps, rigid. Leaf blades flat or involute, obtuse, narrow, to 20cm, grey-green, old sheaths clothing stem; ligules to 1mm, papery, translucent. Panicles narrow-oblong or tapering, interrupted, to 10 x 2cm, tinged with grey or purple; branches very short; spikelets 2–3-flowered, to 0.5cm, glabrous to pubescent, margins silver-grey; glumes acute, awned. Europe, Asia, N America. Z2.

**K.phleoides** (Vill.) Pers. See *Rostraria cristata*.

**K.pyramidata** (Lam.) Beauv. (*K.genevensis* Domin). JUNE GRASS; CRESTED HAIR GRASS.
To 90cm, loosely tufted, sometimes with long, creeping rhizome. Culms robust, glabrous to sparse-pubescent Leaves flat, to 23cm x 3.5mm, green or glaucescent, glabrous or sparse long-white-pubescent at margins; sheaths persistent, sparse-sericeous to dense-pubescent. Panicle pyramidal, to 22 x 3cm; rachis pubescent; spikelets to 8mm; glumes unequal, acute, glabrous or scabrous, green or light brown; lemma acuminate to short-aristate, glabrous to ciliate. Flowers like those of *K.glauca* but with glumes acute. EC Europe. Z2.

**K.vallesiana** (Honck.) Bertol.
Perennial to 40cm, densely tufted. Culms stiff, erect. Leaves to 10cm, flat or involute, glabrous, very narrow, blue-green to silver. Panicles oblong to ovate-oblong, purple to silvery green becoming buff; spikelets 2–3-flowered. W Europe.

**Lagurus** L. (From Greek *lagos*, hare, and *oura*, tail, referring to the fluffy panicles.) HARE'S TAIL Gramineae. 1 species, an annual grass to 60cm. Culms tufted, erect, flexible, branching at base, hairy. Leaf blades arching, flat, linear to narrowly lanceolate, to 20 x 1cm, downy; sheaths loose, hairy; ligules papery, translucent. Panicle spike-like, ovoid to oblong-cylindric, densely and softly hairy, light green or tinged mauve, to 6 x 2cm; spikelets densely overlapping, 1-flowered, to 1cm; glumes to 1cm, unequal, 1-ribbed, aristate, hairy; lemma shorter than glumes, bifid at apex, glabrous, awned from dorsal surface; awn exserted, bent, to 2cm; palea with 2 keels extending into short awns. Summer. Mediterranean. Z9.

CULTIVATION  Grown for the very softly hairy, hare's-tail flowerheads, used fresh or air dried in floral arrangements. Grow in light, sandy, freely draining soils in sun; cut or pull flowerheads for drying before maturity. Seed may be sown *in situ* in spring, or in pots in autumn and overwintered in well-ventilated frost-free conditions for planting out in spring.

*L.ovatus* L.
'Nanus': very dwarf, to 15cm.

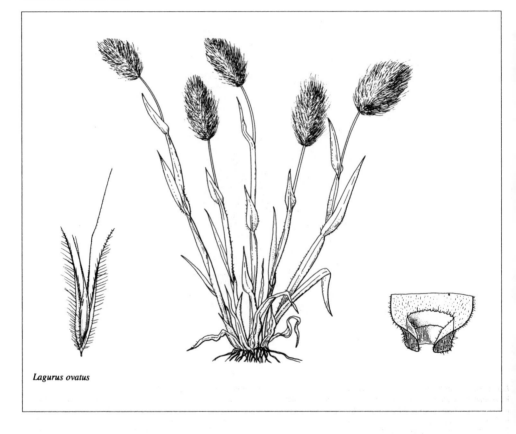

*Lagurus ovatus*

**Lamarckia** Moench. (For the French naturalist J.B.A.P. Monet de Lamarck (1744–1829).) GOLDEN TOP. Gramineae. 1 species, an annual grass to 30cm. Culms loosely tufted, smooth. Leaf blades flat, linear, acute, twisted, to 12.5 × 1cm; sheaths inflated; ligules papery, translucent. Panicle oblong, secund, dense, golden-yellow, often tinged purple, to 7.5 × 2.5cm; branches short, erect; spikelets dimorphic, in groups of 5, laterally compressed, one of group fertile, 2–13-flowered, to 5mm; upper floret asexual, vestigial; glumes equal, acute; lemmas obtuse, 5-ribbed, membranous, aristate; other spikelets sterile, many-flowered, enclosing fertile spikelets, to 8mm; lemmas obtuse, papery, without awns; palea 2-keeled. Summer. Mediterranean. Z7.

CULTIVATION  A beautiful grass with downswept silky plumose inflorescences in shining golden yellow, sometimes flushed purple when mature; these are excellent for drying. Sow seed *in situ* in spring in a light sandy, well-drained soil in full sun. For earlier flowers, sow in autumn and overwinter in frost-free conditions under glass.

*L.aurea* (L.) Moench.

**Leymus** Hochst. (From Greek *elymos*, a kind of millet.) Gramineae. Some 40 species of rhizomatous, perennial grasses. Leaves stiff, flat or rolled, usually blue or grey-green and glaucous, scabrous, pungent. Inflorescence racemose, spike-like, narrow; spikelets borne singly or in pairs, subsessile, adpressed to the rachis axis, to 7-flowered; rachis tough; glumes opposite or overlapping, linear to narrow-lanceolate, coriaceous, to 5-ribbed, acute to short-awned; palea equal to lemma, 2-keeled; lemmas lanceolate, acute to short-awned. N Temperate.

CULTIVATION   *L.arenarius* is often used as a sand-binder with marram grass, *Ammophila arenaria*, but is valuable in gardens as (potentially invasive) ground cover especially in association with other coastal natives. It is one of the finest of the blue grasses, a beautiful foil for pastel-tinted flowers or purple foliage, as are *L.racemosus* and *L.secalinus*. Shear over before flowering for a flush of fresh blue green growth in summer. Cultivate as for *Elymus* but on drier soils with sharper drainage in full sun.

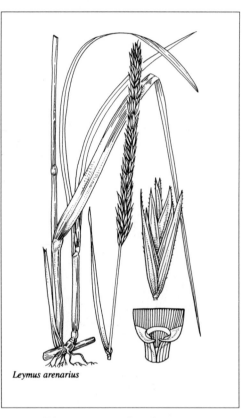

*Leymus arenarius*

**L.angustus** (Trin.) Pilger (*Elymus angustus* Trin.).
Rhizomatous Culms to 1m, clustered. Leaves to 1cm diam. inrolled, scabrous above. Spike to 25cm; spikelets paired, with 2–3 flowers; glumes to 2cm, lanceolate; lemmas to 1.5cm, 5–7-nerved; awn pointed. USSR. Z3.

**L.arenarius** (L.) Hochst. (*Elymus arenarius* L.). BLUE LYME GRASS; BUNCH GRASS; SAND WILD RYE; LYME GRASS; SEA LYME GRASS; EUROPEAN DUNE GRASS.
Culms robust, to 1.5m, upright, glabrous. Leaves flat, to 60 × 1.5cm, metallic blue-grey, margins convolute; ligule truncate. Spike to 35 × 25cm; spikelets paired, oblong to deltoid, to 3.5cm, 4-flowered; glumes narrow-lanceolate, to 2.5cm × 3mm, apex attenuate, acuminate, keeled, glabrous to short-pubescent; lemmas oblong-lanceolate, to 2.5cm, to 7-ribbed, densely pubescent, awns absent. Summer–autumn. N & W Europe, Eurasia. Z4

**L.chinensis** (Trin.) Tzvelev (*Elymus chinensis* (Trin.) Keng; *Agropyron pseudoagropyron* (Griseb.) Franch.).
Rhizome to 3.6m. Culms to 45cm, coarse. Spike to 20cm; rachilla joint smooth; lemma acuminate or awned; callus glabrous. China.

**L.condensatus** Presl & C. Presl. GIANT WILD RYE.
To 2.7m. Rhizome short, thick. Culms robust. Leaves to 75 × 2cm, blue-grey, ribs conspicuous above, scabrous. Spike upright, compact, to 30cm; spikelets in fascicles to 5, to 4-flowered, to 2cm; glumes subulate or awl-shaped, to 1.5cm; lemmas lanceolate, to 13mm, awns absent, or apex mucronate, smooth or slightly scabrous. US (California). Z7. 'Canyon Prince': leaves very silver-blue.

**L.mollis** (Trin.) Hara (*L.mollis* Trin.).
Perennial to 1.3m. Leaves to 1.5cm, broad, as with inrolled margins, rigid, smooth beneath, scabrous above. Spikes to 26cm, dense; glumes to 4mm, lanceolate, long-acuminate, thin, membranous, 3–5-nerved; lemma to 2.5cm, broadly lanceolate, 5–7-nerved, pointed, awnless, soft-pubescent. USSR. Z4.

**L.multicaulis** (Karel. & Kir.) Tzvelev (*Elymus aralensis* Reg.; *E.multicaulis* Karel. & Kir.). ARAL WILD RYE.
Close to *L.arenarius*, differing in its shorter rhizome and more erect leaves. USSR. Z4.

**L.racemosus** (Lam.) Tzvelev (*Elymus racemosus* Lam.; *Elymus giganteus* Vahl; *Elymus glaucus* hort. non Buckley). GIANT BLUE WILDRYE.
To 120cm. Leaves to 30 × 1.5cm, blue-grey, glabrous below, scabrous above. Spikes to 35 × 2cm, ultimately papery, straw-white; apex attenuate; spikelets flattened, in clusters to 6, to 6-flowered; glumes linear-lanceolate, to 2.5cm; lemma to 1.5cm, 7-ribbed, softly pubescent; apex acute, glabrous. Summer– autumn. Eurasia. 'Glaucus' ('Vahl Glaucus' sic.): upright to arching, to 75cm; leaves clear light blue – not distinct from type. Z4. See also note under *Elymus magellanicus*.

**L.secalinus** (Georgi) Tzvelev (*Elymus glaucus* Buckley). BLUE WILDRYE.
Culms 50–150cm, upright, clumped, bent at base. Leaves narrow, to 20 × 1.5cm, convolute, scabrous, blue-green, usually glaucous. Inflorescence spicate, erect to nodding, graceful; spikelets 1–1.2cm, to 6-flowered, adpressed to inflorescence axis; glumes thin, to 1.5cm, to 5-ribbed, apex acute to awned; lemma awned, erect to spreading. W US. Z5.

**Lolium** L. (Latin name for a troublesome weed, *L.tomulentum.*) RYEGRASS; DARNEL. Gramineae. 8 species of annual or perennial grasses to 1m. Rhizomes spreading extensively; culms erect, tufted. Leaf blades flat; sheaths auriculate; ligules obtuse, translucent, papery. Spike simple, flattened or cylindric, to 25cm; rachis scalloped; spikelets solitary, sessile, adpressed to hollows in rachis, opposite, distichous, few- to many-flowered, to 2.5cm; lower glumes reduced, upper glume exserted, stiff; lemmas dorsally convex, obtuse, acute or aristate, 5–7-ribbed, to 5mm. Summer. Eurasia. Z5.

**L.perenne** L. PERENNIAL RYE GRASS.
Tufted perennial to 1m; culms erect or spreading, smooth. Leaves to 30cm, glabrous. Panicle to 30cm, flattened; spikelets oblong, to 14-flowered. Spring and summer. Eurasia, widely naturalized elsewhere. Often used in pasture and turf, the panicles are also attractive and may be used cut, fresh or dried.

**Luzula** DC. (Classical Latin name, possibly from *luciola*, glow-worm, from *lux*, light.) WOOD-RUSH. Juncaceae. 80 species of perennial or, rarely, annual herbs. Leaves mostly basal, grass-like, flat or with a longitudinal groove above, soft, with long curving white hairs. Flowers inconspicuous, bracteolate, in umbel-like, paniculate, corymbose or congested inflorescences; perianth brown or green, sometimes white, with 6 free scarious segments. Fruit a 1-celled capsule; seeds 3, usually with a succulent basal appendage aiding ant dispersal. Cosmopolitan, especially temperate Eurasia. Z6.

CULTIVATION  Those most commonly cultivated are natives of open subalpine woodland, on both dry and moist soils. The rhizomatous and perennial species make effective groundcover in moist and shaded conditions. Their grass-like form provides useful foliage contrasts, especially as they age and sometimes assume attractive parchment colours. All have light soft clusters of brown or creamy flowers in early summer, and leaves conspicuously edged with hair, distinguishing them from *Juncus*.

Plant *L.sylvatica* in moisture-retentive soils on rough ground or in the wild garden, where the cream leaf margins of its cv. Marginata will lighten dark places; its creeping nature is also useful for stabilizing banks of heavy soil. The graceful *L.nivea*, has similar soil requirements to *L.sylvatica*, but will prove less invasive and appreciates a sunny aspect. *L.campestris*, although less decorative, can be used in drier woodland shade and may naturalize on well drained soils. Plant in spring and propagate by division between autumn and spring; also by seed, sown in spring/summer.

**L.albida** (Hoffm.) DC. See *L.luzuloides*.

**L.campestris** (L.) DC. (*L.subpilosa* (Gilib.) V. Krecz.). FIELD WOOD-RUSH.
To 30cm, loosely tufted. Basal leaves to 4mm wide, sparsely ciliate, soft and flat. Inflorescence with 1 sessile and a few pedunculate clusters of 5–12 flowers; peduncles straight, erect, but deflexed in fruit; perianth segments to 4mm, brown; anthers to 6× length of filaments. Capsule to 3mm; seeds to 1.3mm with a basal appendage to half the length of the seed. Europe as far north as Norway.

**L.campestris** ssp. *multiflora* (Retz.) Buchenau. See *L.multiflora*.
**L.campestris** ssp. *occidentalis* V. Krecz. See *L.multiflora*.
**L.cuprina** Rochel & Steud. See *L.luzuloides*.

**L.lutea** (All.) DC.
To 30cm, loosely tufted. Basal leaves short, linear-lanceolate, to 6mm wide, glabrous; sheaths red-brown. Inflorescence erect, condensed into pedunculate clusters of flowers; perianth segmentss to 3mm, straw-coloured, equal; anthers equalling filaments. Capsule 2.5mm, dark brown; seeds to 1.5mm, brown, oblong, basal appendage inconspicuous. S Europe.

**L.luzuloides** (Lam.) Dandy & Willmott (*L.nemorosa* (Pollich) E. Mey.; *L.albida* (Hoffm.) DC.; *L.cuprina* Rochel & Steud.).
To 65cm, loosely tufted. Basal leaves to 6mm wide, flat, light green with long hoary hairs. Inflorescence corymbose, loose or condensed into clusters of 2–10 flowers; perianth segments to 3.5mm, dirty white or pink, the inner segment longer than the outer, acute; anthers to 3× length of filaments. Capsule ovoid; seeds to 1.2mm, dark brown. S & C Europe.

**L.maxima** (Rich.) DC. See *L.sylvatica*.

**L.multiflora** (Retz.) Lej. (*L.campestris* ssp. *multiflora* (Retz.) Buchenau; *L.campestris* ssp. *occidentalis* V. Krecz.). MANY-FLOWERED WOODRUSH.
To 30cm, densely tufted, erect. Basal leaves to 4mm wide, sparsely hairy. Inflorescence umbel-like, with to 10 clusters of to 18 flowers each; flowers sessile or pedunculate; perianth segments to 3.5mm, brown; anthers slightly exceeding filaments. Capsule to 3mm; seeds oblong, to 1.3mm; basal appendage to half length of the seed. Europe, America, Australia.

**L.nemorosa** (Pollich) E. Mey. See *L.luzuloides*.

**L.nivea** (L.) DC. SNOW RUSH; SNOWY WOODRUSH.
To 60cm, loosely tufted. Basal leaves linear, to 30 × 0.4cm, dark green, hairy, flat; stem leaves to 20cm. Panicle loose with to 20 clusters of flowers; perianth segments to 5mm, snowy white, unequal, acute; anthers slightly shorter than filaments. Capsule to 2.5mm, globose; seed 1.5mm, red-brown. Alps, C Europe. 'Little Snow Hare' ('Schneehaeschen'): flower heads very fine, bright white with long hairs. 'Snow Bird': flower heads snow white.

**L.pilosa** (L.) Willd. HAIRY WOOD-RUSH.
To 35cm, tufted. Basal leaves to 10mm wide, flat, sparsely to densely long-hairy. Inflorescence with unequal spreading branches, deflexed in fruit; perianth segments nearly equal, brown with broad hyaline margins; anthers exceeding filaments. Capsule to 4.5mm, pear-shaped, light green; seeds to 1.8mm, pale brown, basal appendage to 1.5mm, hooked. Europe. 'Greenfinch' ('Gruenfink'): an improved selection.

*L.subpilosa* (Gilib.) V. Krecz. See *L.campestris*.

**L.sylvatica** (Huds.) Gaud.-Beaup. (*L.maxima* (Rich.) DC.). GREATER WOODRUSH.
To 80cm, loosely tufted, in large tussocks. Basal leaves to 30 × 2cm, channelled, with a few or many silky hairs. Inflorescence spreading, with many flowers in groups of 2–5; bracteoles lacerate to ciliate; perianth segments to 4mm, brown, the inner segment longer than the outer; anthers to 6× length of filaments. Capsule to 4.4mm; perianth segments ovoid; seeds to 2mm, shiny. S, W & C Europe. 'Aurea': leaves broad, golden-yellow. 'High Tatra' ('Hohe Tatra'): collection from the Tatra Mountains. 'Marginata': habit dense; leaves deep green edged white; spikelets gold and brown, hanging. 'Tauernpass': leaves very broad, low-lying. 'Woodsman' ('Waldler'): leaves very bright, sap green.

**L.cultivars**. 'Botany Bay': young leaves off-white, broad. 'Mount Dobson': hardy; dark brown cymes. 'N.Z. Ohau': habit large, hardy.

**Melica** L. (From Latin *melica in varro*, a kind of vessel or amphora, alluding to the swollen stem bases.) MELIC; MELICK; MELLIC. Gramineae. 70 species of perennial grasses to 150cm. Rhizomes creeping; culms flimsy to stout, erect, clumped, often bulbous at base. Leaf blades linear, flat or involute, arching; sheaths clasping. Panicles open or contracted; spikelets 2- to several-flowered, somewhat laterally compressed; rachilla usually disarticulating above glumes and between fertile florets; glumes unequal, obtuse or acute, membranous, 3–5-ribbed; lower 1–3 florets bisexual; lemma dorsally convex, 5–13-ribbed, papery, with scarious margins, aristate; palea 2-keeled; upper florets vestigial, sterile, rarely with a small bisexual floret enclosed in overlapping lemmas forming a clavate structure. Summer. Temperate regions (except Australia).

CULTIVATION  Charming grasses from woodlands and hedgerows with very finely branched panicles of small but showy florets, often white or silvery and followed by rice-like grains (jet-like in *M.altissima* 'Atropurpurea'). *M.altissima* is perhaps the most imposing species, tolerant of bright sunlight and fairly dry soils. *M.ciliata* and *M.uniflora* are sylvan delights, the first with blue-green leaves, the second a fresh pale green delicately picked out with arching, slender flower heads. Both work well naturalized in semi-wooded situations. Propagate by seed or division.

**M.altissima** L. SIBERIAN MELIC.
To 150cm. Rhizome creeping. Culms erect, stout. Leaf blades acute, mid-green, to 22.5 × 1cm, scabrous with midrib prominent beneath; ligules to 5mm. Panicles loose, interrupted at base, to 25 × 2.5cm; spikelets furled at first then spreading, off-white and showy, oblong, to 1cm; fertile florets 2; fertile lemmas acute, finely scabrous. C & E Europe. 'Alba': spikelets white. 'Atropurpurea': spikelets deep mauve, sweeping downwards and overlapping; grains black. Z5.

**M.ciliata** L. HAIRY MELIC GRASS; SILKY-SPIKE MELIC.
To 75cm. Culms densely tufted, erect, flimsy, grey-green. Leaf blades grey-green, involute and filiform or flat and broader, acute, to 17.5cm, grooved, ligules to 2mm, split. Panicles dense, nodding, cylindric, 15 × 1cm, silky, pale green or tinged purple; spikelets subsessile, elliptic-oblong, to 0.5cm; glumes ovate, acute, to 0.8cm; lemma narrowly ovate, acute, to 0.5cm, silky white-ciliate. Europe, N Africa, SW Asia. Z6. 'Alba': florets white.

**M.nutans** L. MOUNTAIN MELIC; NODDING MELIC.
To 60cm. Culms clumped or solitary, slender, tetragonal in section. Leaf blades to 20 × 0.5cm, green, pubescent above; sheaths tubular; ligules obtuse, to 0.1cm. Panicles secund, loose, nodding to 15cm; spikelets to 1cm, elliptic-oblong, obtuse, purple or maroon; fertile florets 2–3, on thread-like stalks to 1.5cm; glumes elliptic, obtuse, papery; fertile lemmas dorsally convex, obtuse, finely scabrous; palea elliptic, keels winged. Europe, N & SW Asia. Z6.

**M.transsilvanica** Schur.
As for *M.ciliata* except to 1m; leaves flat, to 0.6cm wide, midrib prominent; panicle shorter; glumes distinctly unequal, upper twice as long as lower. Eurasia. Z6.

**M.uniflora** Retz. WOOD MELIC.
To 60cm. Culms loosely, tufted. Leaf blades acute to 20×0.8cm, pubescent above, flat, fresh pale green; sheaths tubular, mucronate at apex, sparsely hairy or glabrous.

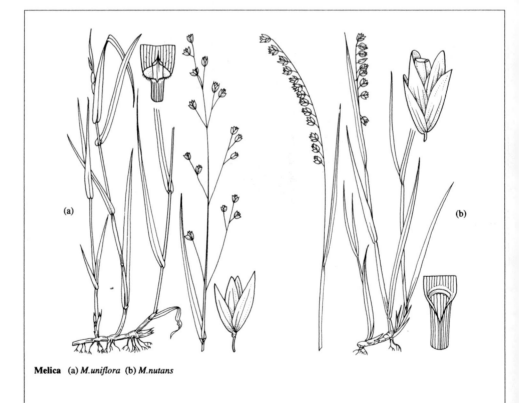

**Melica**  (a) *M.uniflora*  (b) *M.nutans*

Panicles lax, sparsely and finely branched, to 20 × 10cm; spikelets obovate, to 0.8cm, purple or tinged brown, on threadlike stalks to 0.5cm; fertile floret solitary; glumes equal; fertile lemmas obtuse, smooth. Europe, SW Asia.

'Alba' ('Albida'): very pale green form with white spikelets. 'Aurea': leaves tinted gold. 'Variegata': very small; leaves longitudinally striped with cream and green; more often culti- vated than type. Z7.

**Mibora** Adans. (Name invented by Michel Adanson (1727–1806).) EARLY SAND-GRASS; SAND BENT. Gramineae. 1 species, an annual grass to 3cm, rarely to 10cm. Culms very slender. Leaves tufted, arching, glabrous grey-green; blades flat or rolled, to 2.5cm, very narrow; sheaths overlapping; ligules obtuse, translu- cent, papery. Raceme spike-like, secund, erect to drooping, to 2cm, tinged red or purple; spikelets subsessile, distichous, overlapping, laterally compressed, oblong, obtuse, 1-flowered, to 3mm; glumes equal, 1-veined, membranous; lemmas shorter than glumes, 5-ribbed, densely pubescent; palea equals lemma. Late winter–spring. Mediterranean, coastal W Europe. Z7.

CULTIVATION  *M.minima* is one of the smallest of grasses, one of the earliest to emerge and in bloom in early winter; on moist soils, it retains its bright foliage well into the season and may produce a second flush of flowers in autumn under favourable conditions. *M.minima* is sometimes used to provide shade at the roots of terrestrial orchids and in the rock garden. Grow in damp but light sandy soils in light shade or in sun. Propagate by seed; *M.minima* may self-sow where conditions suite.

*M.minima* (L.) Desv. (*M.verna* P. Beauv.).

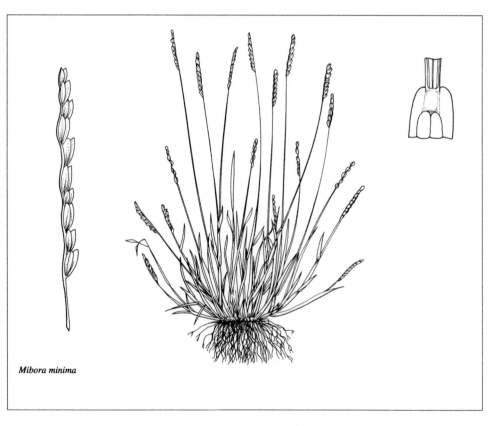

*Mibora minima*

**Milium** L. (Latin *milium*, millet.) Gramineae. 6 species of annual or perennial grasses to 2m. Culms in loose tussocks, flimsy to stout. Leaf blades flat; ligules papery, translucent. Panicles loose or contracted; spikelets single-flowered, somewhat longitudinally flattened; glumes subequal, obtuse, 3-ribbed; lemma shorter than glumes, elliptic, dorsally convex or flattened, 5-ribbed, stiff, shining, not awned; palea obtuse, 2-ribbed, not keeled, rigid. Summer. Eurasia, N America. Z5.

CULTIVATION   *M.effusum*, found in oak and beech woodland frequently on damp, heavy, humus-rich, lime soils, has been widely planted in English woodlands as a food source for game birds. Grown as ornamentals for flowers and foliage, the golden and variegated forms are beautiful ground cover for semi-shade or in the dappled sunlight of the woodland garden. The fresh spring leaves of *M.e.* 'Aureum' are also useful in floral arrangements. It is a cool-season grower, becoming nearly dormant in midsummer in hot, humid climates. Hardy to temperatures of –20°C/–4°F and below. Plant in fertile, moist but well drained soil. Propagate by seed and division, *M.e.* 'Aureum' comes true from seed. *Milium* will self-sow but seldom to the point of nuisance.

***M.effusum*** L. WOOD MILLET.
Perennial to 180cm. Leaf blades 30 × 1.5cm, glabrous, undulate; sheaths smooth; ligules obtuse, to 1cm. Panicles to 30 × 20cm, lax, lanceolate to ovate or oblong, nodding, finely branched, light green or tinged purple; branches in whorled cluster, flexuous; spikelets to 5mm. 'Aureum' ('Bowles' Golden Grass', this name often used as a popular, rather than cultivar name): leaves and spikelets wholly suffused golden yellow to lime green; comes true from seed. 'Variegatum': leaves bright green with white longitudinal stripes; weak-growing. Z6.

*M.multiflorum* Cav. See *Oryzopsis miliacea*.

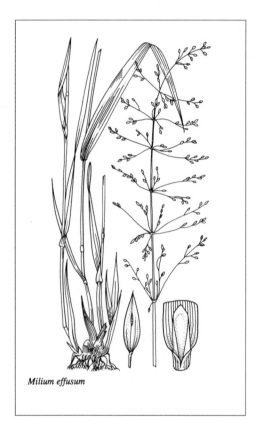

*Milium effusum*

**Miscanthus** Anderss. (From Greek *mischos*, stalk, and *anthos*, flower, referring to the stalked spikelets.) Gramineae. 17 species of perennial grasses, culms reed-like, strong, clumped. Leaf blades narrow, not rolled but somewhat folded, gracefully arching; ligules short, membranous. Panicle terminal, oblong-elliptic or obtusely pyramidal, composed of arching, spike-like, pubescent racemes; rachis continuous; spikelets in pairs, equal, with a ring of long hairs at base, stalks of pairs unequal, 2-flowered; upper floret hermaphrodite, lower sterile; glumes equal, stiff, awnless; lemmas papery, translucent; lemma of fertile floret softly aristate, shorter than lemma of sterile floret. Old World Tropics, S Africa, E Asia.

CULTIVATION   To the uninitiated, it must sometimes seem as if *Miscanthus* were all there is to ornamental grasses, with so many cultivars available. Fortunately this is not the case; but their popularity is well deserved, for the *Miscanthus* is an all-round superb performer and an excellent choice for a multitude of landscape purposes. Use varies with the different species and cultivars; however, most are suited for specimen use, for massing, or for screening. *M.sinensis*, the Eulalia or Japanese Silver Grass, hails from slopes in lowlands and mountains of Japan, also the southern Kuriles, Korea, China, Ryukus and Taiwan. Its height varies from 1.5m to 3m. The leaves of the typical form are up to 2.5cm wide. They are medium green in colour with a prominent white midrib. There are innumerable cultivated selections of the species, many of which have variegated foliage.

New growth begins in mid-spring. The foliage begins to have some effect in the garden by late spring or early summer. *Miscanthus* is a warm-season grower, and most growth takes place in the hot summer months. Typical fall foliage colour is yellow, but it varies with cultivars. The leaves of most are dry and dormant by mid winter. The typical form of the species produces its reddish inflorescences in the early autumn in the mid-Atlantic region. The flowering time for the various cultivars ranges from early summer to early autumn. Although the flowers of some selections open with a reddish cast and others open silver, all take on a silver-white appearance upon drying. *Miscanthus* flowers are superb for cut or dried arrangements, and frequently remain attractive in the landscape long into winter. Many of the more recently developed cultivars are earlier flowering. Although this potentially extends the blooming season, cultivars which flower earlier generally shed their seeds earlier, and their flower heads are not fluffy and full in winter.

*M.floridulus* is almost 1m taller than *M.sinensis* and the large inflorescences have a strong central stalk. For a number of years authors and growers used this name to refer to the commonly cultivated clone that has both the tall stature and the clump- forming habit of this species and the awnless florets typical of *M.sacchariflorus*. This clone is now more properly known simply as *Miscanthus* 'Giganteus'. Currently *M. floridulus* may be best represented by plants sold under the cultivar name 'Nippon Summer'.

*M.oligostachyus* is another Japanese native far smaller than *M.sinensis* with a more open habit and relatively thin leaves. Uncommon in western cultivation, variegated forms are frequently grown in Japan. Probable western cultivars include 'Herbstfeuer' (similar to 'Purpurascens' but shorter and slower-growing) and 'Purpurascens' (usually listed under *M.sinensis*, it is most closely allied to *M.oligostachyu*s) with good red-orange fall foliage colour. This grows to 1.5m, is narrowly upright and never needs staking. It will perform satisfactorily even in light shade, in which case the fall foliage colours will be various pastels. The flower heads are more erect than in most *Miscanthus*, and appear relatively early, in tones of pink-flushed oatmeal.

*M.sacchariflorus*, the Silver Banner Grass, originates from wet places in lowlands of Japan, also Manchuria, Ussuri, Korea and northern China. It is an upright-arching grass to 2.5m, spreading rapidly by rhizomes. The leaves are to 1.25 cm wide, medium green with white midrib, turning yellow in autumn. The flowers appear in late summer, opening silver and becoming fluffy-white upon drying. The inflorescences are narrower and more upright than those of *M.sinensis*. They remain attractive through most of winter. Useful for cut or dried arrangements. Its general culture is similar to that of *M.sinensis*. It prefers full sun and tolerates damp or wet soils. It is well suited for colonizing large sites, but can be difficult to control in smaller gardens. The running nature is less pronounced on heavy soils.

*M.transmorrisonensis*, recently introduced from Taiwan, is closely related to *M.sinensis* and sometimes considered synonymous with it by taxonomists. Horticulturally, it is quite distinct, with narrow green foliage rarely topping 1m and flowers held high above the foliage in midsummer on long graceful stems. The foliage often stays green into midwinter. It may succumb to cold injury.

These are generally plants for full sun only, however some cultivars are more shade-tolerant than others. Siting in shade promotes lax growth and the attendant need for staking and tying. Even in full sun, some cultivars are notoriously floppy when they attain full size. They tolerate a wide range of soils, and will easily grow in moist or even wet soils, yet most are reasonably drought-tolerant. Although plants growing on sandy soils may grow fuller if fertilized, feeding is unnecessary on normally fertile garden soils, and may cause over- lax growth. Clumps eventually tend to die out in the centre, and may require lifting and dividing after 5–7 years. Annual maintenance generally consists of cutting back by mid-spring before new growth begins, or in late autumn if winter interest is not a consideration. The species and some of the earlier-flowering cultivars have a definite tendency to self-sow. Whilst some have naturalized in the US and in Europe, *Miscanthus* seeding is easily controlled in the garden.

Until recently this genus could be said to be nearly disease-free; however, a recently introduced mealybug, *Pilococcus miscanthi* now poses a serious threat to *Miscanthus sinensis* and possibly to other species. There is little information currently available on this mealybug's tenure in the United States, although it seems to have been introduced at least 3 or 4 years ago. It is thought to be of oriental origin. It seems to have only one generation per year, overwintering as adult females. The eggs hatch and crawlers appear in late spring. This mealybug is big and prolific, and can reduce even the even the largest clump of *Miscanthus* to a stunted, chlorotic mess

**Miscanthus** (a) *M. sinensis* (a1) leaf tip of typical plant (a2) spikelet (a3) raceme (a4) 'Giganteus'
(b) *M. sacchariflorus* (b1) panicle (b2) culm tip and leaves

that will cease to flower. By autumn the lower culms and insides of the leaf sheaths may be white-caked with mealybugs. The mealybug attacks all parts of the plant including the roots, so above-ground mechanical or chemical methods are not sufficient for control. Unfortunately, the only proven cure at this time is fumigation with ethyl bromide or Vapam. Entomologists have suggested that systemic insecticides work if applied in spring after the eggs hatch. For those not fond of using pesticides in the garden, the best strategy is probably to remove and destroy any plants that show infestation and to be cautious when purchasing new plants. It may be hoped that a natural predator will be found. The species are easily grown from seed. The cultivars, most of which are clonal, should be propagated by division. Division is the safest in spring, but may succeed in an autumn followed by a mild winter. The same holds true for transplanting times.

**M.floridulus** (Labill.) Warb. (*M.formosanus* A. Camus; *M.japonicus* Anderss.).
Robust, tufted, evergreen perennial. Culms stout, to 2.5m, glabrous. Leaves elongate, to 3cm diam., light glaucous green, scabrous at margins, pubescent above towards base; ligules truncate, to 2mm; sheaths glabrous. Panicles erect, pyramidal, to 50cm, white, puberulent-scaberulous, with axillary tufted-pubescence; racemes many, to 20cm, slender, branched at base, subsessile; spikelets obliquely patent, lanceolate, to 3.5mm, acute, with tufted-pubescence to 6mm at base; glumes herbaceous, glabrous; awn to 1cm. Summer. Ryukyus, Taiwan, Pacific Is. Z6. The name *M.floridulus* is sometimes misapplied to *M.*'Giganteus'.

**M.formosanus** A. Camus. See *M.floridulus*.
**M.**'Giganteus'. See under *M.sinensis*.
**M.japonicus** Anderss. See *M.floridulus*.
**M.oligostachyus** Stapf. and cvs. See CULTIVATION above.

**M.sacchariflorus** (Maxim.) Hackel. AMUR SILVER GRASS; SILVER BANNER GRASS.
To 3m. Rhizome creeping, roughly bumpy. Leaf blades spreading, linear, acute, flat, stiff, to 90 × 3cm, smooth, margins rough, midrib very pale green to silver-white; sheaths beige. Panicles finely hairy, to 40 × 13cm, very pale green, tinged red or purple; racemes many, to 30cm, flimsy; spikelets narrow, to 5mm, dull beige, with long silky hairs twice length of spikelet from base. Late summer–autumn. Asia; escaped in US. 'Aureus': small, not vigorous, to 1.5m; leaves striped gold. 'Robustus': habit large, vigorous, to 2.2m 'Variegatus': leaves narrowly striped white. Z8.

**M.sieboldii** Honda. See *M.tinctorius*.

**M.sinensis** Anderss. EULALIA; JAPANESE SILVER GRASS.
To 4m. Rhizome short, not stoloniferous. Culms in large clumps, rigid, erect, smooth. Leaf blades mostly basal, erect or arching, linear, acute, flat, to 120 × 1cm, glabrous to sparsely hairy above, typically blue-green; sheaths stiff, rolled, minutely serrate; ligule silky-haired. Panicles obpyramidal, branching, to 40 × 15cm, pearly white to light grey, tinged purple-brown or maroon; racemes erect or spreading, to 20cm; spikelets lanceolate to 5mm, with a ring of long, silky, white or purple hairs at base; upper, fertile lemma with awn to 1cm. Summer. E Asia. Z6. var. **condensatus** is limited mostly to coastal areas near seashores in Japan, but also occurs at higher elevations in Japan and also in China, Indochina and the Pacific islands. Some taxonomists consider it a separate species, *M.condensatus*. For garden purposes, it is nearly identical to the typical form of the species, only slightly taller and more robust. A selection with distinctly copper-red flowers has been sold in the US as *M.sinensis* 'Condensatus'.
CULTIVARS. 'Adagio': a diminutive selection similar to 'Yaku Jima'; it is reputed, however, to be more even in its flower production. Z7. 'Autumn Light': leaves green, undistinguished. 'Cabaret': the boldest of all the variegated selections. The 2.5cm-wide leaves are cream-white in the centre with narrow dark-green margins. The copper-red flowers appear in early autumn. Flowering stems turn pink in autumn. Very sturdy and upright, to 2.5m tall in flower. Individual stems occasionally revert to all green foliage. Slightly less cold-hardy than the type. Z7. 'Condensatus': plants sold under this name represent a vegetatively propagated selection of *M.sinensis* var.*condensatus*, which is slightly larger than the type and more coastal in its range. Z6. 'Cosmopolitan': variegated foliage that is, in effect, the reverse of 'Cabaret', with a

dark green centre and cream-white margins. The leaves exceed 2.5cm in width, and a mature specimen is over 3m tall in flower. It flowers more freely and slightly earlier than 'Cabaret', opening copper-red at the beginning of autumn. Occasionally individual stems revert to solid green. Z7. 'Ferner Osten' ('Far East'): leaves green, flowers red-bronze. 'Giganteus' (GIANT MISCANTHUS): frequently listed as belonging to *M.sinensis*, but intermediate between *M.floridulus* and *M.sacchariflorus*. Its origin is obscure, and it is probably best referred to simply as *M.* 'Giganteus'. Upright in form, with maximum height approaching 3.3m. The 2.5cm-wide solid green leaves are pendent, giving the overall effect of a large fountain. Frequently sheds its lower leaves by late summer. Flowers very late or sometimes not at all in cooler years. Z4. 'Goldfeder' (GOLDEN VARIEGATED MISCANTHUS): originated as a sport of 'Silberfeder'. Leaves attractively gold-striped. A fine garden form, but slow-growing. 'Goliath': early-flowering, tall. Leaves green. 'Gracillimus' (MAIDEN GRASS): one of the oldest and perhaps the best known of all the *Miscanthus* cultivars, grown primarily for the fine texture of its narrow green foliage and for the gracefully arching overall form. One of the last to flower, its copper-red inflorescences open in mid-autumn. Seedlings may be narrow-leaved, and this cultivar has often been grown from selected seedlings over the years. Plants purchased as 'Gracillimus' should have narrow leaves but they may not all be identical. Sometimes requires staking or tying. Z5. 'Grosse Fontane' ('Big Fountain'): leaves green.'Graziella': one of several largely green-leaved early flowering cultivars developed by Ernst Pagel, 'Graziella' is more upright and slightly wider-leaved than 'Gracillimus'. The flowers open more silver than red in colour, and are extremely fluffy and white when dry. The foliage often, but not always, turns rich orange in late autumn. Z5. 'Hinjo' (DWARF ZEBRA GRASS) similar to 'Zebrinus' but much smaller.
'Kleine Fontane' ('Little Fountain'): leaves green. 'Malepartus': early- flowering. Slightly coarser in texture than 'Gracillimus'. The flowers are extraordinarily fluffy and white when dry. Z5. 'Morning Light': distributed for a time under the name 'Gracillimus Variegatus'. Habit as graceful as that of 'Gracillimus', leaves clean cream-white-margined perhaps the most beautiful and elegant of all the *Miscanthus*. Flowers open copper-red in autumn. Slightly lower-growing than 'Gracillimus', it never needs staking or tying. Z5. 'Nippon': diminutive, suited to smaller spaces. Leaves green. Flowers red-bronze. 'Purpurascens': see under *M.oligostachyus* in CULTIVATION above. Z4. 'Sarabande': very similar to 'Gracillimus' but narrower-leaved and finer textured overall. Very cold-hardy. Z5. 'Silberfeder': the cultivar name translates as 'Silver Feather', and refers to the inflorescences, which emerge silver. There is good separation between the top of the foliage and the point on the stems where the flowers begin. The foliage is 2cm wide and medium green. Somewhat lax-stemmed, and may flop even in full sun. Z5. 'Silberpfeil' ('Silver Arrow'): promoted as being more upright than 'Variegatus', but it is nearly impossible to distinguish from it. 'Silberspinne': flowers red-bronze; leaves green, mid-sized. 'Silver Shadow': purported to be a diminutive form of 'Variegatus' but indistinguishable. 'Strictus' (PORCUPINE GRASS): has been grown for almost a century (sometimes as 'Zebrinus Strictus'). The gold-banded foliage is similar to 'Zebrinus'; this cultivar, however, holds its leaves more erect, like porcupine quills, increasing the effectiveness of the variegation and

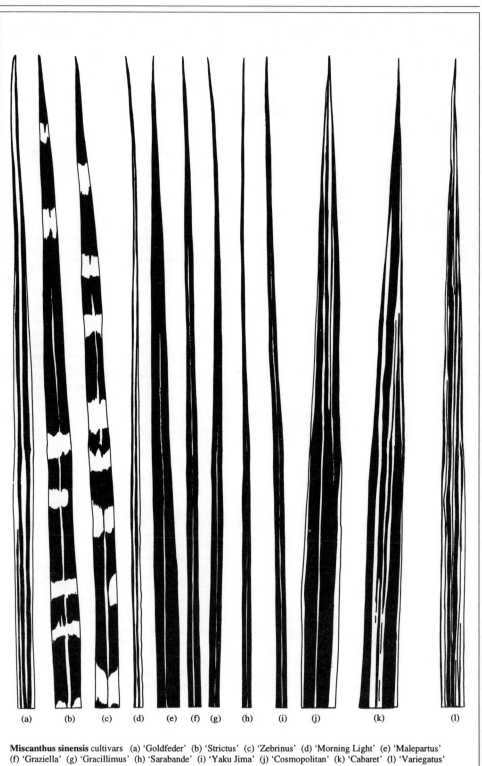

**Miscanthus sinensis** cultivars  (a) 'Goldfeder'  (b) 'Strictus'  (c) 'Zebrinus'  (d) 'Morning Light'  (e) 'Malepartus'
(f) 'Graziella'  (g) 'Gracillimus'  (h) 'Sarabande'  (i) 'Yaku Jima'  (j) 'Cosmopolitan'  (k) 'Cabaret'  (l) 'Variegatus'

making for a more spiky effect. The stems are quite sturdy and hardly ever need staking. 3m tall at maturity. Flowers in early to mid autumn. 'Variegatus': the cream-white striped foliage of this older cultivar still provides the most dramatic overall 'white' effect of all the *Miscanthus*, unfortunately mature specimens almost always need staking. Flowers in early to mid autumn. Z6. 'Wetterfahne' ('Weathervane'): leaves green. 'Yaku Jima': many of the species that grow naturally on the Japanese island of Yaku Jima (Yakushima) are smaller than their counterparts on the larger islands. This vegetatively propagated cultivar is representative of the diminutive types on Yaku Jima. With narrow green leaves, it looks like a miniature version of 'Gracillimus'. Mature height of the foliage is about 1m, with flowers about 30cm above in early autumn. Sometimes some of the flower stalks fail to emerge from the foliage. The cultivar 'Adagio' is said to be an improvement on this. Ideal for smaller gardens. Z5. 'Zebrinus' (ZEBRA GRASS): the lax green leaves have irregularly spaced horizontal bands of yellow variegation. Flowers copper in early autumn. Although the foliage is more graceful and the overall effect less formal than 'Strictus', it usually requires staking. Z6. 'Zebrinus Strictus': see 'Strictus'.

**M.tinctorius** (Steud.) Hackel (*M.sieboldii* Honda).
Tufted perennial. Culms to 1m. Leaves flat, broadly linear, to 40 × 1.2cm, scabrous on margins, with slender midrib, mostly glabrous, pilose towards base; ligules semi-rotund or truncate, to 3mm; sheaths sometimes sparsely pilose. Inflorescence erect with a very short (i.e. barely exserted) axis; racemes 2–12, to 15cm; spikelets to 6mm, acute, tufted-pubescence at base to 3mm; awn absent; glumes coriaceous, acute, long-pilose, rather brown, the first bidentate, scabrous on the 2 ribs above. Autumn. Japan. 'Nanus Variegatus' ('Variegatus'): a loosely spreading clump to 30 × 45cm of rather dwarf, bamboo-like growth, the leaves yellow-green with white stripes. Inflorescences produced in autumn, feathery, off-white to buff, disproportionately large. Z6.

**M.transmorrisonensis** Hayata. See under CULTIVATION above.

*M.zebrinus* (Beal) Nak. ex Matsum. See *M.sinensis* 'Zebrinus'.

**Molinia** Schrank. (For J.I. Molina (1740–1829), writer on the natural history of Chile.) Gramineae. 2–3 species of perennial grasses to 120cm. Culms tufted, lax; basal internode bulbous. Leaves flat; ligule a row of hairs. Panicles narrow-lanceolate, loose; spikelets laterally compressed, 1–4-flowered; glumes subequal, acute, papery, translucent, lower glumes 1-nerved, upper glume 3-nerved; lemma obtuse, 3-nerved, convex on dorsal surface, glabrous; palea equals lemma. Eurasia. Z5.

CULTIVATION  *Molinia* occurs on open moorland, in moist, humus-rich infertile, base-poor soils; in the wild, plants act as secondary host to the ergot fungus, which infects cultivated rye. The species is used in waterside and other informal naturalistic plantings, the variegated and coloured cultivars in the border or containers – all are attractive when in bloom, especially in the purple-flowered forms. The typical plant seldom exceeds 1m tall; ssp. *arundinacea* is far taller with a strong, sculptural form and lofty flower heads that play in the wind. Grow in moisture-retentive, humus-rich acid or neutral soils in part-shade – they may suffer in long, hot, humid summers. Propagate by division in early autumn and spring, species also by seed.

*Molinia caerulea*

*M.altissima* Link. See *M.caerulea* ssp. *arundinacea*.
*M.arundinacea* Schrank. See *M.caerulea* ssp. *arundinacea*.

**M.caerulea** (L.) Moench. PURPLE MOOR-GRASS.
To 120cm. Leaf blades to 45 × 1cm, deciduous, green becoming yellow in autumn; ligules densely hairy. Panicles dense, spike-like to 40 × 10cm, purple to olive-green; spikelets to 1cm; lemmas acute or obtuse, to 5mm; anthers purple-brown; stigmas mauve. Summer–autumn. 'Dauerstrahl' ('Ramsey'): green-leaved and arching. 'Edith Buksaus': to 1m; culms strong; panicles dark, long-lasting. 'Heidebraut' ('Heatherbride'): culms erect to divergent, soft straw yellow, to 1.5m; spikelets tinted yellow, forming a glistening cloud. 'Moorhexe' ('Bog Witch'): narrowly upright; stems slender, to 50cm; flowers dark. 'Moorflamme': to 70cm; leaves tinted purple in autumn; inflorescence dark. 'Nana Variegata': dwarf form of 'Dauerstrahl', to 80cm. 'Strahlenquelle' ('Fountain Spray'): green-leaved with arching-pendent flower stems.

'Variegata': tufted, compact; stems slender and upright, bright ochre, to 60cm; leaves dramatically striped dark green and cream-white; panicle tinted purple; slow-growing. ssp. **arundinacea** (Schrank) H. Paul (*M.altissima* Link; *M.arundinacea* Schrank; *M.litoralis* Host.). TALL PURPLE MOOR GRASS. To 2.5m. Leaves green turning yellow in autumn to 12mm diam. Panicle with long, patent to erecto-patent branches; lowest lemma long-acute, to 6mm. W Europe. 'Altissima': a synonym of ssp. *arundinacea*. 'Bergfreund' ('Mountain's Friend'): leaves strong yellow in autumn; panicles brown. 'Fontäne' ('Fountain'): stems inclining, forming a fountain. 'Karl Foerster': tall, leaves to 80cm, arching. 'Skyracer': to 2.5m, leaves to 1m, clear gold in autumn. 'Transparent': spikelets sparse, giving the whole infloresence a light, spacious, transparent quality. 'Windspiel' ('Windplay'): stems slender, swaying; plumes gold-brown, dense.

*M.litoralis* Host. See *M.caerulea* ssp. *arundinacea*.

**Muhlenbergia** Schreb. (for Gotthiff Heinrich Ernest Muhlenberg (1753–1815), botanist). Gramineae. 125 species of annual or perennial grasses. Culms usually tufted and mound-forming. Ligule membranous (rarely briefly so and hair-fringed). Panicle open, contracted or spiciform; spikelets typically lanceolate, lightly laterally compressed, rounded or keeled, often hairy; glumes 0–1(–3)-nerved, characteristically subequal and shorter than lemma, but sometimes exceeding it, usually entire, occasionally awned or tridentate; lemma usually firmer than glumes, 3-nerved, entire or bidenticulate, usually with a terminal or subterminal awn. New World, especially southern US and Mexico; c8 species in Southern Asia.

CULTIVATION    Those listed here are handsome perennials from dry, open grasslands; their stems are slender, mounding and topped with narrow, usually parchment-coloured flower spikes. The foliage is usually very fine, massed and weeping. *M.rigens* is most often grown, a California native ideal for hot arid regions. It has proved hardy as far north as Maryland, USA, but is perhaps best treated as a container plant, moved outside from spring to autumn in zones 8 and under. The erect flower spikes persist well into winter, making this an excellent fall grass, especially on thin, dry soil in mild areas.

Other *Muhlenbergia* species of interest include *M.dumosa*, the Bamboo Muhly Grass, a clumping, graceful, bamboo-like native of Arizona and Northern Mexico with arching culms to 2m and shimmering foliage; *M.emersleyi*, the Bull Grass, native to Texas, Arizona and Mexico; *M.filipes*, the Purple Muhly Grass from N Carolina and the Gulf Coast, producing a cloud of purple panicles; *M.lindheimeri* from Mexico and N Texas, with clumps of blue-grey foliage; *M.mexicana* from the US and Mexico, with grey-tinted foliage and closely branched, narrow panicles, and *M.pubescens* the Hairy or Blue Muhly, a striking blue-green grass from Central Mexico.

All require perfectly drained sandy or rocky soil, full sunlight and protection from prolonged, hard frost. Propagate by division in spring, or by seed.

*M.rigens* (Benth.) Hitch. MUHLY GRASS; DEERGRASS.
Culms tufted, erect, 70–180cm tall, rather slender; sheaths smooth or scabrous, covering the nodes; blades scabrous, 5–30 × 0.5–0.8cm, grey-green, involute, with long slender point. Panicles slender, mostly spicate, 20–60cm, whip-like, drying a pale straw colour and remaining on plant long into winter, strongly erect; glumes 2–3mm long, scarcely keeled, acute to obtuse; lemma scaberulous, sparsely pilose below, slightly exceeding glumes, 3-nerved upward, awnless. Summer–autumn. California. Z8.

**Neyraudia** Hook. f. Gramineae. Some 6 species of large, reed-like perennial grasses. Culms filled with pith. Leaves linear. Inflorescence an open, many-flowered panicle; spikelets 4–8-flowered, stipitate. Inflorescence rachilla jointed midway between florets, glabrous below joint, pubescent above; glumes persistent, membranous, unequal; lemma ovate-lanceolate, apex bilobed, awn recurved, margins long soft-pubescent. Old World tropics. Z9.

CULTIVATION    As for *Thysanolaena*.

*N.reynaudiana* (Kunth) Keng ex A. Hitchc. BURMA REED.
Culms to 3m. Leaves flat, to 100 × 4cm, midrib green, acute; leaf sheaths ciliate or glabrous, ligule a dense ciliate fringe.    Panicle pendent, to 80cm; spikelets to 9mm, short-stipitate, awn flat, recurved. S Asia.

**Oplismenus** P.Beauv. (From Gk *hoplismenos*, armed, referring to the awned spikelets.) Gramineae. Some 5 species of trailing annual or perennial grasses. Culms creeping and rooting, leafy. Leaves lanceolate to ovate, flat; ligules very short, ciliate. Inflorescence unilateral, few to several, spicate racemes on a common axis; spikelets paired, 2-flowered, laterally flattened, lower spikelet reduced, abscising at maturity; upper flower hermaphrodite, lower flower sterile or male; glumes subequal, to half length of spikelet, awned, awns sticky; lower lemma acute to awned, upper lemma crested, acute, flexible. Tropics, subtropics. Z9.

CULTIVATION    Found in forest shade in tropical zones, *Oplismenus* species are foliage grasses grown in temperate areas in hanging baskets and as low edging in the warm glasshouse or conservatory. *O.hirtellus* 'Variegatus' the most commonly grown in these situations and an excellent plant for glasshouse bedding and underplanting. Grow in any moderately fertile and moisture-retentive potting mix in bright indirect light or filtered light and water plentifully when in growth. Maintain a minimum temperature of 15°C/60°F. *O.undulatifolius* is the hardiest species, grown in favoured gardens outdoors, tolerating several degrees of short-lived frost. Propagate by division of rooted stems.

*O.bromoides* (Lam.) P.Beauv. See *O.burmannii*.

*O.burmannii* (Retz.) P.Beauv. (*O.bromoides* (Lam.) Palib.).
Annual, glabrous to pubescent, to 50cm. Culms leafy, slender, procumbent. Leaves lanceolate to elliptic, to 6 × 8cm, acuminate. Inflorescence to 10cm; racemes to 8, tightly clustered, bristled, erect or spreading; spikelets oblong-lanceolate, to 0.3cm; glumes awned, to 1cm. Summer–winter. Tropics. 'Albidus': leaves white with a pale green median stripe.

*O.compositus* (L.) P.Beauv. BASKET GRASS.
Creeping, procumbent perennial, resembling *O.hirtellus* but culms more robust, spikelets longer. Leaves narrow-elliptic to ovate, to 15 × 2.5cm. Inflorescence to 30cm; spikelets borne in fascicles to 10, to 10cm; glumes subequal, lanceolate to lanceolate-oblong. Africa, Asia, Polynesia.

**O.hirtellus** (L.) P.Beauv. (*O.imbecillicus* Roem. & Schult.;*O.loliaceus* (Lam.) HBK. BASKET GRASS; IDIOT GRASS. Evergreen perennial, to 90cm+. Culms procumbent, flimsy, becoming erect, rooting at nodes. Leaves narrow-lanceolate to ovate, to 5 × 1.3cm, acuminate, pubescent, dark green; sheaths glabrous to pubescent. Inflorescence to 15cm; racemes to 10, densely arranged, to 2.5cm; spikelets lanceolate-oblong to oblong, to 0.3cm, glabrous to sparsely pubescent; lower glume awned, awn to 1cm; upper glume short-

awned. Summer–winter. Tropical America, Africa, Polynes 'Variegatus' ('Vittatus'): leaves striped white, sometin tinted pink.

**O. imbecillicus** Roem. & Schult. See *O.hirtellus*.
**O.loliaceus** (Lam.) HBK. See *O.hirtellus*.

**O.undulatifolius** (Ard.) P.Beauv.
Differs from *O.hirtellus* in leaves small, ovate, strongly und late to puckered. Tropics & Subtropics.

**Oryza** L. (From Gk *oryza*, for the rice plant and its grain.) RICE. Gramineae. Some 19 species of annual perennial rhizomatous grasses. Culms flimsy to robust. Leaves linear, flat; ligules subcoriaceous to paper Inflorescence a panicle, loose or dense; spikelets laterally compressed, 3-flowered, upper flower fertile, low flowers sterile; lodicules 2; glumes obscure; sterile lemma lobed, attached to base of fertile flower, fertile lemm keeled, rigid, awned; palea 3–7-ribbed, equal to lemma, tightly enclosing grain; stamens 6. Tropical Asi Africa. Z10.

CULTIVATION The staple food of up to 50 percent of the world's population, many cultivars of rice have be developed to suit a wide diversity of climates and soil types; most are adapted to tropical and subtropical cond tions and to the wet system of cultivation used in tropical lowlands; at higher altitudes, cultivars adapted to d systems are more commonly used. In the wet system, flooding usually occurs after germination; seed may al be broadcast directly into shallow water or, in some regions, introduced as sprouts into wet mud. Rice is cap ble of germination either in the soil, or on the soil surface underwater; it is seldom successful sown in soil ar underwater. It is grown as an ornamental for the dark foliage (in some cultivars) and for the inflorescenc attractive if dried once the seed grains have swollen, although in cool temperate zones this may only occur grown under glass.

Treated as a frost-tender annual, *O.sativa* may be introduced to pond margins after the danger of frost ha passed, thus ensuring the necessary constantly moist soil. It is sometimes cultivated in collections of econom crops. Under glass, grow in clay pans or in shallow (c15–20cm/6–8in.) fibreglass trays with a drain hole, whic will allow periodic flooding and draining. Use a heavy and fertile loam-based medium, preferably with a hig proportion of clay and silt, and sow seed thinly in late winter/early spring; cover with about 2cm3/4in. of coars sharp sand and submerge to a depth of 2.5cm/1in. of water. Alternatively, germinate seed in moist soil an transplant sprouts into pans of soil covered by about 2.5cm of water. Grow in full sunlight and with moderate high humidity; the application of a balanced general fertilizer is beneficial. The optimum water temperature i between 20–30°C/68–86°F; temperature above and below this range adversely affect growth. As growth pr ceeds the depth of water is gradually increased to 15–30cm/6–12in. Provided the soil is kept continuously mois flooding is not strictly necessary; periodic drainage and re-flooding will reduce algal infestation and the harmf build-up of salts in the water. Rice will take between 90–260 days to mature, depending on cultivar, climate c environmental conditions, and from flowering to seed-ripening takes about 30 days, with a range of 15–60 days Water should be drained off about a month before harvest to allow ripening; at this stage, the roots will hav formed a dense mat at or near the soil surface, sometimes covered by green algae; although unsightly, this is se dom harmful. Rice is usually grown as an annual, but in favourable conditions a second ratoon crop is some times taken from the re-growth of stubble after the primary harvest.

**O.sativa** L.
Annual, to 180cm. Culms stout, upright, arching. Leaves elongate, to 150 × 2.5cm, usually smaller. Panicle arching to pendent, to 45cm; spikelets to 1cm; palea scabrous,

mucronate to long-awned; lemma scabrous. SE Asia 'Nigrescens': leaves dark purple. var. **rufipogon** (Griff Watt. Awns long, red; ornamental.

**Oryzopsis** Michx. (From Gk *oryza*, rice, and *opsis*, appearance, due to its similarity to *Oryza*.) RICE GRASS Gramineae. Some 35 species of perennial grasses. Culms clumped. Leaves flat to rolled; ligules membranous Inflorescence paniculate; spikelets stipitate, 1-flowered; flower narrow-lanceolate to ovate; lodicules to 3; callu short, obtuse; glumes persistent, papery; lemma leathery to rigid, brittle, convex, shorter than glumes, 5-ribbed awn deciduous, straight; palea leathery, 2-ribbed, acute. N Hemisphere, temperate and subtropics. Z8.

CULTIVATION *Oryzopsis* species are grown for their light, open flowering panicles and slender graceful habit. Grov in full sun in any moderately fertile, moisture-retentive soil. Propagate by division or by seed *in situ* in spring.

**O.hymenoides** (Roem. & Schult.) Ricker. SILKGRASS; INDIAN MILLET.
To 60cm. Culms clumped. Leaves slender, margins inrolled. Panicle to 15cm, branches flexible, spreading; glumes to 6mm; lemma to 3mm, fusiform, straight, pilose, caducous, awn to 3mm. SW US, N Mexico.

**O.miliacea** (L.) Asch. & Schweinf. (*Milium multiflorum* Cav. *Piptatherum multiflorum* (Cav.) P.Beauv.). SMILO GRASS.
To 1.5m. Culms erect to arching, loosely clumped, rigid smooth. Leaves flat, to 30 × 1cm, smooth; ligule very short Panicle very loose and open, linear to oblong, pendent, to 30cm; spikelets short-stipitate, numerous, to 4mm, occasionally tinged purple; glumes distinctly striped green; lemmas smooth, awn to 8mm. Summer–autumn. Mediterranean.

**Otatea** (McClure & E.W. Sm.) Cald. & Söderstr. (A corruption of the Aztec name for these bamboos.) Gramineae. 2 species of delicate, tender bamboos forming open clumps. Culms medium-sized, nearly solid, sulcate above, glabrous; sheaths white-green, with few or no auricles or bristles; branches 3 at first, eventually many. Leaves long and narrow, obscurely tessellated, very variably hairy, margins slightly scaberulous on one margin; sheaths practically without auricles or bristles. Mexico to Nicaragua. Z9.

CULTIVATION This clump-forming bamboo carries exceptionally graceful foliage on slender culms that arch almost to the ground. It requires plentiful water and feed in buoyant, semi-shaded conditions. Frost-tender, it should be attempted outdoors only in the most sheltered positions and the most favoured locations. It is otherwise a fine tub or border plant for the cool glasshouse or conservatory.

*O.acuminata* (Munro) Cald. & Söderstr. (*O.acuminata* ssp. *aztecorum* (McClure & E.W. Sm.) Guzman; *Arthrostylidium longifolium* (Fourn.) Camus; *Yushania aztecorum* McClure & E.W. Sm.). MEXICAN WEEPING BAMBOO.
Culms 2–8m × 2–4cm, ultimately curving gracefully, with white powder below the nodes; sheaths glabrescent, the upper sheaths deciduous, the lower disintegrating *in situ*. Leaves 7–16 × 0.3–0.5cm, numerous, pendulous. Range as for the genus.

**Panicum** L. PANIC GRASS. (From a Lat. name for millet.) Gramineae. Some 470 species of annual or perennial grasses. Leaves threadlike to linear-ovate. Inflorescence paniculate to racemose, open to contracted; spikelets symmetric to laterally compressed, awnless, abscising at maturity, 2-flowered, lower flower male, upper flower hermaphrodite; lower glume truncate to awned, equal to or shorter than spikelet, upper glume as long as spikelet; lower lemma as for upper glume, leathery to rigid, margins involute, apex obtuse to acute or apiculate, upper lemma convex, becoming coriaceous. Pantropical to temperate N America. Z5.

CULTIVATION *Panicum* species are grown for the large intricately branched flowerheads which are used for drying; *P.virgatum*, a fairly hardy perennial species, is noted for its generosity of bloom and the summer blue foliage colours of some cultivars, or the russet tones in autumn which persist into winter. Grow in any moderately fertile well-drained soil in sun. Propagate annuals by seed sown in early spring under glass and set out after danger of frost is passed, perennials by division.

*P.boscii* Poir.
Culms to 70cm, glabrous or minutely puberulent. Leaf sheaths glabrous; leaves to 12×3cm, spreading, sparsely ciliate at base, glabrous. Panicle to 12cm; spikelets papillose. E US.

*P.capillare* L. OLD WITCH GRASS; WITCH GRASS.
Annual, to 90cm. Culms clumped, upright to spreading, slender to robust. Leaves undulate, linear to narrow-lanceolate, to 30 × 1.4cm, stiffly pubescent; sheaths stiffly pubescent. Panicles open, to 45cm+, green to purple; branches very fine, spreading; spikelets stipitate, to 3mm; lower glume to 2mm, upper glume, lower lemma acuminate, equal. Summer–autumn. S Canada, US. Diffuse inflorescences useful for bouquets, interior decoration, etc. With their massed and thread-like branchlets and minute spikelets, they resemble a cloud. var. *occidentale* Rydb. Spikelets slightly larger than in type.

*P.clandestinum* L. (*Dichanthelium clandestinum*). DEER TONGUE GRASS.
Upright-open, densely tufted, semi-evergreen, rhizomatous perennial. Culms scabrous to papillose-hispid below the nodes, 50–80cm. Leaves light grey- to dark green, dense, coarse; blades spreading or finally reflexed, acuminate, 10–20 × 1.5–3cm, usually glabrous above and beneath, scabrous toward the apex, margins scabrous and papillose-ciliate below; sheaths papillose-hispid or glabrous, except the ciliate margins; ligule a pubescent ring of short hairs at the summit. Panicles 8–18cm, long-exserted, broadly oval, many-flowered, branches flexuous in distant fascicles, with ascending short branchlets in axils bearing spikelets; spikelets obovate-oblong, 2–3mm. N America. Z4. Valued for its fine foliage and the inflorescences, the axillary branches of which bear closed and self-fertile florets. The whole becomes richly tawny in autumn.

*P.crus-galli* L. See *Echinochloa crus-galli*.
*P.germanicum* Mill. See *Setaria italica*.
*P.glaucum* L. See *Setaria glauca*.
*P.italicum* L. See *Setaria italica*.

*P.maximum* Jacq.
Rhizomatous perennial to 3m, densely tufted. Culms erect, 3–4-noded, simple or sparsely branched. Leaf sheaths firm, lower compressed, otherwise terete; leaves linear, glabrous or softly pubescent, with scabrous margins; ligule membranous, very short, ciliolate. Panicle erect or nodding, to 45cm, occasionally glabrous, otherwise villous; spikelets to 5cm, oblong, subobtuse to acute, glabrous, pale green or tinged purple; glumes very dissimilar, obscurely nerved, lower rounded or slightly acute, hyaline, with 1–3 veins or unveined, upper membranous, 5-veined, same shape and size as spikelet. Tropical Africa, naturalized US.

*P.miliaceum* L. MILLET; BROOM CORN MILLET; HOG MILLET; INDIAN MILLET.
Annual, to 120cm. Culms robust, erect, or decumbent at base. Leaves linear to linear-lanceolate, to 40 × 2cm; sheaths short, stiffly pubescent. Panicle open to contracted, to 30cm, nodding, green to purple, branches rigid; spikelets stipitate, ovate to elliptic, to 6mm; upper flowers white or yellow to dark brown; lower glume to 4mm, apex sharp, acuminate to acute; upper glume, lower lemma long-acute. C, S & E Europe, Asia. 'Violaceum': inflorescence purple-tinged.

*P.palmifolium* Koenig. See *Setaria palmifolia*.
*P.plicatile* Hochst. See *Setaria plicatilis*.
*P.plicatum* Willd. non Lam. See *Setaria palmifolia*.
*P.plicatum* hort. See *Setaria plicatilis*.
*P.plicatum* Lam. non Willd. See *Setaria plicata*.
*P.polystachyum* HBK. See *Echinochloa polystachya*.
*P.purpurascens* Raddi. See *Brachiaria mutica*.
*P.ramosum* L. See *Brachiaria ramosa*.
*P.sanguinale* L. See *Digitaria sanguinalis*.
*P.spectabile* Nees ex Trin. See *Echinochloa polystachya*.
*P.subquadriparum* Trin. See *Brachiaria subquadripara*.
*P.sulcatum* Aubl. See *Setaria sulcata*.
*P.teneriffae* R. Br. See *Tricholaena teneriffae*.
*P.tonsum* Steud. See *Rhynchelytrum repens*.
*P.variegatum* hort. See *Oplismenus hirtellus* 'Variegatus'.

**P.virgatum** L. SWITCH GRASS.
Perennial, narrowly upright, to 180cm. Culms clumped, flimsy to robust, purple to glaucous green; rhizomes robust, creeping. Leaves linear, flat, erect, to 60cm × 1.4cm, usually glabrous, green, sometimes glaucous blue, becoming vivid yellow, bronze or burgundy in autumn. Panicle open, to 50 × 25cm, resembling a fine-textured cloud, lasting well into winter; branches spreading, feathery, stiff to drooping; spikelets short-stipitate, elliptic-ovate, to 6mm; lower glume to 5mm; upper glume slightly longer than lower lemma. Summer–autumn. C America to S Canada. 'Cloud Nine': to 2m; leaves blue-green. 'Hänse Herms' (RED SWITCH GRASS):

to 1.25m; habit weeping; leaves plum to rich burgundy in autumn; flowers suffused red. 'Heavy Metal': leaves stiffly erect, pale metallic blue, yellow in autumn. 'Rotbraun' ('Red Bronze'): to 80cm; leaves light brown, flushed red at tips, rich autumn colour. 'Rotstrahlbusch': leaves vivid red in autumn. 'Rubrum': to 1m; leaves flushed red, bright red in autumn; seedheads rich brown, in clouds. 'Squaw': leaves tinted red in autumn. 'Strictum' (TALL SWITCH GRASS): narrowly upright, to 1.2m. 'Warrior': tall, strong-growing; leaves tinted red-brown in autumn.

**P.viride** L. See *Setaria viridis*.

**Paspalum** L. (From Greek *paspalos*, millet.) Gramineae. Some 330 species of glabrous to sparsely pubescent annual or perennial grasses. Culms flimsy to robust. Leaves narrow-linear to lanceolate or ovate, flat to rolled; ligules membranous. Inflorescence terminal, composed of more or less digitately arranged, secund racemes; rachis winged; spikelets 2-flowered, solitary or paired, hemispherical to ovate, in 2–4 rows, awnless; lower flower sterile, upper flower hermaphrodite; lower glume absent or inconspicuous, upper glume and lower lemma equal spikelet, upper lemma usually obtuse, tough, involute. New World Tropics.

CULTIVATION  Native to forest margins, savannahs and damp habitats, *Paspalum* species are valued as ornamentals for their branched flower spikes, in which the individual florets are arranged in exceptionally neat and symmetrical formation along the raceme and, in *P.ceresia*, are clothed in long silver hairs. *P.ceresia* is tolerant of light, short-lived frost but may be grown in the cool glasshouse; *P.dilatatum* survives temperatures to −15°C/5°F. *P.notatum* is used as a coarse, low-maintenance, warm-season lawn grass. Grow in a sheltered position in a light, moderately fertile, porous soil in sun. Propagate by division or seed in early spring with gentle heat under glass.

**P.ceresia** (Kuntze) Chase (*P.membranaceum* Lam.; *P.elegans* Roem. & Schult.).
Perennial, to 75cm. Culms flimsy, ascending. Leaves linear-lanceolate, to 20 × 1cm, flat, glabrous, glaucous. Racemes to 4 per inflorescence, ascending, to 8cm; spikelets solitary, lanceolate to oblong, to 3mm, enveloped by silver hairs; ligules very short; anthers vivid yellow. Summer. Tropical S America. Z9.

**P.dilatatum** Poir. DALLIS GRASS; PASPALUM; WATER GRASS.
Perennial, to 180cm. Culms clumped, erect or ascending. Leaves linear, to 45 × 1.3cm. Racemes to 5, to 11cm; rachis winged; racemes borne on a central axis, to 20cm; spikelets elliptic to ovate, to 4mm, green tinged yellow; upper glume slightly pubescent; lower lemma glabrous. Summer. S America, naturalized throughout US south of New Jersey. Z8.

**P.elegans** Roem. & Schult. See *P.ceresia*.
**P.membranaceum** Lam. See *P.ceresia*.

**P.notatum** Fluegge. BAHIA GRASS.
Perennial. Culms to 50cm. Leaves flat or folded. Racemes to 7cm, recurved-ascending, mostly paired; spikelets to 2mm, ovate to obovate, smooth and glossy. Mexico, W Indies, S America; introduced US. Z8.

**P.racemosum** Lam. (*P.stoloniferum* Bosc).
Annual, to 90cm. Culms flimsy to robust, nodes dark brown. Leaves lanceolate to ovate, to 15 × 2.5cm glabrous; sheaths inflated. Inflorescence densely arranged; racemes to 80, beige to purple; spikelets solitary, elliptic, to 3cm. Autumn. Peru. Material under this name is often *P.elegans*. Z8.

**P.stoloniferum** Bosc. See *P.racemosum*.

**Pennisetum** Rich. ex Pers. (From Latin *penna*, feather, and *seta*, bristle, referring to the plume-like bristles of some species.) FOUNTAIN GRASS. Gramineae. Some 80 species of rhizomatous or stoloniferous annual or perennial grasses, to 340cm. Culms flimsy to robust, clumped. Leaves flat; ligule ciliate. Inflorescence terminal to axillary, spicate, cylindric to globose; spikelets clustered, to 4 per cluster, lanceolate to oblong, sessile to short-stipitate, 2-flowered, enclosed by an involucre of bristles; upper flower hermaphrodite, lower flower sterile or male; lower lemma absent or equal to spikelet, upper lemma blunt to acute, more robust than lower. Tropics, subtropics, warm temperate.

CULTIVATION    Natives of woodland and savannah throughout the tropics, many species are valued as ornamentals for their feathery and beautifully coloured panicles; in *P.* 'Burgundy Giant' the whole plant is richly coloured, as are several cultivars of *P.alopecuroides* and *P.setaceum*. Tender species are reliably perennial only in essentially frost-free zones. In cool temperate zones, most are amenable to treatment as annuals or to pot or tub cultivation (displayed outdoors in spring and summer, overwintered in a cool glasshouse). Grow in light porous soils in full sun. Propagate annuals by seed, perennials by division. *P.* 'Burgundy Giant' is propagated by stem cuttings under mist in late summer.

*P.alopecuroides* (L.) Spreng. (*P.japonicum* Trin.). CHINESE PENNISETUM; SWAMP FOXTAIL GRASS; FOUNTAIN GRASS.
Perennial, to 1.5m. Culms clumped, slender, upright. Leaves to 60cm × 1.2cm, glabrous, scabrous; sheaths flattened. Inflorescence cylindric to narrow-oblong, to 20 × 5cm, yellow-green to dark purple; axis minutely warty, to 3cm, hispidulous; spikelets solitary or in pairs, lanceolate, to 8mm. Summer–autumn. E Asia to W Australia. var. *purpurascens* (Thunb.) Ohwi. CHIKARA-SHIBA. Spikelet bristles dark purple. var. *viridescens* (Miq.) Ohwi. Spikelet bristles pale green. The foliage of *P.alopecuroides* turns golden-yellow in autumn. The flower heads resemble large foxtails or bottle-brushes. Pink to purple in colour, they arch out from the mounded foliage and form a fountain-like effect. In the landscape the flowers begin to deteriorate by early winter. They are superb for cut arrangements. Of easy culture in full sun or light shade on most soils. Prefers good drainage but can be grown in heavy soils. Will flower more fully and for a longer period with adequate moisture but is attractive even under droughty conditions once established. The species in general has a tendency to self-sow. Usually this is at a low, manageable level; however, the wider-leaved autumn blooming cultivars such as 'Moudry' and 'National Arboretum' are particularly fertile and can become serious pests by seeding into lawns adjacent to flower beds and borders. If these are grown near turf, it is advisable to cut the flower-heads back before they go to seed.
'Cassian' (CASSIAN'S FOUNTAIN GRASS): named for Cassian Schmidt. To 60cm. Strong golden autumn foliage tones. 'Caudatum' (WHITE FLOWERING FOUNTAIN GRASS): often sold as *P.caudatum* but is more correctly considered a cultivar of *P.alopecuroides*. The flowers are mostly white. 'Hameln' (DWARF FOUNTAIN GRASS): about 1m in height to the top of the flowers, which is at least 30cm lower than typical forms. It is the most commonly available low-growing cultivar, and is a reliable performer. The flower heads are slightly smaller than the typical plant and greenish-white. Has little tendency to self-sow. Often used very succesfully in groundcovering masses. The autumn color is an attractive golden-yellow. 'Little Bunny' (MINIATURE FOUNTAIN GRASS): only 15cm tall to the top of the flowers. 'Moudry' (LATE BLOOMING FOUNTAIN GRASS): quite unlike the most common examples of the species. The leaves are relatively wide, dark green, and very glossy. They form a neat, dense mound 60cm high. The flower heads are dark purple, emerging on stiff stalks in autumn. In some years, the flowers do not emerge fully from the foliage. This cultivar can be traced back to seed introduced from Japan by the US National Arboretum. Populations of plants very similar to 'Moudry' are common in parts of Japan, although they have not been clearly distinguished taxonomically. These wide-leaved plants come mostly true from seed, and although they are quite attractive, care must be taken or they will become a self-seeding nuisance in the garden. It is advisable to remove flower heads before seed ripens, especially if plants are grown near lawns. 'National Arboretum': very similar in appearance to 'Moudry'; however, the flower-heads are supposedly held out further from

the foliage in a more consistent manner. 'Paul's Giant' (GIANT FOUNTAIN GRASS): an outstanding new selection discovered by Paul Skibinski and introduced by Longwood Gardens. Produces a magnificent fountain effect, to 2m tall at maturity. Autumn foliage turns a rich gold. Effective through winter. 'Weserbergland': intermediate in height between 'Hameln' and the typical species form. Flowers greenish-white. Z6.

*P.americanum* (L.) Schum. See *P.glaucum*.
*P.asperifolium* (Desf.) Kunth. See *P.setaceum*.
*P.atrosanguineum* hort. See *P.setaceum*.
*P.caudatum* hort. See *P.alopecuroides* 'Caudatum'.
*P.cenchroides* hort. non Rich. See *Cenchrus ciliaris*.
*P.ciliare* (L.) Link. See *Cenchrus ciliaris*.
*P.compressum* R. Br. See *P.alopecuroides*.
*P.cupreum* A. Hitchc. ex L.H. Bail. See *P.setaceum*.

**P. incomptum** Nees ex Steud. SPREADING FOUNTAIN GRASS.
Rapidly spreading, upright to 120cm. Leaves linear green to grey-green. Racemes are much more slender than in *P.alopecuroides*, held nearly upright, cream-white in overall appearance late summer. China & HImalayas. Z6. Spreading too aggressively to be manageable in a mixed flower border, it is relatively early flowering, has good autumn and winter presence, and can be just the right choice for a mass planting in difficult sites.

*P.japonicum* Trin. See *P.alopecuroides*.

**P.latifolium** Spreng. (*Gymnothrix latifolia* (Spreng.) Schult.). URUGUAY PENNISETUM.
Perennial, to 270cm. Culms robust, erect, branched towards apex, branches bearing inflorescences, nodes pubescent. Leaves lanceolate, base attenuate, to 75 × 5cm, scabrous. Inflorescence compact, pendent, to 9 × 1.8cm, spikelets to 5mm, lanceolate, solitary, borne on threadlike pedicels; bristles 1–2× length of spikelets. Summer–autumn. Brazil, Peru, Argentina. Z9.

*P.longistylum* Vilm. non Hochst. See *P.villosum*.

**P.macrostachyum** (Brongn.) Trin.
Resembles *P.setaceum* but leaves to 2.5cm wide, inflorescence more compact, tinged purple-brown; bristles not plumed. 'Burgundy Giant' (GIANT BURGUNDY FOUNTAIN GRASS): upright, clump-forming, often to over 2m. The boldest-textured of the fountain grasses, with bronze-burgundy leaves over 2.5cm wide. It is not winter-hardy and must be treated as an annual. Flowers are produced from midsummer to autumn. Does not set viable seed, and must be propagated by division or stem cuttings rooted in sand and held indoors over winter. Plants should be set out in late spring. The parentage is uncertain. This cultivar was named at Longwood Gardens in cooperation with Marie Selby Botanical Gardens, from material obtained from Selby.

*P.macrostachyum* Freis. non (Brongn.) Trin. See *P.setaceum*.

**P.macrourum** Trin. (*Gymnothrix caudata* Schräd.).
Perennial, to 180cm. Culms flimsy to robust, erect, clumped. Leaves linear, flat or rolled, to 60 × 1.3cm, scabrous.

**Pennisetum**  (a) *P.setaceum*  (b) *P.alopecuroides* 'Hameln'  (c) *P.orientale*  (d) *P.alopecuroides*

Inflorescence cylindric, compact, erect or inclined, to 30 × 2cm, pale brown to purple, with bristles equal to or longer than spikelet; axis covered with tiny warts; spikelets solitary, lanceolate, to 6mm. Summer– autumn. S Africa. Z7.

**P.orientale** Rich. HARDY ORIENTAL FOUNTAIN GRASS.
Rhizomatous perennial, to 90cm. Culms clumped, slender, decumbent, erect to spreading; nodes pubescent. Leaves narrow-linear, apices straight to flexuous, to 10 × 0.4cm, grey-green, slightly rough. Inflorescence loose, to 14 × 2.5cm, pure white becoming pearly pink; axis pubescent with bristles to twice length of spikelets; spikelets in fascicles of 2–5, lanceolate, to 6mm. Summer–autumn. C, SW Asia to NW India. Z7.

**P.rueppelianum** Hochst. See *P.setaceum*.
**P.ruppelii** Steud. See *P.setaceum*.

**P.setaceum** (Forssk.) Chiov. (*P.asperifolium* (Desf.) Kunth; *P.macrostachyum* Fres. non (Brongn.) Trin.; *P.rueppelianum* Hochst.; *P.ruppelii* Steud.). TENDER FOUNTAIN GRASS.
Perennial (often treated as an annual in cool temperate gardens), to 90cm. Culms erect, stiff, slender, clumped, occasionally pubescent beneath inflorescence Leaves narrow-elongate, erect, flat or rolled, to 30 × 0.3cm, rigid, very scabrous. Inflorescence erect to inclined, plumed, to 30 × 3cm, tinged pink to purple; axis scabrous; bristles unequal, loosely pubescent near base, to 2.5cm, one to 3cm, more robust; spikelets to 0.6cm, solitary or in group to 3. Summer. Tropical Africa, SW Asia, Arabia. 'Cupreum Compactum' (COMPACT PURPLE FOUNTAIN GRASS): as red-leaved as 'Rubrum' but reaches a maximum height of 60cm. It usually does not set seed and should be propagated by division. Does not survive prolonged low temperatures. The commercial name is contrary to nomenclatural rules. 'Rubrum' ('Cupreum', 'Atropurpureum', 'Purpureum') (TENDER PURPLE FOUNTAIN GRASS): red-purple leaves up to 1.5cm wide and flower-heads more than 30cm long. It does not survive prolonged low temperatures. Usually does not set seed and should be propagated by division. It is possible that these purple-red leaved forms do not properly belong to *P.setaceum*. Treated as half hardy from plants held over in a glasshouse or purchased annually, this dramatic colour form can be stunning as a specimen or in groups or masses. Z9.

**P.spicatum** Roem. & Schult. See *P.glaucum*.
**P.typhoides** (Burm.) Stapf & C. Hubb. See *P.glaucum*.

**P.villosum** R. Br. ex Fries. (*P.longistylum* Vilm. non Hochst.). FEATHERTOP; TENDER WHITE-FLOWERED FOUNTAIN GRASS.
Perennial or annual in cool temperate cultivation, to 60cm. Culms loosely clumped, erect to ascending, pubescent below inflorescence Leaves flat or folded, to 15 × 0.6cm. Inflorescence cylindric to subglobose, compact, plumed, to 11 × 5cm, white, sometimes tinged tawny brown to purple; bristles spreading, variably pubescent below middle, to 7.5cm; spikelets lanceolate, to 1.5cm; acute. Summer–autumn. NE Tropical Africa (mts). Z8. The flowers are almost pure white, often grown for cut-flower use.

**Phalaris** L. (Classical Greek name for a kind of grass.) Gramineae. Some 15 species of annual or perennial grasses. Culms flimsy to robust. Leaves flat, often long-acute; ligules thin. Inflorescence a compact panicle, spike-like to ovoid; spikelets ovate, to 3-flowered, short-stipitate, laterally compressed, awnless; flowers 1–3, 1 flower hermaphrodite, fertile, the others sterile, reduced to subulate lemmas; glumes equal, conspicuously keeled; fertile lemma keeled, leathery, shiny, acute. N Temperate (Mediterranean, S America, California).

CULTIVATION Small reed-like grasses valued for their tolerance of a range of soil situations, dry and wet, and grown for attractive seed heads for drying (*P.canariensis*) and for the brightly variegated foliage (*P.arundinacea* cultivars), most perennial species are also cold-tolerant to between −15 and −20°C/5 to −4°F. *P.arundinacea* and *P.a.* 'Picta' are noted for their invasive tendencies (of great value in making impenetrable groundcover), providing useful colour contrast in the foliage border and in wet soils at stream and pondside; they can be confined where necessary to planting in a sunken, bottomless half barrel. *P.arundinacea* 'Picta' (Gardener's Garters), is perhaps the first variegated ornamental grass to be cultivated, its long history in gardens underwritten by its almost unkillable nature. In the American prairie States it has become a serious weed. *P.a.* 'Feesey' has brighter colour and is less invasive, as is *P.a.* 'Dwarf's Garters'. Propagate perennials and cultivars by division, annuals from seed sown *in situ* in spring, or earlier under glass.

**P.aquatica** L. (*P.stenoptera* Hackel; *P.tuberosa* L.; *P.nodosa* Murray). TOOWOMBA CANARY GRASS; HARDING GRASS.
Perennial, to 1.5m. Stem loosely or densely clumped, slender to robust, internodes swollen near base of stem. Leaves to 30 × 1cm, glabrous, rough to smooth, green sometimes tinged blue, narrow-acuminate; ligules to 8mm. Inflorescence cylindric to ovoid-cylindric, compact, to 11 × 1cm, pale green or tinged purple; spikelets elliptic-oblong, to 8mm; glumes lanceolate, winged above; wing entire; fertile lemma ovate, to 6mm, tough, downy. Summer–autumn. S Europe, Mediterranean. Z8.

**P.arundinacea** L. (*Typhoides arundinacea* (L.) Moench; *Digraphis arundinacea* (L.) Trin.; *Baldingera arundinacea* (L.) Dumort.). REED CANARY GRASS; RIBBON GRASS.
Rhizomatous perennial, to 1.5m. Culms robust, erect, or bent at base, glaucous. Leaves to 35 × 1.8cm, glabrous, ligule to 1cm; sheaths smooth. Inflorescence narrow, to 17cm; spikelets oblong, to 4mm; glumes narrow-lanceolate, to 6mm; fertile lemma narrow-oblong, sterile lemmas villous, to 2mm. Summer–autumn. Eurasia, N America, S Africa. 'Dwarf's Garters': as 'Picta' but dwarf, to 30cm. 'Luteopicta': small; leaves striped golden-yellow. 'Feesey' ('Mervyn Feesey'; 'Feesey's Form'): small, less invasive; leaves light green, boldly striped white; inflorescence tinted pink. 'Picta'

(GARDENER'S GARTERS): invasive; leaves striped white, usually predominantly on one side of leaf. 'Streamlined': leaves mainly green, edged white. 'Tricolor': leaves white-striped with a strong pink flush. Z4.

**P.canariensis** L. CANARY GRASS; BIRDSEED GRASS.
Annual, to 120cm. Culms flimsy to robust, solitary to clumped. Leaves linear to linear-lanceolate, to 25 × 1.3cm, glabrous, scabrous. Inflorescence ovoid to ovoid-cylindric, compact, erect, to 6 × 2cm; spikelets obovate, to 1 × 0.4cm; glumes oblanceolate, winged above, abruptly acute; fertile lemma narrow-ovate, to 6mm, pubescent, shiny yellow when mature, sterile lemmas to 5mm+. Summer–autumn. W Mediterranean. Z6.

**P.minor** Retz.
Annual, to 120cm. Culms flimsy, solitary or clumped, erect or ascending. Leaves to 15 × 0.6cm, glabrous; ligule to 6mm. Inflorescence ovoid to cylindric, compact, to 7 × 1.6cm, pale green and white; spikelets to 6mm; glumes narrow-oblong, with a toothed wing; fertile lemma narrow-ovate, to 3mm, grey, downy. Summer. Mediterranean to NW France. Z6.

*P.nodosa* Murray. See *P.aquatica*.
*P.stenoptera* Hackel. See *P.aquatica*.
*P.tuberosa* L. See *P.aquatica*.

**Phalaris**   (a) *P.arundinacea* (b) *P.canariensis* (c) *P.minor*

**Pharus** P. Browne. (From Greek *pharos*, a wide cloth or covering, referring to the broad leaf blades.) Gramineae. Some 5 species of perennial grasses. Leaves linear to oblong, sometimes ciliate, petiole-like base and sheath twisted; ligule papery, glabrous. Inflorescence a delicate panicle; spikelets awnless; male spikelet terminal; female spikelet 1-flowered; glumes 2, membranous, sometimes abscising, shorter than flower, entire; palea 2-ribbed; lemma cylindric, chalky, becoming leathery, to 5-ribbed or more, minutely pubescent, margins inrolled; male spikelets smaller than female. Summer. Tropical America. Z9.

CULTIVATION As for *Hyparrhenia*.

**P.latifolius** L.
To 90cm. Leaves oblanceolate to narrowly-obovate, to 25 × 10cm; 'petiole' to 10cm. Panicle to 30cm, loose; female spikelets to 18mm; glumes tinged brown or purple; lemma tinged pink; male spikelets to 4mm, glabrous. 'Vittatus': leaves striped white, tinged pink.

**Phleum** L. (From Greek *phleos*, a type of reed with a compact inflorescence.) Gramineae. Some 15 species of annual or perennial grasses. Leaves flat. Inflorescence paniculate, spicate, cylindric to subglobose; spikelets axis elongated, laterally flattened, abscising; glumes equal, longer than flower, membranous, conspicuously keeled, mucronate to awned; lemma to 7-ribbed, keeled, shorter than glumes; apex awnless or mucronate. N, S temperate regions, S America. Z5.

CULTIVATION Occasionally grown for its soft, densely flowered cylindrical inflorescences which dry and dye well, *P.pratense* is an attractive addition to the wild flower meadow. Grow in any ordinary soil in sun. It is grown for pasture and is a common cause of hayfever. *P.bertolinii* is a pasture and turf grass. In addition to those described below, *P.paniculatum* Huds. and *P.subulatum* Asch. & Gräbn. may be worthy of cultivation. Sow seed of *P.subulatum* in spring *in situ*, perennials also by division.

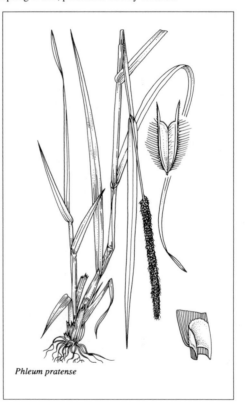

*Phleum pratense*

*P.boehmeri* Wibel. See *P.phleoides*.
*P.nodosum* L. See *P.pratense*.
*P.phlaroides* K. Koch. See *P.phleoides*.

**P.phleoides** (L.) Karst. (*P.boehmeri* Wibel; *P.phlaroides* K. Koch).
Perennial to 60cm. Culms clumped, slender, erect. Leaves flat or rolled, to 13 × 0.4cm. Inflorescence narrow cylindric, to 10 × 0.6cm, tinged green to purple; spikelets narrow-oblong, mucronate, with rough and bristly keel. Summer. NW Africa, Europe, N Asia.

**P.pratense** L. (*P.nodosum* L.; *P.pratense* var. *nodosum* (L.) Huds.). TIMOTHY; CAT'S TAIL; MEADOW CAT'S TAIL.
Perennial to 1.5m. Culms clumped, robust, ascending, swollen at base. Leaves to 45 × 1cm, glabrous; ligules papery, blunt, to 0.6cm. Inflorescence compact, cylindric, to 30 × 1cm; spikelets oblong, compact, to 0.4cm, 1-flowered; glumes narrow-oblong, keel ciliate, awned; awn to 0.2cm; palea equalling lemma. Summer. C, N, W Europe.

*P.pratense* var. *nodosum* (L.) Huds. See *P.pratense*.

**Phragmites** Adans. (From Greek *phragma*, fence or screen, referring to the screening effect of many plants growing together along streams.) REED. Gramineae. Some 4 species of rhizomatous perennial reed grasses, to 3m or more. Culms robust. Leaves linear, flat, attached to stem, blades deciduous; ligules papery, ciliate. Inflorescence terminal a large, plumed panicle; spikelets stipitate, to 13-flowered rachis laterally compressed; lowest flower male or sterile; callus plumed; glumes shorter than lowest lemma, to 5-ribbed; palea longer than lemma; lemma membranous, to 3-ribbed, glabrous, entire. Cosmopolitan, Tropics to temperate regions.

CULTIVATION   Occurring in marsh, fen and riverside habitats in temperate and tropical zones, *P.australis* is an elegant perennial, valued for the soft and showy flowering panicles, which retain their beautiful metallic sheen even on drying, and for the golden russet autumn colours. The species is notable for its aggressive invasiveness and tolerance of extremely low winter temperatures to –20°/–4°F and below (*P.a.* ssp. *altissimus* is slightly less hardy). *Phragmites* thrives in deep, moisture-retentive soils, but is generally suitable only for the larger landscape unless confined in containers, or as a marginal or submerged aquatic. Propagate by division.

*P.australis* (Cav.) Trin. ex Steud. (*P.communis* Trin.;*P.vulgaris* (Lam.) Crépin; *Arundo phragmites* L.). COMMON REED; CARRIZO.
Culms robust, to 3.5m. Leaves to 60 × 5cm+, apex attenuate, arching, mid to grey-green, margins scabrous. Inflorescence plumose, soft, oblong to ovoid, erect to pendent, loose to compact, silky grey , tinged pearly pink-brown to purple, then brown, to 45cm; spikelets lanceolate, to 17mm, 2–10-flowered; glumes lanceolate, persistent, papery, to 6mm; fertile lemmas enveloped in white hairs, to 8mm. Summer–autumn. Cosmopolitan. 'Flavescens': inflorescence tinted yellow. 'Humilis': dwarf. 'Rubra': inflorescence tinted crimson.

'Striatopictus': less vigorous, leaves striped pale yellow. 'Variegatus': leaves striped bright yellow, fading to white. ssp. **altissimus** Clayton (*P.communis* var. *giganteus* (Gay) Husnot). To 6m; panicles to 40cm; glumes tridentate. Z5.

*P.communis* Trin. See *P.australis*.
*P.communis* var. *giganteus* (Gay) Husnot. See *P.australis* ssp. *altissimus*.
*P.*'Giganteus'. See *P.australis* ssp. *altissimus*.
*P.maxima* Chiov. See *P.australis*.
*P.vulgaris* (Lam.) Crépin. See *P.australis*.

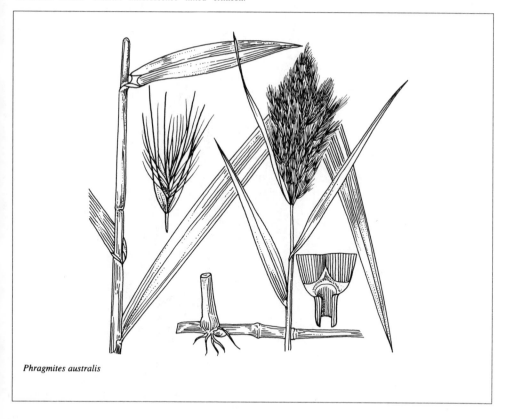

*Phragmites australis*

**Phyllostachys** (a) *P.aurea* (b) *P.aureosulcata* (c) *P.aureosulcata* 'Spectabilis' (d) *P.bambusoides* 'Castillonis' (e) *P.bambusoides*

**Phyllostachys** Sieb. & Zucc. (From Greek *phyllon*, leaf, and *stachys*, spike, for the 'leafy' inflorescence.) Gramineae. Some 80 species of medium and large bamboos, readily recognized by their grooved culms and branching habit. Rhizomes running, but in colder climes and less favourable conditions the culms stay in close clumps appearing pachymorph. Culms hollow and grooved, or at least flattened on alternate sides where the branches emerged; lower down they may be terete or, when young, flexuous; sheaths caducous, except sometimes at the base, with or without auricles or scabrous bristles; nodes with varying amounts of white powder below; branches typically two, unequal, often with a third rune between which may soon drop off, branchlets soon numerous. Leaves medium-sized or small, narrow-lanceolate, tessellate, one third green, two thirds glaucous beneath, glabrous above but often with a few hairs beside the midrib beneath, margins varyingly scaberulous; sheaths soon caducous, with or without auricles and bristles. China, India, Burma.

CULTIVATION   Spectacular bamboos suited to specimen and grove planting on rich damp soils, sheltered from harsh winds and prolonged exposure to hard frosts. *Phyllostachys* includes some of the most ornamental bamboos – their culms yellow, yellow-striped, purple, jade green, tortoiseshell or black, the culm sheaths sometimes beautifully marked. They are all runners and are used in milder climates to create dense stands and even living fences. Elsewhere, these plants tend to form clumps slowly and are treated as choice specimens, often planted in containers. All require heavy feeding when active. Propagate by large divisions in spring, transferring straight to the desired site and misting and watering freely until established. For general cultivation see BAMBOOS.

*P.*'Allgold' (sensu Chao). See *P.sulphurea*.

**P.angusta** McClure.
Resembles *P.flexuosa*, but with culms straight, not flexuous, little white powder, yellow-green not black; sheaths much paler, striped and hardly dotted, with paler ligules. Z8.

**P.arcana** McClure.
Culms somewhat ribbed, sometimes curved at base, with white powder, sometimes blackening with age; sheaths striped with no or few dots, no auricles or bristles, ligule broad, convex; nodes prominent. Z8.

**P.aurea** (Carr.) A. & C. Riv. FISHPOLE BAMBOO; GOSAN-CHIKU; HOTEI-CHIKU.
Culms 2–10m × 2–5cm, thick-walled, glabrous, green later brown-yellow with a little waxy powder below the nodes where, on most culms, are markedly cup-shaped swellings, the lowest internodes very short and asymmetrical; sheaths lightly spotted and streaked, glabrous except for a narrow fringe of white hairs at the base, lacking auricles and bristles, blades long and narrow, ligule very short, ciliate; sheath scars with short white hairs; branches rather erect. Leaves 5–15 × 0.5–2cm; sheaths usually with small auricles and bristles, ligule downy. SE China, naturalized Japan. 'Albovariegata': culms slender; leaves striped white. 'Holochrysa': culms yellow, sometimes striped green; leaves occasionally striped. 'Violascens': culms to 6m, gouty, swollen, green thinly striped purple or yellow in time, ultimately bruised grey-purple to violet; nodes prominent; branches short, dense; leaves to 12cm, glossy above, glaucous beneath. Sometimes treated as a form of *P.bambusoides*, *P.viridiglaucescens* or as a species in its own right. Z6.

**P.aureosulcata** McClure. YELLOW-GROOVE BAMBOO.
Culms 3–10m × 1–4cm, sometimes markedly flexuous below, somewhat rough below the prominent nodes with white waxy powder at first, yellow-green, the grooves fading to yellow; sheaths striped, with few or no spots, sometimes hairy at base with bristles and auricles, ligule broad, blades rather broad; branches erect. Leaves 5–17 × 1–2.5cm; sheaths with or without auricles and bristles. NE China. 'Aureocaulis' ('Spectabilis'): culms yellow with a green groove. Z6.

**P.bambusoides** Sieb. & Zucc. (*P.mazellii* A. & C. Riv.; *P.quilioi* (Carr.) A. & C. Riv.; *P.reticulata* sensu K. Koch). GIANT TIMBER BAMBOO; MADAKE; KU-CHIKU.
Culms 3–30m × 1.5–20cm, arising late, noticeably erect, stout, fairly thick-walled, green, glabrous with no or very little white powder; sheaths large, thick, rough and rather heavily marked, glabrescent, mostly with 1–2 small auricles and kinked bristles, ligule narrow, ciliate, blades mostly short and rather narrow. Leaves noticeably large and broad, 9–20 × 2–4.5cm, their stalks longer than in other species, (to 8mm), sheaths usually with auricles and prominent bristles to 1.2cm, ligule medium-sized, hairless. Native in China and perhaps Japan, widely grown elsewhere. 'Albovariegata': leaves poorly marked white. 'Castillonis': differs from 'Holochrysa' in having green grooves to the culms. 'Castillonis Inversa': a sport from 'Castillonis' with green culms and yellow grooves. 'Holochrysa' ('Allgold'): smaller and more open, with golden-yellow culms, sometimes with green stripes. 'Marliacea': internodes strangely wrinkled. Z7.

**P.bissetii** McClure.
Culms medium-sized, slightly scabrous, powdery at first; sheaths lightly striped, unspotted, mostly with auricles and bristles, blades narrow. Leaves large, sheaths in auricles. Z5.

**P.dulcis** McClure.
Culms tall, thick, strongly tapered, ribbed, striped, hairless, with abundant powder; sheaths smooth, pale, striped, with a few spots, auricles and bristles, ligule broad, rounded, blade crinkled, narrow; nodes prominent. Leaves sometimes pubescent beneath; sheath with auricles and bristles. Z8.

**P.edulis** (Carr.) Houz. (*P.heterocycla* (Carr.) Matsum.; *P.heterocycla* f. *pubescens* (Mazel ex Houz.) Muroi in Sugimoto; *P.mitis* auctt. non A.&C. Riv.; *P.pubescens* Mazel ex Houz.). MOSO BAMBOO; MOUSOU-CHIKU.
Culms 3–27m × 4–30cm, very thick, strongly tapered, grey-velvety when young, ultimately green or almost orange, with white powder below the nodes, curved near the base, where the nodes are much closer together; sheaths thick with much brown mottling, ciliate and strewn with erect hairs, ligule long, ciliate, blades small, glabrous, auricles large, sometimes absent, bristles prominent. Leaves 5–12 × 0.5–2cm, numerous; sheaths with no or poorly developed auricles and bristles. China, introduced Japan. 'Bicolor': culms stiped yellow. f. **heterocycla** (Carr.) Muroi. TORTOISESHELL BAMBOO; KIKKOUCHIKU; NABESHIMANA. A distortion with lowest internodes of some culms short, bulging on alternate sides. Z7.

*P.elegans* McClure. See *P.viridiglaucescens*.
*P.fastuosa* (Marliac  ex  Mitford)  Nichols.  See *Semiarundinaria fastuosa*.

**P.flexuosa** (Carr.) A. & C. Riv.
Culms 2–10m × 2–7cm, slender, often flexuous to the top, the outermost spreading to strongly arching, green-yellow, finally almost black, glabrous with white powder below the nodes; sheaths acuminate, smooth with some dark markings, lacking auricles and bristles, ligule dark, ciliate, blades narrow. Leaves 5–15 × 1.5–2cm, sheaths lacking auricles except in first young shoots, medium-sized, downy. China. Z6.

*P.heterocycla* (Carr.) Matsum. See *P.edulis*. f. *heterocycla*.
*P.heterocycla* f. *pubescens* (Mazel ex Houz.) Muroi. See *P.edulis*.
*P.henonis* Mitford. See *P.nigra* var. *henonis*.

**P.humilis** Munro.
Culms small, 3–5m, rough, dark at first, later pale green; sheaths papery, grey-white, striped, ciliate with white hairs,

**Phyllostachys**  (a) *P.flexuosa*  (b) *P.sulphurea*  (c) *P.makinoi*  (d) *P.nidularia*  (e) *P.sulphurea* var. *viridis*

**Phyllostachys** (a) *P.nigra* var. *henonis* (b) *P.nigra* var. *nigra* (c) *P.nigra* f. *punctata* (d) *P.viridiglaucescens* (e) *P.nigra* 'Boryana'

auricles and, sometimes, bristles, ligule long. Mature leaves small, glabrous, with or without bristles. Z8.

**P.kumasasa** (Zoll. ex Steud.) Munro. See *Shibataea kumasasa*.

**P.makinoi** Hayata.
Very tall and vigorous; culms to 18m × 9cm, widely spaced, very thick at base, rough, glaucous for some time, with white powder; sheaths spotted, glabrous, ligule long, dark, lacking auricles or bristles. Leaf sheaths with auricles and bristles. Z8.

**P.mazellii** A. & C. Riv. See *P.bambusoides*.

**P.meyeri** McClure.
Strong grower; culms with little white powder; sheaths spotted and blotched with white hairs at the base, without auricles or bristles, blade long and narrow. Leaf sheaths with few or no auricles or bristles. Z8.

**P.mitis** sensu A. & C. Riv. See *P.sulphurea* var. *viridis*.
**P.mitis** auctt. non A. & C. Riv. See *P.edulis*.
**P.nevinii** Hance. See *P.nigra* var. *henonis*.

**P.nidularia** Munro.
Culms 3–5–10m × 0.5–4cm, somewhat flexuous to arching, nearly solid below, slightly ribbed, yellow-brown, glabrescent, with white powder below the nodes; sheaths streaked, ciliate, with brown hairs near the base, ligule short, ciliate, blades often flared at the base and merging into the broad smooth clasping auricles, bristles absent or few and small, nodal ridges prominent, sheath scars fringed with brown hairs. Leaves 6–14 × 0.9–2cm, rather broad and stubby, sheaths with small or no auricles or bristles. N & C China. Z7.

**P.nigra** (Lodd. ex Lindl.) Munro. BLACK BAMBOO; KURO-CHIKU.
Culms 3–10m × 1–4cm, rather thin with white waxy powder below the nodes, green at first, typically turning shining jet black in the second or third year; sheaths glabrescent, unspotted except toward the apex, auricles and bristles prominent, ligule tall, blades broad, crinkled; nodes prominent; branches rather erect. Leaves 4–13 × 0.8–1.8cm, bright green, thin, glabrous with poor or no auricles or bristles, ligule small, downy. E & C China, widely cultivated elsewhere. The typical plant, **var. nigra**, has glossy, ultimately pure black culms; an excellent selection is 'Othello', the culms very black even when young and tightly clumped. 'Boryana' (f. *boryana* (Mitford) Muroi): culms green blotched brown. var. **henonis** (Mitford) Stapf ex Rendle (*P.henonis* Mitford; *P.nevinii* Hance; *P.puberula* (Miq.) Nak.). The condition most often encountered in the wild: culms all green, later yellow-green, downy and rough when young, foliage abundant. f. **punctata** (Bean) Mak. Culms variously marked dark purple-brown, often mistaken for young examples of the type. Z7.

**P.nuda** McClure.
Robust; culms thick-walled, ribbed, glabrous, sometimes geniculate below, with much white powder at first; sheaths rough, striped, the lower blotched, ligule prominent, rounded, without auricles or bristles; nodes rather prominent. Leaf sheaths without auricles or bristles. Z8.

**P.propinqua** McClure.
Differs from *P.meyeri* in lacking the white hairs at the base of the shining culm sheath. Z8.

**P.puberula** (Miq.) Nak. See *P.nigra* var. *henonis*.
**P.pubescens** Mazel ex Houz. See *P.edulis*.
**P.quilioi** (Carr.) A. & C. Riv. See *P.bambusoides*.
**P.reticulata** sensu K. Koch. See *P.bambusoides*.

**P.rubromarginata** McClure.
Culms slender, hairless, habit rather open, sheaths unspotted and unstriped, the lower edged red above when fresh, ligules fringed red, without auricles or bristles, blades narrow. Leaves rather broad; sheaths with red ligules. Z8.

**P.ruscifolia** (Sieb. ex Munro) Satow. See *Shibataea kumasasa*.

**P.sulphurea** (Carr.) A. & C. Riv. (*P.*'Allgold' sensu Chao). OUGON-KOU CHIKU; ROBERT OUGON-CHIKU.
Culms 4–12m × 0.5–9cm, increasingly yellow with age, sometimes striped green, lower nodes usually minutely pitted, hairless, with a little white powder below them; sheaths thick, ultimately spotted and blotched brown, glabrous, usually lacking auricles or bristles; branches erect. Leaves 6–16 × 1.5–2.5cm, rather narrow, sometimes striped; sheaths on young culms with auricles and bristles. E China. var. *viridis* R.A. Young (*P.mitis* sensu A. & C. Riv.; *P.viridis* (R.A. Young) McClure; *P.viridis* var. *robertii* Chao & Renvoize; *P.viridis* 'Robert Young'). KOU-CHIKU. The larger and more usual wild plant with green culms which never reach a good size in cooler climes. Z7.

**P.tranquillans** (Koidz.) Muroi. See × *Hibanobambusa tranquillans*.

**P.violascens** auct. See *P.aurea* 'Violascens'.

**P.viridiglaucescens** (Carr.) A. & C. Riv. (*P.elegans* McClure.)
Culms 4–12m × 1–5cm, arising early, often curved at the base, smooth, glabrous, with white waxy powder below the nodes; sheaths rough, well marked with dark spots and blotches, glabrescent with 0–2 auricles and bristles, ligule long, blades narrow; nodes fairly prominent; branches spreading. Leaves 4–20 × 0.6–2cm, sheaths usually with auricles and bristles. E China. Z7.

**P.viridis** (R.A. Young) McClure. See *P.sulphurea* var. *viridis*.
**P.viridis** var. *robertii* Chao & Renvoize. See *P.sulphurea* var. *viridis*.
**P.viridis** 'Robert Young'. See *P.sulphurea* var. *viridis*.

**P.vivax** McClure.
A vigorous species of open habit. Culms 3–25m × 2.3–12.5cm, ribbed, pale grey, glabrous, with white waxy powder below the nodes; sheaths with many spots and blotches, glabrous, lacking auricles or bristles, ligule small, blades narrow, crinkled; nodes prominent, somewhat asymmetrical. Leaves 7–20 × 1–2.5cm, sheaths lacking auricles or bristles. E China. Z8. 'Aureocaulis': culms striped green and deep yellow.

**Pleioblastus** Nak. (From Greek *pleios*, many, and *blastos*, buds, alluding to the branches which are borne several per node.) Gramineae. Some 20 species of dwarf to medium-sized bamboos with short or far-running rhizomes . Culms erect, almost always hollow, typically with 3–7 branches spreading from the upper nodes, in some species with 1–2 branches from low down, all leafing from the top downwards; sheaths fairly persistent. Leaves scaberulous, tessellate, sheaths with smooth white bristles or none, the apex horizontal in all those here except *P.simonii*. China, Japan.

CULTIVATION    A genus rich in small or medium-sized bamboos ideal for the smaller garden, containers, companion planting and medium-height groundcover. Some are very small (e.g. the dwarf and ferny *P.pygmaeus*), others graceful with narrow weeping leaves (*P.gramineus*, *P.linearis*) and many of the remainder possess leaves variegated with white, cream, yellow, lime or gold, the finest among them including the popular *P.variegatus*, a well-mannered miniature with white-striped leaves, and *P.auricomus* f. *chrysophyllus*, pure gold and a superb contrast plant. Their general culture is as for *Sasa*, although they are seldom so invasive and almost always more attractive.

**P.akebono** (Mak.) Nak. (*P.argenteostriatus* f. *akebono* (Mak.) Muroi; *P.pygmaeus* var. *distichus* 'Akebono'; *Arundinaria variegata* var. *akebono* Mak.; *Nippocalamus chino* var. *akebono* (Mak.) Mak.).
Culms slender, 20–50 × 1–2cm, glabrous throughout; branches erect, 1–2 from low down. Leaves 5–7cm×1–1.5mm in 2 rows, mostly suffused white or yellow. Not yet traced in the wild, but comes true from seed. Z7.

**P.angustifolius** (Mitford) Nak. See *P.chino* f. *angustifolius*.

**P.argenteostriatus** (Reg.) Nak. (*P.chino* var. *argenteostriatus* (Reg.) Mak.; *Arundinaria argenteostriata* (Reg.) Vilm.; *Nippocalamus argenteostriatus* (Reg.) Nak.; *Sasa argenteostriata* (Reg.) Camus).
Differs from the more frequently encountered *P.variegatus* in being sturdier and more upstanding with culms to 1m. Leaves 1.4–2.2 × 1.0–2.1cm, hairless; sheaths hairless except at the base, nodes pubescent. at first. Unknown in the wild; cultivated in Japan.Z7.

**P.argenteostriatus** f. *akebono* (Mak.) Muroi. See *P.akebono*.
**P.argenteostriatus** f. *pumilus* (Mitford) Muroi. See *P.humilis* var. *pumilus*.

**P.auricomus** (Mitford) D. McClintock (*P.kongosanensis* 'Auricomus'; *P.viridistriatus* (Reg.) Mak.; *Arundinaria auricoma* Mitford; *Arundinaria viridistriata* (Reg.) Mak. ex Nak.; *Sasa auricoma* (Mitford) Camus). KAMURO-ZASA.
Culms 1–3m × 2–4mm, sometimes purple-lined, softly hairy; sheaths rather persistent, shorter than the internodes, ciliate and markedly downy when young; nodes prominent with white waxy powder below; branches usually 1–2 from low down. Leaves 12–22 × 1.5–3.5cm, softly hairy throughout, especially when young, brilliant yellow with green stripes of various breadths; sheaths downy, ciliate, purple, bristles few or none. *P.kongosanensis*, a later name, probably describes the wild original. Cultivated and naturalized in Japan. f. *chrysophyllus* Mak. Leaves entirely golden-yellow. Z7.

**P.chino** (Franch. & Savat.) Mak. (*P.maximowiczii* (A. & C. Riv.) Nak.; *Arundinaria chino* Franch. & Savat.; *Nippocalamus chino* (Franch. & Savat.) Nak.).
Rhizome sometimes invasive. Culms 2–4m × 0.5–1cm, sometimes purple, glabrous with white wax below nodes; sheaths glabrescent, lacking bristles; nodes hairless; branches 3 or more, spreading. Leaves 12–25 × 1–3cm, green throughout, sometimes slightly downy beneath, or ciliate; sheaths often tinted purple above, ciliate with bristles. C Japan. f. *angustifolius* (Mitford) Muroi & H. Okamura in Sugimoto (*P.angustifolius* (Mitford) Nak.). Leaves to 1.3cm across, often downy beneath, striped white in varying degrees, occasionally entirely green. f. *gracilis* (Mak.) Nak. (*P.vaginatus* (Hackel) Nak.; *P.chino* 'Elegantissimus'; *Arundinaria chino* var. *argenteostriata* f. *elegantissima* Mak.). Leaves to 0.6cm across, glabrous, usually variegated white. Z6.

**P.chino** var. *argenteostriatus* (Reg.) Mak. See *P.argenteostriatus*.
**P.chino** var. *viridis* f. *pumilus* (Mitford) S. Suzuki. See *P.humilis* var. *pumilus*.

**P.distichus** (Mitford) Muroi & H. Okamura. See *P.pygmaeus* var. *distichus*.
**P.fortunei** (Van Houtte) Nak. See *P.variegatus*.
**P.gauntlettii** auct. See *P.humilis* var. *pumilus*.

**P.gramineus** (Bean) Nak. (*Arundinaria graminea* (Bean) Mak.).
Differs from *P.linearis* in its glabrous culm sheaths; culms 2–5m × 0.5–2cm, glabrous with white waxy powder below the nodes; sheaths glabrescent, lacking bristles; branches many, from the upper parts of the culms. Leaves 15–30 × 0.8–2cm, numerous, narrow, long-acuminate, glabrous, pendulous, somewhat twisted toward apex, sheaths glabrous, often lacking bristles, ligule long, rounded. Japan, E China. Z7.

**P.humilis** (Mitford) Nak. (*Arundinaria humilis* Mitford; *Nippocalamus humilis* (Mitford) Nak.; *Sasa humilis* (Mitford) Camus; *Yushania humilis* (Mitford) W.C. Lin).
Culms 1–2m × 2–3mm, glabrous; sheaths glabrous, purple when young; nodes glabrous with white waxy powder below; branches 1–3, low down. Leaves 10–25 × 1.5–2.3cm, glabrescent; sheaths slightly downy, with or without bristles, ligule minute, downy. C Japan. var. **pumilus** (Mitford) D. McClintock (*P.argenteostriatus* f. *pumilus* (Mitford) Muroi; *P.chino* var. *viridis* f. *pumilus* (Mitford) S. Suzuki; *Sasa pumila* (Mitford) Camus; *Arundinaria pumila* Mitford; *Arundinaria gauntlettii* auct.). More robust and compact, differs most clearly in the nearly always conspicuously bearded scars of the upper culm sheaths. Leaves a fresher green.

**P.kongosanensis** 'Auricomus'. See *P.auricomus*.

**P.linearis** (Hackel) Nak. (*Arundinaria linearis* Hackel).
A graceful medium-sized bamboos with long narrow leaves; it differs from *P.gramineus* in its hairy culm sheaths with a truncate ligule and flatter, rather shorter leaves C Japan. Z7.

**P.maximowiczii** (A. & C. Riv.) Nak. See *P.chino*.

**P.pygmaeus** (Miq.) Nak. (*Arundinaria pygmaea* (Miq.) Mitford; *Nippocalamus pygmaeus* (Miq.) Nak.; *Sasa pygmaea* (Miq.) Rehd.). DWARF FERN-LEAF BAMBOO; KE-OROSHIMA-CHIKU.
Stems very small and slender, 10–20cm × 1mm, typically solid and flattened above, glabrous with white powder. Sheaths sometimes hairy or ciliate; nodes sometimes pubescent; branches 1–2, low down. Leaves 2–4cm × 0.2–0.5cm, in 2 close ranks, usually slightly downy, especially beneath; margins sometimes withering papery white in hard winters. Unknown in the wild; cultivated in Japan. var. **distichus** (Mitford) Nak. (*P.distichus* (Mitford) Muroi & H. Okamura; *Arundinaria disticha* Pfitz.; *Nippocalamus argenteostriatus* var. *distichus* (Mitford) Nak.; *Sasa disticha* (Mitford) Camus). Much more frequently grown than the type, often misnamed *P.pygmaeus*, and sometimes under cultivar names such as 'Minezuzme' 'Orishimazasa' or 'Tsuyuzasa'. All selections have taller stems, to 1m; sheaths, nodes and leaves usually hairless; leaves 3–7x0.3–1cm, to 8 pairs arranged in 2 close ranks atop the stems. A vigorous runner with very sharp, pointed rhizomes. Z6.

**Pleioblastus** (a) *P. simonii* (b) *P. pygmaeus* var. *distichus* (c) *P. chino* 'Elegantissimus'

**Pleioblastus** (a) *P. auricomus* (b) *P. variegatus* (c) *P. humilis* (d) *P. humilis* var. *pumilus*

*P.pygmaeus* var. *distichus* 'Akebono'. See *P.akebono*.

*P.shibuyanus* 'Variegatus'. See *P.variegatus*.

**P.simonii** (Carr.) Nak. (*Arundinaria simonii* (Carr.) A. & C. Riv.; *Nippocalamus simonii* (Carr.) Nak.). SIMON BAMBOO; MEDAKE.

Rhizomes usually not far-running. Culms 3–8m × 1.2–3cm, stout, thin-walled, glabrous; sheaths rather persistent, eventually standing away as the branches develop, sometimes slightly hairy toward the base, ligule prominent, ciliate; nodes with white waxy powder below; branches finally numerous from higher up the culm. Leaves 13–27 × 1.2–2.5cm, often half-green, half-glaucous beneath, some of the earliest white-striped at first; sheaths glabrous, usually with a sloping apex, ligule downy. C & S Japan. f. **variegatus** (Hook. f.) Muroi (*P.simonii* var. *heterophyllus* (Mak. & Shirasawa) Nak.; *Arundinaria simonii* var. *albostriata* Bean; *Arundinaria simonii* var. *striata* Mitford). Less robust than the type with leaves variable, some broad, some narrow, some white-striped. Z6.

*P.simonii* var. *heterophyllus* (Mak. & Shirasawa) Nak. See *P.simonii* f. *variegatus*.

*P.vaginatus* (Hackel) Nak. See *P.chino* f. *gracilis*.

**P.variegatus** (Sieb. ex Miq.) Mak. (*P.fortunei* (Van Houtte) Nak.; *P.shibuyanus* 'Variegatus'; *Arundinaria fortunei* (Van Houtte) Nak.; *Arundinaria variegata* (Sieb. ex Miq.) Mak.; *Nippocalamus fortunei* (Van Houtte) Nak.; *Sasa variegata* (Sieb. ex Miq.) Camus). DWARF WHITE-STRIPED BAMBOO; CHIGO-ZASA.

Culms 20–75 × 0.5–2cm; sheaths minutely downy, ciliate, lacking bristles; nodes glabrous with white waxy powder below; branches 1–2, erect, mostly from low down. Leaves 10–20 × 0.7–1.8cm, downy chiefly below, dark green with cream stripes of varying breadth; sheaths purple-lined, minutely downy, ciliate, bristles few or none. Probably a selection of the green-leaved var. **viridis** (Mak.) D. McClintock (*P.shibuyanus* Nak.). Widely cultivated in Japan; unknown in the wild. Z7.

*P.viridistriatus* (Reg.) Mak. See *P.auricomus*.

**Poa** L. (Classical Greek name for pasture grass.) MEADOWGRASS; BLUE GRASS; SPEAR GRASS. Gramineae. Some 500 species of mostly perennial, sometimes dioecious grasses, of variable habit. Culms slender to robust. Leaves narrow, folded to flat or bristled, basal sheaths occasionally thickened or flattened, apex blunt to hooded, ligules membranous. Inflorescence paniculate, open to compact; spikelets stipitate, 2- to several flowered, upper flower rudimentary; rachilla glabrous; glumes acute, persistent, upper glume 3-ribbed; palea keeled, rough to ciliate or smooth; lemmas protruding from glumes, awnless, herbaceous to papery, keeled, keels to 7-ribbed, base glabrous to ciliate; stamens 3; ovary glabrous; hilum round. Cool temperate regions.

CULTIVATION  *Poa* includes a number of species valued for their blue foliage, especially bright in *P.colensoi* and *P.labillardieri*. *P.pratensis* and *P.nemoralis*. The blue grasses or meadow grasses, are amongst the most commonly used of cool-season turf grasses. Several have attractive inflorescences. Grow in sun in any porous moderately fertile garden soil, with good moisture retention for the New Zealand species. Propagate by division or seed.

*P.abbreviata* R. Br.
Habit very dwarf, to 2cm. Leaves dark green. Spikelets minute, soft.

*P.alpina* L. ALPINE MEADOW GRASS; BLUE GRASS.
Perennial, tufted. Culms to 50cm, glabrous, smooth, somewhat thickened at base. Leaves to 5×0.3cm, linear, short-acuminate, thick, flattened; ligules to 4mm. Panicles to 7cm, ovoid, dense, shortly branched; spikelets to 1cm, with 3–7 fls; glumes broad, almost equal; lemmas glabrous, downy on keel and marginal veins. C Asia, USSR. var. *vivipara* (L.) Tzvelev. Spikelets replaced by numerous plantlets. Z4.

*P.annua* L.
Annual or biennial. Culms to 30cm, smooth, creeping or erect. Leaves to 3.5mm diam., narrowly linear, flat, smooth; ligules to 2mm. Panicles to 7cm, pyramidal, loose; spikelets to 5.5mm, closely spaced, with 3–7 flowers; glumes unequal, the lower 1-veined, the upper 3-veined; lemmas subobtuse, sparsely pubescent on keel and marginal veins. Europe, US. Z5.

*P.bulbosa* L. BULBOUS MEADOW GRASS.
Perennial, densely tufted. Basal sheaths swollen, bulb-like. Culms to 55cm, erect, glabrous, terete. Leaves to 1.5mm diam., narrow-linear, tightly involute; ligules to 2mm, acute to rounded, hyaline. Panicles to 6cm, oblong, compact; branches scabrous, short; spikelets to 6mm, green or on keel and marginal veins, obscurely veined. Eurasia, N Africa; naturalized US. Z5.

*P.caesia* Sm. See *P.glauca*.

*P.chaixii* Vill. (*P.sudetica* Haenke). BROAD-LEAVED MEADOW GRASS; FOREST BLUEGRASS.
Perennial, to 120cm. Culms clumped, erect, robust. Leaves to 45 × 1cm, flat or folded, bright green; sheaths flattened, keeled. Panicles ovoid to ovoid-oblong, open, to 25 × 11cm; spikelets ovate to oblong, to 6mm, to 4-flowered; lemmas lanceolate-oblong, to 4mm, glabrous. Spring–summer. Europe, SW Asia. Z5.

*P.colensoi* Hook. f.
Perennial, to 25cm. Culms erect or arching, flimsy, smooth, ridged, blue-green. Leaves to 16cm × 0.2cm, threadlike, intensely blue-green, margins involute. Panicle lax, to 5cm; spikelets blue, becoming brown-tinged; glumes oblong-ovate. New Zealand. Z7.

*P.curvula* Schräd. See *Eragrostis curvula*.
*P.fertilis* Host. See *P.palustris*.

*P.glauca* Vahl (*P.caesia* Sm.). GLAUCOUS MEADOWGRASS.
Glaucous perennial, to 40cm. Culms upright, stiff. Leaves to 8cm × 0.4cm, glabrous, glaucous blue. Panicle erect, stiff, to 10 × 4cm, sometimes tinted purple; spikelets ovate to oblong,

to 6mm, to 6-flowered; glumes 3-ribbed; lemma oblong, obtuse, to 4mm, keels downy. Summer–autumn. N Eurasia, northern N America. Z5.

*P.iridifolia* Hauman.
Perennial, to 90cm. Culms clumped, robust. Leaves flat or folded, erect, to 50 × 1.8cm, glabrous; lower sheaths conspicuously keeled, flattened; ligules to 3mm. Panicles ovoid, erect to pendent, to 30 × 7cm, green; spikelets densely arranged, oblong to oblong-lanceolate, to 1cm, to 5-flowered; lemmas lanceolate to oblong, to 4mm, base pubescent, keels downy. Summer. Uruguay, Argentina. Z8.

*P.labillardieri* Steud.
Perennial to 1m, forming mounds. Leaves to 30cm or more, slender, bright glaucous blue to grey-green, filiform. Panicles to 20cm, loose, tinged purple or green, with hair-like stalklets. Australia, New Zealand.

*P.mexicana* Hornem. See *Eragrostis mexicana*.

*P.nemoralis* L. (*Agrostis alba* L.). WOOD MEADOWGRASS.
Perennial, to 90cm. Culms clumped, erect to spreading. Leaves to 12 × 0.3cm, glabrous; ligules very short, membranous. Panicles ovoid to cylindric, to 20cm; branches spreading; spikelets lanceolate to ovate, laterally flattened, to 5-flowered; glumes persistent, subequal, 3-ribbed, keeled. Summer. Europe, temperate Asia, NE America. Grown for ornament and cover, will not tolerate mowing. Z5.

*P.palustris* L. (*P.fertilis* Host; *P.serotina* Ehrenb.). SWAMP MEADOWGRASS.
Perennial, to 1.5m. Culms clumped, flimsy to robust. Leaves to 20 × 0.4cm, scabrous; ligules to 5mm. Panicles ovoid to oblong, to 30 × 15cm, tinged yellow-green or purple; spikelets ovoid to oblong, to 5mm, to 5-flowered; lemmas narrow-oblong, keeled, keels short pubescent, tipped brown-yellow. Summer. N temperate. Z5.

*P.pilosa* L. See *Eragrostis pilosa*.
*P.plumosa* Retz. See *Eragrostis tenella*.

*P.pratensis* L. KENTUCKY BLUE GRASS; JUNE GRASS; SMOOTH MEADOW GRASS.
Perennial, loosely tufted. Culms to 90cm, ascending, smooth, terete. Leaves to 4mm diam., narrow-linear, flattened, smooth or scabrous; ligule to 1.5mm, rounded to truncate. Panicles to 20cm, pyramidal, purple-tinted; branches scabrous, lowest in whorls of 3–5; spikelets to 6mm, ovate, green, with 2–5 flowers lanceolate, pubescent on keel and marginal veins. Eurasia, N Africa. Z3.

*P.serotina* Ehrenb. See *P.palustris*.
*P.sudetica* Haenke. See *P.chaixii*.
*P.trichodes* Nutt. See *Eragrostis trichodes*.

**Poa**  (a) *P. bulbosa*  (b) *P. annua*  (c) *P. alpina*  (d) *P. alpina* var. *vivipara*

**Poa**   (a) *P.chaixii*   (b) *P.palustris*   (c) *P.nemoralis*   (d) *P.glauca*

**Pogonatherum** P.Beauv. (Greek, bearded with bristles). Gramineae. 3 species, delicate low-growing, clump-forming to spreading bamboo-like grasses with many very slender culms. Inflorescence axillary, a single raceme with fragile rachis born on a flexuous peduncle; spikelets laterally compressed, the callus obtuse with an involucral beard; lower glume cartilaginous, strongly convex, obtuse; upper glume slenderly awned; upper lemma oblong, bifid with a long slender awn. Tropical Asia. Z9.

CULTIVATION  *P.saccharoideum*, Bamboo Grass, is obviously, if rather tautologously, a bamboo-like grass, found on damp, mossy rocks and steep banks in subtropical and tropical regions of E Asia. A delightful minia-ture, it forms a low thicket or mop of fragile, leaning culms dense with narrow, bright green leaves. It favours moist soils rich in garden compost, leafmould or composted bark, semi- to rather deep shade, a minimum tem-perature of 10°C/50°F and frequent misting. That said, it gains in popularity as a verdant survivor of bad condi-tions in the home, office and interior landscapes (good groundcover for the last) and is also reported to have withstood short, light frosts. It makes an excellent indoor bonsai. Plants damaged by cold or wanting vigour will regenerate readily if cut hard back. Propagate by division.

*P.saccharoideum*  Beauv. (*P.crinitum* Trin.; *P.paniceum* Trin.). BAMBOO GRASS.
Culms 15–40cm, simple or branched, tufted, slender, leafy. Leaves 1.8–2cm, linear-lanceolate, bright fresh green, base ciliate. Spikes 1–2.5cm; spikelets to 0.8cm., exceeded by basal hairs. India, China, Malaysia. 'Bambino': said to be a compact selection suitable for container cultivation.

**Polypogon** Desf. (From Greek *polys*, many, and *pogon*, beard, referring to the setaceous panicles.) Gramineae. Some 18 species of annual or perennial grasses, differing from *Agrostis* in its deciduous spikelets. Culms often decumbent, slender, upright or ascending. Leaves flat, scabrous. Inflorescence a dense, bristly, spicate panicle; spikelets short stipitate, 1-flowered, soon abscising; glumes narrow, equal, enclosing flower, papery to leathery, scabrous, 1-ribbed, entire to 2 lobed, awned; lodicules absent; palea inconspicuous; lemma membranous, truncate to acute, 5-ribbed; awn straight to sharply bent or absent, shorter than those of the glumes. Summer–autumn. Warm Temperate. Z8.

CULTIVATION    Inhabitants of damp pasture and salt marshes, *Polypogon* species are grown for their inflorescences, used in fresh flower arrangements and, in the garden, in drift plantings. The panicles of *P.monspeliensis* are exceptionally silky and fine. Sow seed *in situ* in spring in any moderately fertile, well drained soil in sun.

**P.fugax** Nees ex Steud. (*P.littoralis* hort.). Annual to 60cm. Culms loosely clumped, upright or ascending. Leaves to 20 × 1cm, glabrous, scabrous above; ligules to 0.5cm. Inflorescence cylindric, compact, to 15 × 3cm, silky; spikelets oblong, to 0.3cm, tinged green to purple; glumes emarginate, rough, awns to 0.3cm. Warm temperate Asia, NE Africa.

*P.littoralis* hort. See *P.fugax*.

**P.monspeliensis** (L.) Desf. ANNUAL BEARD GRASS; BEARD GRASS; RABBIT'S FOOT GRASS.
Annual to 60cm. Culms solitary or clumped, slender. Leaves to 15 × 0.8cm, glabrous; ligules to 1.5cm. Inflorescence narrow-ovoid to cylindric, to 15 × 3cm, with fine bristles, tinged light green to yellow-green then brown, very dense and silky; spikelets narrow oblong, to 0.3cm; glumes obtuse, emarginate, short-ciliate, rough below, awns to 0.8cm, glumes to twice length of lemmas; lemmas shiny, awned. Cosmopolitan in Europe, naturalized N America.

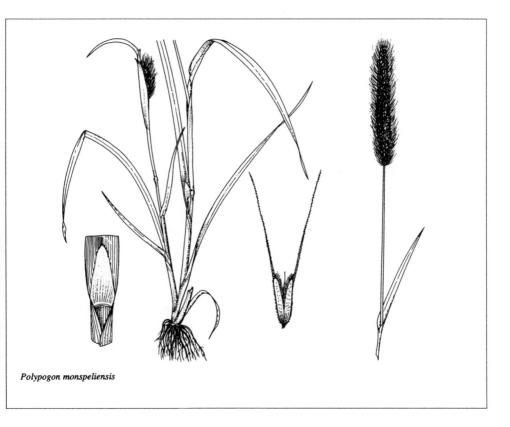

*Polypogon monspeliensis*

**Pseudosasa** Mak. ex Nak. (From Greek *pseudos*, false, and *Sasa*.) Gramineae.
6 species of tall or short bamboos with running rhizomes (clumped in colder climates). Culms erect; sheaths longer than the internodes, bristles white and smooth but often absent; nodes not prominent, with white waxy powder below, glabrous; branches 1–3 from the upper nodes. Leaves tessellate, glabrous; sheaths glabrous, almost always without bristles. China, Japan, Korea.

CULTIVATION    Tall or small, robust bamboos with short rhizomes which soon consolidate to form dense thickets. *P.japonica* is among the most commonly planted bamboos. In recent years, many examples of this species have been diminished by flowering: flowering plants should be cut back and fed to restore vigour. *Pseudosasa japonica* is hardy to –15°C/5°F and will resist high winds with little damage to foliage. It favours damp, rich soils and will even endure near-saturated conditions. Plant young material at 1m intervals to provide a dense hedge after 3–4 years. *P.amabilis*, the Lovely Bamboo, is rather more tender than *P.japonica*, suffering in harsh winds and prolonged frosts. It requires a sheltered, warm situation in zone 8. This is the celebrated producer of long, straight, rigid tonkin canes, or tsingli, with many uses. Introduced to the US for commercial production by F.A. McClure.

**P.amabilis** (McClure) Keng f. (*Arundinaria amabilis* McClure). TONKIN CANE.
Culms 5–13m × 2–7cm, stout, very straight and more or less bare below, arching above, bristly when young, rather thick-walled; sheaths fairly persistent, bristly-hairy, often with a shaggy midline with bristles, ligule fringed; branches usually 3. Leaves 10–35 × 1.2–3.5cm, often of two different sizes on the same plant, scaberulous; sheaths ciliate, usually with bristles. S China; widely cultivated throughout SE Asia. Z8.

**P.japonica** (Sieb. & Zucc. ex Steud.) Mak. ex Nak. (*Arundinaria japonica* Sieb. & Zucc. ex Steud.; *Sasa japonica* (Sieb. & Zucc. ex Steud.) Mak.; *Yadakea japonica* (Sieb. & Zucc. ex Steud.) Mak.). ARROW BAMBOO; METAKE.
Culms 3–6m × 1–2cm, stiff, thin-walled, glabrous, branching in the second year and eventually arching above when well-grown; sheaths long-persistent, coarse, with scattered hairs at first and few or no bristles, ligule blunt; nodes often oblique; branches 1 per node. Leaves 20–36 × 2.5–3.5cm, rough above, one-third green, two-thirds glaucous beneath; sheaths lacking bristles. A rugged species useful for shelter and in shade; distinguished from the similar *Sasamorpha borealis* in the white powder and the bicoloured undersides to the leaves. Japan, Korea. Z6.

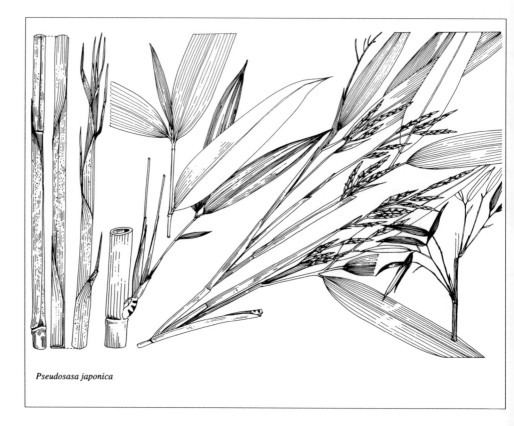

*Pseudosasa japonica*

**Pycreus** P. Beauv. (An anagram of *Cyperus*.) Cyperaceae. 70 species of perennial herbs. Stems terete or 3-angled, leafy. Leaves grasslike, linear, usually equalling stem; sheaths often tinged purple or red, or fibrous. Inflorescence an umbel, subtended by leaflike bracts which exceed it in length; spikes sessile or stalked; spikelets few- to many-flowered; flowers hermaphrodite, minute, spirally arranged, subtended by a scale-like, caducous glume; sepals and petals absent; stamens 2; style 2-branched. Fruit a nut, compressed, 2-angled. Cosmopolitan. Close to *Cyperus*.

CULTIVATION A hardy *Cyperus*-like herb with very fine foliage and yellow to ochre inflorescences. It should be naturalized in bog gardens and the margins of ponds and lakes.

*P.congestus* (Vahl) Hayek. See *Cyperus congestus*.

*P.filicinus* (Vahl) T. Koyama (*Cyperus filicinus* Vahl; *Cyperus nuttallii* Eddy). NUTTALL'S CYPERUS. Annual or short-lived perennial, 10–35cm. Stems slender, tufted. Leaves equalling or exceeding stem. Bracts unequal, 1–2 very much exceeding the others; umbel compound, ses-sile or with 1–2 rays; spikelets linear-lanceolate, 3mm, compressed, golden yellow to brown; glumes oblong, 2–4mm, pointed, bristle-tipped, midrib green. Fruit brown, 1–1.5mm. Summer. N America. Z7.

*P.longus* (L.) Hayek. See *Cyperus longus*.


**Qiongzhuea** Hsueh et Yi. Gramineae. Several medium-sized bamboos. Rhizomes amphipodial. Culms erect, internodes terete or slightly tetragonal, often compressed on branching side, glabrous, not glaucous, solid or nearly solid on lower internodes, culm wall rather thick, nodes not swollen to strongly swollen; culm sheaths thick, papery. Primary branches usually 3 per node, becoming many-branched as secondary branches develop, fastigiate to spreading. Leaf blades lanceolate to narrow-lanceolate with distinct tessellation. China.

CULTIVATION An unusual, more or less clump-forming bamboo with slender, glossy green culms which bulge like a spinning-top at the nodes. The foliage, light and graceful, is carried on spreading branches. Will tolerate semi-shade and, reportedly, temperatures as low as −15°C, much assisted by the protection of surrounding shrubs or woodland.

*Q.tumidinoda*. Hsueh et Yi.
Culms 2.5–6m × 1–3cm; lower culm dull to glossy olive or sap green, internodes very slender, nodes strongly swollen. Primary branches 3 per node, with lateral branches 1–4, twigs slender, to 20 × 0.2cm, spreading. Leaves slender, 3–4 on twigs, blades green above, grey-green beneath, glabrous. NE Yunnan and SW Sichuan. Z6

**Rhynchelytrum** Nees. RUBY GRASS. (From Greek *rhynchos*, beak, and *elytron*, scale, referring to the beaked upper glume and lower lemmas.) Gramineae. Some 14 species of annual or perennial grasses. Culms clumped. Leaves linear to thread-like; ligules short-ciliate. Inflorescence a showy panicle with racemose branches, compact to open; spikelets short-stipitate, laterally flattened, keeled, 2-flowered, upper flower male; upper glume laterally swollen, 5-ribbed, awns present or absent, lower glume inconspicuous; upper lemma laterally flattened, lower lemma tapering to a narrow beak. Tropical Africa to SE Asia. Z8.

CULTIVATION   A savannah native grown for its richly coloured and highly ornamental panicles, *R.repens* is tolerant of several degrees of short-lived frost and is easily grown in sun, in any moderately fertile, porous garden soil. It seldom achieves more than a quarter of its potential height in cultivation. Propagate by seed sown under glass in early spring, or by division.

**R.repens** (Willd.) C. Hubb. (*R.roseum* (Nees) Stapf & C. Hubb.; *Tricholaena rosea* Nees; *Tricholaena repens* (Willd.) A. Hitchc.). NATAL GRASS; RUBY GRASS.
Short-lived, deciduous perennial or annual, to 120cm. Culms flimsy, upright or decumbent to ascending. Leaves to 30×1cm, flat, long-acute. Inflorescence cylindric to ovoid, compact to loosely arranged, to 20 × 10cm; spikelets to 4mm, densely silky-pubescent, hairs white, tinged purple to pink or red; pedicels slender. Summer– autumn. Tropical Africa. Sometimes offered as *R.nerviglume.*

*R.roseum* (Nees) Stapf & C. Hubb. See *R.repens.*

**Rhynchospora** Vahl. Cyperaceae. Around 200 species of perennial herbs. Stem round or 3-angled, leafy. Leaves grass like, linear. Inflorescence a scapose panicle; spikelets 2–3-flowered, round in section, subtended by a collar of leaf-like bracts; flowers very small, hermaphrodite, borne in the axils of spirally arranged, scale-like bracts (glumes); sepals and petals represented by 5–13 rough bristles; stamens 2–3, stigmas 2. Fruit a 2-sided or 3-angled nut with a persistent style base. More or less cosmopolitan.

CULTIVATION    An attractive perennial sedge resembling a small papyrus with bright white inflorescence bracts. It may be grown outdoors in areas where little or no frost occurs, planted in a moist, acid soil in part shade or full sun. Elsewhere, it will thrive used as a glasshouse or conservatory pool plant, its pot partly submerged. It may be moved in summer to the margins of outdoor ponds. It has no absolute requirement for saturated soils and will grow quite happily in pots of damp, ericaceous compost, making a subtly pleasing house- or conservatory plant.

**R.nervosa** (Vahl) Böckeler (*Dichromena nervosa* Vahl). Tufted or rhizomatous, 8–150cm. Stems erect or arching. Leaves 8–50 × 0.2–1cm, mostly basal, glabrous above, hairy or glabrous beneath. Bracts 3–8, exceeding inflorescence, linear-lanceolate, base triangular, apex finely tapering to a point, brilliant white on upper surface at base, 4–22cm. C America, Caribbean, Northern S America. ssp. *ciliata* (*Dichromena ciliata* Pers.).    Bracts white for 10–35cm; glumes with red striations. Z9.

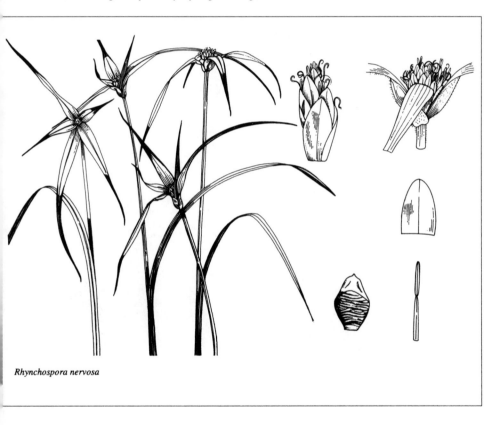

*Rhynchospora nervosa*

**Rostraria** Trin. Gramineae. Some 10 species of annual grasses. Culm bases occasionally swollen. Leaves linear; ligule membranous. Inflorescence paniculate; branches rough; spikelets laterally flattened, to 10-flowered, with one rudimentary sterile floret; rachilla glabrous to short-pubescent; glumes unequal to subequal, keeled, persistent; lemmas membranous, obtuse to acute, longer than glumes, 5-ribbed, keeled, awned. Summer. Temperate Europe, Asia, N Africa. Z6.

CULTIVATION A hardy grass grown for its tail-like inflorescences; it should be sown *in situ* in spring and may naturalize. General requirements as for *Phleum*.

**R.cristata** (L.) Tzvelev (*Koeleria phleoides* (Vill.) Pers.; *Lophochloa cristata* (L.) Hylander; *Lophochloa phleoides* (Vill.) Rchb.). CAT TAIL GRASS. Annual to 45cm. Culms solitary to loosely clumped, erect or bent. Leaves flat, pubescent, to 18 × 0.8cm, long-acute. Inflorescence cylindric to narrowly oblong, compact, to 12 × 1.5cm, pale green; spikelets to 0.8cm, to 8-flowered; glumes acute, glabrous to pubescent; lemma 5-ribbed, awn to 0.3cm, straight.

*Saccharum ravennae*

**Saccharum** L. (From Greek *sakchar*, sugar.) PLUME GRASS. Gramineae. Some 37 species of clumped or rhizomatous perennial grasses. Culms stout, hard, cane-like, jointed, green to violet, coarse to glossy, ringed and
woody with vestigial adventitious roots at base. Leaves flat, linear-lanceolate to ligulate, in 2 ranks, basally
sheathing. Inflorescence terminal, paniculate, often plumed, branches racemose; rachis brittle; spikelets lanceolate, paired, 2-flowered, lower flower sterile, reduced to a lemma, one spikelet sessile, one stipitate; callus conspicuously bearded; glumes equal, as long as spikelet, base with silky hairs; lemmas and palea membranous,
upper lemma bilobed or entire, awn short or absent, lower glume flat to convex, bi-keeled. Old World Tropics,
warm temperate zones. Z9 unless specified.

CULTIVATION   Occurring predominantly in the tropics and subtropics by riversides and in the rich soils of valley
bottoms, *Saccharum* species are large vigorous grasses valued for their strong form, foliage and for the soft
dense inflorescences. The pulp of *S.arundinaceum* is used in paper manufacture; *S.officinarum* is the sugar cane
of commerce. *S.ravennae* has a strong form, much resembling a smaller, finer Pampas Grass; it creates a beautiful shimmering effect, although it may fail to thrive and flower in cool, dull and wet climates. *S.spontaneum* is
used as hedging or screening in the Mediterranean; *S.officinarum* is sometimes grown in botanic collections of
economic plants, but easily merits a place as a large ornamental for conservatories, glasshouses and subtropical
gardens. Most need warm glasshouse protection in cool temperate zones, but may survive in sheltered sites,
although foliage will invariably be damaged by frosts and a semi-herbaceous habit may develop. *S.ravennae* and
the North American Plume Grasses are basically cold-hardy, however. Their silvery, plume-like inflorescences
are suitable for cutting and their foliage often turns attractive shades in fine autumns following long hot summers. Under glass grow in beds or tubs of rich, damp, loam-based medium. Well grown plants may become
invasive or overcrowd the glasshouse.

*S.alopecuroides* (L.) Nutt. (*Erianthus alopecuroides* (L.)
Ell.). SILVER PLUME GRASS.
Differs from *S.contortum* in its silvery panicles. Autumn. SE
US. Z8.

*S.arundinaceum* Retz.
To 3.6m. Culms clumped, robust, upright. Leaves linear, to
180 × 5cm, dark green, midrib stout, margins scabrous.
Panicle cylindric, to 90cm, compact to diffuse, silky pubescent, white, tinged silver or pink; spikelets lanceolate, to
4mm, enveloped by white to light grey hairs, to 3 × as long as
spikelets. India to Malay Is.

*S.barberi* Jesw. See *S.officinarum*.

*S.bengalense* Retz. (*S.munja* Roxb.; *S.sara* Roxb.; *S.ciliare*
Anderss.).
To 5m. Culms stout, clumped, erect. Leaves to 2m × 2.5cm,
glaucous, scabrous. Panicles cylindric, compact, to 90cm,
pearly white sometimes tinged purple; spikelets lanceolate, to
6mm, with silver-grey hairs. Summer. N & NW India,
Pakistan, to Iran.

*S.brevibarbe* (Michx.) Pers. (*Erianthus brevibarbis* Michx.)
Differs from *S.contortum* in the absence of twisted awns.
Foliage turns uniformly bronze in autumn. US. Z8.

*S.ciliare* Anderss. See *S.bengalense*.

*S.contortum* (Ell.) Nutt. (*Erianthus contortus* Ell.) BENT AWN
PLUME GRASS.
Clump-forming, narrowly upright, 2–3m. Leaves to 2cm
wide, green becoming purple, bronze or orange-red in
autumn, this tinge redder and persisting in winter. Panicle
held 0.75m above leaves, narrow, strictly erect, tinted red at
first, becoming paler upon drying; awns twisted. Autumn. E
US. Z7.

*S.giganteum* (Walt.) Pers. (*Erianthus giganteus* (Walt.)
P.Beauv.) SUGAR CANE PLUME GRASS; GIANT PLUME GRASS.
To 3m. Similar to *S.ravennae* with large, feathery panicles.
Late summer. US (NY). Z7.

*S.munja* Roxb. See *S.bengalense*.

*S.officinarum* L. SUGAR CANE.
To 6m. Culms upright or ascending, to 5cm + diam. Leaves
linear-lanceolate, to 180 × 5cm and over, erect to pendent,
margins spiny, rough. Panicle pyramidal, to 90cm+, compact
to open, brittle; spikelets lanceolate to oblong-lanceolate, to
4mm, enveloped by silky white pubescence, to 1cm. Summer.
Tropical SE Asia, Polynesia. 'Violaceum': stems and leaves
purple to violet.

*S.procerum* Roxb.
Resembles *S.arundinaceum*, but larger. Panicles larger, more
lax, more silky; spikelets to 4mm, shorter than internodes of
inflorescence axis. NE India, Burma. Z8.

*S.ravennae* (L.) L. (*Erianthus ravennae* (L.) P. Beauv.)
HARDY PAMPAS GRASS.
Resembles *Cortaderia*. To 4m. Culms robust, glabrous,
smooth. Leaf sheaths tight, lowest hirsute, otherwise
glabrous; leaves linear, lower over 90cm, upper to 60cm long,
white-pubescent, rough throughout, turning brown in autumn
with tints of purple and orange. Panicle to 60cm, erect, dense
or somewhat loose and lobed, silvery white tinged grey or
purple; racemes sessile; sessile spikelet 9cm, lanceolate,
acuminate; glumes subequal, membranous, the lower acuminate, upper lanceolate, acuminate, minutely mucronate. N
Africa, Mediterranean. Z7. 'Purpurascens': panicles tinted
purple.

*S.sara* Roxb. See *S.bengalense*.
*S.sinense* Roxb. See *S.officinarum*.

*S.spontaneum* L.
Culms over 3m. Leaf-sheaths tight, terete, smooth, glabrous;
leaves 12.5cm, linear, tapering, glaucous, glabrous or occasionally hirsute above the ligule, margins very scabrous with
rigid midrib; ligules short, membranous, brown, longer ciliate
from the back. Panicle oblong, over 45 × 12.5cm, dense,
somewhat contracted, silvery; spikelets to 7.5cm, lanceolate;
glumes subequal, lanceolate, lower acuminate, sparsely ciliate, upper acute, 1-nerved, mucronate, densely ciliate; anthers
2.5cm, yellow. Tropical Africa.

**Sasa** (a) *S.nipponica* (b) *S.kurilensis* (c) *S.tsuboiana* (d) *S.palmata*

(a) *Sasa veitchii*  (b) *Sasamorpha borealis*  (c) *Sasaella ramosa*

**Sasa** Mak. & Shib. (From *zasa*, Japanese name for the smaller bamboos.) Gramineae. About 40 small or medium-sized bamboos; rhizomes running. Culms terete, ascending, glabrous; sheaths persistent, shorter than the internodes, bristles scabrous, spreading or absent; blades very small; nodes with white waxy bloom below; branches 1 or none, about as thick as the culms. Leaves large and broad, mostly thick, tessellate, denticulate, one quarter green, three quarters glaucous beneath; sheaths with or without bristles, ligule downy. Japan, Korea, China. In the past, many species were included in this genus. It is now more clearly defined and some of the segregate species may be found under *Indocalamus, Pleioblastus, Pseudosasa, Sasaella* and *Sasamorpha*.

CULTIVATION Small or medium-sized bamboos with creeping rhizomes, found in damp hollows and woodlands. Most species spread rapidly to form open thickets of slender erect culms topped by a loose canopy of large spreading leaves borne on short branches. *S.palmata* may attain considerable heights and should not be planted where it may become invasive or overshadow smaller subjects. *S.veitchii* is a shorter, denser plant, particularly its semi-dwarf form loosely described as forma *minor*, and is an excellent choice for evergreen, medium-high groundcover or a low loose hedge. Its broad, dark green leaves are handsomely ribbed and wither partially in their first winter, developing decorative, papery white margins.

*Sasa* species are hardy to –20°C/–4°F and fare best in partial shade on damp soils rich in humus. Divide in early spring, taking large divisions and causing minimal rhizome and root disturbance. Divisions of fewer than 5–6 culms seldom succeed. Establish new plants in sheltered standing bed conditions in pots of high-fertility sandy medium, misting foliage regularly; plant out once a solid root mass has formed (usually after a year).

*S.albomarginata* (Franch. & Savat.) Mak. & Shib. See *S.veitchii*.

*S.argenteostriata* (Reg.) Camus. See *Pleioblastus argenteostriatus*.

*S.auricoma* (Mitford) Camus. See *Pleioblastus auricomus*.

*S.borealis* (Hackel) Nak. See *Sasamorpha borealis*.

*S.cernua* auctt., non Mak. See *S.palmata*.

*S.disticha* (Mitford) Camus. See *Pleioblastus pygmaeus* var. *distichus*.

*S.humilis* (Mitford) Camus. See *Pleioblastus humilis*.

*S.japonica* (Sieb. & Zucc. ex Steud.) Mak. See *Pseudosasa japonica*.

**S.kurilensis** (Rupr.) Mak. & Shib.
A variable species. Culms stout, 1–4m × 2–3mm, rather thick-walled, sheaths almost hairless, disintegrating untidily, bristles rare; nodes glabrous, not prominent. Leaves 15–30 × 1.2–4.8cm, dark shining green above, scabrous mostly on one margin only; sheaths minutely downy, bristles few or none, ligule short. Japan, Korea. 'Shimofuri': leaves striped white. Z7.

*S.masumuneana* (Mak.) Chao & Renvoize. See *Sasaella masumuneana*.

**S.nipponica** Mak. (Mak. & Shib.). MIYAKO-ZASA.
Culms 50–80cm, slender, unbranched, often purple, with rather prominent nodes, glabrous, as are culms and sheaths. Leaves 8–20 × 1.4–3cm, small, thin, typically pubescent beneath, scaberulous, edged white in winter; sheaths usually with bristles. Japan. Has a certain resemblance to the widely branched *Sasaella ramosa*. Z7.

**S.palmata** (Burb.) Camus (*S.paniculata* Mak. & Shib.; *S.cernua* auctt., non Mak.; *S.senanensis* auctt., non (Franch. & Savat.) Rehd.).
Rhizome rampant. Culms stout, 1.5–4m × 2–5mm, often sloping, usually soon streaked purple (f. *nebulosa* (Mak.) Suzuki);

sheaths finally ragged, usually with no bristles, ligule broad. Leaves 25–40 × 5–9.5cm, well tapered, glabrous, bright shining green even in winter, the midrib yellow, margins scaberulous, stalks green-yellow; sheaths rarely with bristles, ligule short. Japan. Z7.

*S.paniculata* Mak. & Shib. See *S.palmata*.

*S.pumila* (Mitford) Camus. See *Pleioblastus humilis* var. *pumilus*.

*S.purpurascens* var. *borealis* (Hackel) Nak. See *Sasamorpha borealis*.

*S.pygmaea* (Miq.) Rehd. See *Pleioblastus pygmaeus*.

*S.ramosa* (Mak.) Mak. & Shib. See *Sasaella ramosa*.

*S.senanensis* auctt., non (Franch. & Savat.) Rehd. See *S.palmata*.

*S.tessellata* (Munro) Mak. & Shib. See *Indocalamus tessellatus*.

**S.tsuboiana** Mak.
Rhizome very shortly running. Culms 1–2m × 1–2mm; sheaths glabrous, often tinted purple, with bristles, ciliate; nodes glabrous, not prominent. Leaves 12–26 × 2.4–4.6cm, glabrous; sheaths glabrous, soon without bristles, ligule short. C Japan. Z6.

*S.variegata* (Sieb. ex Miq.) Camus. See *Pleioblastus variegatus*.

**S.veitchii** (Carr.) Rehd. (*S.albomarginata* (Franch. & Savat.) Mak. & Shib.). KUMA ZASA.
Rhizome gently running. Culms 80–150 × 2–3cm, purple-lined, glaucous, glabrescent, lower internodes short; sheaths long-persistent, sometimes decaying untidily, hairy at first, with bristles. Leaves 15–25 × 3.2–6cm, broad, lanceolate-ovate, shortly tapered, glabrous, soon developing a broad, papery white margin, midrib yellow, stalks yellow-purple; sheaths often purple, glabrous, ciliate, bristles soon falling, ligule short, dark. Japan. Z7.

**Sasaella** Mak. (Diminutive of *Sasa*.) Gramineae. Some 12 species of bamboos, differing from *Sasa* in having thinner culms, erect and unbent at the base, with longer, more horizontal branches and small, narrower, thinner leaves, the bristles typically scabrous only at the base. Japan. Z7.

CULTIVATION As for *Sasa*.

*S.glabra* (Nak.) Nak. ex Koidz. See *S.masumuneana*.

*S.masumuneana* (Mak.) Hatsusima & Muroi in Sugimoto (*S.glabra* (Nak.) Nak. ex Koidz.; *Arundinaria atropurpurea* Nak.; *Sasa masumuneana* (Mak.) Chao & Renvoize). SHIYA-ZASA. Culms 0.5–2m × 3–4mm, glabrous; sheaths glabrous, or ciliate, with a few bristles; nodes glabrous. Leaves 10–19 × 1.5–3.5cm, glabrous; sheaths purple-lined, ciliate, with bristles. f. *albostriata* (Muroi) D. McClintock. Leaves striped white maturing yellow; more often grown than the type. f. *aureostriata* (Muroi) D. McClintock. Leaves striped yellow.

*S.ramosa* (Mak.) Mak. (*Arundinaria vagans* Gamble; *Sasa ramosa* (Mak.) Mak. & Shib.). AZUMA-ZASA. A variable rampant spreader. Culms 1–1.5m × 3–4mm, sometimes purple, glabrous; sheaths glabrous, lacking bristles; nodes prominent, glabrous. Leaves 10–20 × 1.4–3cm, palpably pubescent beneath, the margins scaberulous, withering white; sheaths glabrous, ciliate, often with bristles.

**Sasamorpha** Nak. (From its resemblance to *Sasa*, with which it is sometimes combined.) Gramineae. About 6 species of medium-sized, usually glabrous bamboos, differing from *Sasa* in having less rampant rhizomes, erect culms, the sheaths longer than the internodes; nodes flattened, lacking bristles and rarely with any white powder. Far East. Z7.

CULTIVATION As for *Sasa*.

*S.borealis* (Hackel) Nak. (*S.purpurascens* (Hackel) Camus; *Sasa purpurascens* var. *borealis* (Hackel) Nak.; *Arundinaria borealis* (Hackel) Mak.; *Sasa borealis* (Hackel) Nak.). Culms 1–3m × 5–8mm, glabrescent or glabrous, well branched above; sheath pubescent, disintegrating untidily; nodes pubescent. Leaves 10 × 20 × 2–3cm, scaberulous, paler and concolorous beneath, usually tinted purple toward the base of the stalk and the near glabrous sheath. Liable to be mistaken for *Pseudosasa japonica*. Japan, China.

*S.purpurascens* (Hackel) Camus. See *S.borealis*.

*S.tessellata* (Munro) Nak. See *Indocalamus tessellatus*.

**Scirpoides** Ség. (*Holoschoenus* Link). Cyperaceae. 1 species, a grass-like perennial herb, 30–90cm. Roots fleshy; rhizomes creeping, woody. Stems leaf-like (i.e. grassy) in clusters, thick at base. Leaves mostly bladeless, blades when present round, grooved, flat at tip, margin rough; sheaths green at first, often becoming hard or fibrous, spotted red, ligule absent. Inflorescence a compact umbel; bracts leaflike, unequal, the longer bracts 10–40cm, exceeding inflorescence, the shorter bracts variable, deflexed; spikelets blunt, stalked, 2–3.5 × 1–2mm; flowers hermaphrodite, minute, spirally arranged, subtended by a scale-like glume; glumes 1–1.5mm, blunt, bristle-tipped, keel green; stamens 3; style very short, 3-branched. Fruit a 3-angled nut, around 1mm. Europe, SW Asia. Previously included in *Scirpus*. Z7.

CULTIVATION As for *Schoenoplectus*.

*S.holoschoenus* (L.) Soják (*Scirpus holoschoenus* L.; *Scirpus romanus* L.; *Holoschoenus vulgaris* Link; *Holoschoenus australis* (Murr) Rchb.).

**Schizachyrium** Nees. Gramineae. Some 100 species of perennial grasses differing from *Andropogon* in their solitary, terminal racemes. Cosmopolitan.

CULTIVATION   *S.scoparium* hails from prairies, open woods, dry hills over much of central and eastern North America. It is a clump-forming perennial with blue-grey or green foliage turning copper-red in autumn, rich golden-orange by winter. The inflorescences have a striking translucence seen *en masse* in winter sunshine; they may also be used cut, fresh or dried. It requires full sun and will tolerate poor, infertile soils. Used to best advantage when naturalized in large drifts on slopes in the wild or meadow gardens, the Little Bluestem nonetheless appreciates being mown to a height of 10cm in early spring. Propagate by seed or division in spring.

*S.littorale* (Nash) C. Bickn. See *S.scoparium*.

**S.scoparium** (Michx.) Nash. (*Andropogon scoparius* Michx.)
LITTLE BLUESTEM; BUNCHGRASS.
Culms 80–150cm, clumped, slender, strong, erect, branched above. Leaves green or glaucous often tinted purple bronze, to 0.5cm across, blades glabrous, usually weakly downy at junction with sheath. Racemes to 15cm; peduncle very slender largely enclosed by sheaths, rachis flexuous, weakly hairy; sessile spikelet to 7mm, scabrous, long-awned, stalked spikelet short-awned. N America. 'Blaze': a seed cultivar developed for forage, it has, however, exceptional orange winter colour. var. **littorale** (Nash) Gould (*S.littorale* (Nash) C. Bickn.; *Andropogon littoralis* Nash). Stems narrower, crowded, often decumbent at base. Lower leaf sheaths blue-green, glaucous. Inflorescence villous. E N America. 'Aldous': culms and sheaths intensely blue; comes true from seed. Z4.

*Schizachyrium scoparium*

**Schoenoplectus** (Rchb.) Palla. (From Greek *schoinos*, rush, and *pleko*, plait, alluding to the mat-forming rhizomes of some species.) Cyperaceae. 80 species of annual or perennial, grasslike, rhizomatous herbs. Rhizomes often creeping. Stems terete or 3-angled, not leafy. Leaves reduced to sheaths or absent, or basal and sometimes submerged. Inflorescence atop stem but subtended by 1–2 bracts with the lowest stem-like itself, erect, so that inflorescence appears lateral; spikelets sessile or stalked. Flowers hermaphrodite, minute, spirally arranged, subtended by a scale-like glume; sepals and petals absent or represented by 6 bristles; stamens 3; style persistent in fruit, 2–3-branched. Fruit a nut, 3-angled or biconvex, smooth or ridged. Cosmopolitan. Close to and sometimes included in *Scirpus*.

CULTIVATION  The true Bulrushes or Clubrushes (as opposed to *Typha*, the Reed Maces or Cat-tails also known by these names) are clump-forming aquatic or marginal rushes with negligible leaves and tall, slender, green stems resembling those of *Juncus* and bearing clusters of minute brown flowers. In their typical states, they provide a muted backdrop or fringe for water and bog gardens. They can also be used in the pond itself, placed in submerged containers where they will break up hard margins and otherwise unrelieved surfaces and impart a naturalistic grace. The variegated cultivars of *S.lacustris* are among the most striking plants for the water garden. All prefer a neutral to acid substrate. Propagate by division.

**S.hudsonianus** (Michx.) Palla (*Scirpus hudsonianus* (Michx.) Fern.). 1.5–4m.
Rhizome sometimes creeping. Stems slender, rough, leafless. Basal leaf sheaths bristle-tipped. Glumes brown, around 2mm, slightly emarginate; bristles 6, wavy, far exceeding fruit . Fruit obovoid. Summer. N America. Z5.

**S.lacustris** (L.) Palla (*Scirpus lacustris* L.). BULRUSH; CLUBRUSH.
Perennial, 1–3m. Rhizome stout, creeping. Stems to 15mm across, terete. Sheaths brown, membranous. Spikelets in a dense head, 5–8mm, red-brown; glumes 3–4mm, emarginate; stigmas 3. Fruit 2.5mm, brown-grey. Europe, Asia, Africa, Northern S America. ssp. **tabernaemontani** (C. Gmel.) Löve & D. Löve (*Scirpus tabernaemontani* C. Gmel.). Stems usually less than 160cm, glaucous. Rays in axils of lower bracts. Glumes red-papillose; stigmas 2. Fruit biconvex. 'Albescens': stems white, longitudinally and narrowly striped green. 'Zebrinus': stems banded creamy white. Z4.

**S.validus** (Vahl) Löve (*Scirpus validus* Vahl). GREAT BULRUSH.
0.5–1m. Rhizome stout, scaly. Stem 8–25mm wide, thick at base. Lower sheaths with hyaline margins; leaf blades if present to 8cm. Inflorescence bracts solitary, round, shorter than inflorescence rays 1–6cm, thin, flexible; spikelets solitary or in tight clusters, 5–10mm, ovate; glumes nearly round, dorsally pubescent, equalling fruit, style 2-branched. Fruit dull black, plano-convex, 1–1.5mm. Summer. N America. Z7.

**Schoenus** L. (From Greek *schoinos*, rush, referring to the resemblance to the true rush (*Juncus*).) Cyperaceae. Some 80 species of perennial or, rarely, annual herbs. Stems simple, not jointed, tufted, few-leaved. Leaves basal or sub-basal, narrowly linear, shortly ligulate, often reduced to sheaths. Inflorescence terminal, capitate, 1–10 spikelets; involucral bracts 1–2; flowers small, bisexual, uppermost staminate; glumes distichous; perianth setae 3–5, rarely absent; stamens 3; styles caducous, stigmas 2–3. Fruit consisting in trigonous or subglobose achenes, somewhat sunken in the rhachilla. Australia, S Asia, Africa.

CULTIVATION  A hardy rush for damp, acid situations in the bog garden or the water margin. The form usually cultivated is strongly tinted purple-red or maroon throughout, acquiring coppery tints in winter.

**S.pauciflorus** Hook. f.
To 90cm. Stems very slender, pale, leafy at base, angled, grooved. Leaves reduced to red-tinted basal sheaths. Inflorescence a short lateral panicle, spikelets lanceolate, 6.5mm, few, slender, 3–4-flowered, dark brown; glumes acuminate, setae 6, capillary; style very long. Achenes narrow-oblong, glabrous, dark brown, lined with impressed dots. New Zealand (North Is.). Z6.

**Scirpus** L. (The Latin name.) Cyperaceae. Around 100 species of rhizomatous, perennial, grasslike herbs. Stem erect, jointed, solitary or in tufts, leafy. Leaves broadly linear, grasslike, with a membranous ligule. Inflorescence a terminal panicle with many branches (rays), subtended by leaf- or bristle-like bracts; spikelet stalked or in clusters on rays; flowers hermaphrodite, minute, spirally arranged, subtended by a scale-like bract (glume); sepals and petals represented by 6 rough bristles; stamens 3; style 3-branched. Fruit a nut, 3-angled or biconvex, smooth. Cosmopolitan. Here treated in a narrow sense; in this work, *Scirpoides*, *Isolepis*, *Bolboschoenus* and *Schoenoplectus* are treated as separate genera.

CULTIVATION   Quietly attractive rushes with grassy foliage and freely branching heads of minute green, brown or black flowers gathered into spikelets and glomerules. They are more *Cyperus*-like than the Bulrushes (*Schoenoplectus*), also once included in this genus, and their inflorescences seen in quantity have a pleasing airy appearance. They will naturalize readily on damp, preferably acid soils in light woodland, bog and shallow water. They are especially useful in creating wild gardens and recreating habitats.

**S.atrovirens** Willd.
Caespitose perennial; stems slender or stout, to 1.5m. Principal blades to 18mm wide, mostly on the lower half of the stem. Inflorescence once or usually twice branched; spikelets ovoid or short-cylindric, 2–8mm long, densely crowded in subglobose glomerules; scales broadly elliptic or obovate, the body obtuse or acute, the pale midvein prolonged into a conspicuous mucro usually 0.5–0.8mm long. Fruit very pale to white, compressed-trigonous, obovate, 0.8–1.2mm; bristles pale and inconspicuous, straight or nearly so, shorter to barely longer than the achene, or rudimentary, or lacking. N America. Z5. var. **atrovirens**. Scales dark green, becoming black in age, 1.3–1.6mm long; sheaths and lower parts of the blades conspicuously cross-septate, the ribs in the blades 0.25–0.4mm apart; spikelets usually more than 5mm long. var. **georgianus** (Harper) Fern. Scales smaller; sheaths and blades not cross-septate or only rarely so, the ribs in the blades usually 5 or 6 per mm. of width; spikelets usually 2–3mm long. var. **pallidus** (Gray) Britt. (*S.pallidus* Gray). Spikelets very numerous in large glomerules often 1cm in diam.; scales at first pale brown, becoming dark brown.

*S.cernuus* Vahl. See *Isolepis cernua*.

**S.cyperinus** (L.) Kunth. WOOL GRASS. 1–1.5m.
Stem subterete. Leaves narrow-linear, ridged, equalling or exceeding stem. Bracts 3–5, exceeding inflorescence, 15–30cm; rays somewhat drooping; spikelets 3–10cm, numerous, oval to round, clustered; glumes blunt, red-brown; bristles exceeding glumes making spikelets appear woolly. Fruit short-pointed. Late summer. Eastern N America. Z7.

*S.cyperoides* L. See *Cyperus cyperoides*.
*S.filiformis* Savi non Burm. f. See *Isolepis cernua*.
*S.holoschoenus* L. See *Scirpoides holoschoenus*.

*S.hudsonianus* (Michx.) Fern. See *Schoenoplectus hudsonianus*.
*S.lacustris* L. See *Schoenoplectus lacustris*.
*S.maritimus* L. See *Bolboschoenus maritimus*.
*S.pallidus* Gray. See *S.atrovirens* var. *pallidus*.
*S.palustris* L. See *Eleocharis palustris*.
*S.parvulus* Roem. & Schult. See *Eleocharis parvula*.
*S.prolifer* Rottb. See *Isolepis prolifera*.
*S.romanus* L. See *Scirpoides holoschoenus*.

**S.rubrotinctus** Fern.
Stems solitary, stout, to 1m. Sheaths suffused with red at base and usually at the opening; blades 8–15mm wide. Inflorescence repeatedly branched with numerous ascending rays; spikelets ovoid to cylindric, 2.5–6mm, sessile and glomerate; scales broadly round-ovate, 1.2–1.5mm long, obtuse or acute and minutely mucronulate, with narrow, pale midnerve and black-tinted sides. Fruit plano-convex or biconvex, pale to nearly white, obovate, sharply beaked, 0.9–1.2mm; bristles usually 4, straight, pale, as long as to half as long against as the achene. N America. Z5.

*S.setaceus* L. See *Isolepis setacea*.

**S.sylvaticus** L. WOOD CLUB RUSH.
30–120cm. Stems 3-angled, smooth. Leaves to 2cm wide, margins rough. Bracts unequal, the long bracts just equalling inflorescence, shorter bracts bristle-like; panicle lax, to 15cm; spikelets green-brown, 3–4mm; glumes 1.5–3mm; bristles as long as or exceeding fruit. Fruit 1mm, yellow, dull. Europe to Siberia. Z5.

*S.tabernaemontani* C. Gmel. See *Schoenoplectus lacustris* ssp. *tabernaemontani*.
*S.validus* Vahl. See *Schoenoplectus validus*.

**Scirpus and Schoenoplectus**    (a) *Schoenoplectus validus*  (b) *Scirpus rubrotinctus*  (c) *Scirpus atrovirens*

**Semiarundinaria** Mak. ex Nak. (From its similarity to *Arundinaria*.) Gramineae. 10 or more species of tall bamboos. Rhizomes typically running, but tightly clumped in colder climes. Culms terete or with the upper internodes grooved or flattened on one side; sheaths soon falling but sometimes left hanging for a while, bristles absent or scarce, blades narrow; branches rather short, 3–7 developing from the base upwards, the lowest nodes branchless. Leaves small, tessellate, scaberulous, (chiefly on one margin); sheaths with scabrous bristles. The genus has been postulated as a cross between *Phyllostachys* and *Pleioblastus*. Far East.

CULTIVATION   Tall, slow-growing bamboos forming open thickets and suited to use as specimens in sheltered gardens or in tubs in the cold greenhouse or conservatory. The new shoots of most species are a striking plumblack and the culms ultimately glossy green, flushing maroon or chestnut in bright sunlight. They are very erect and tend to lose the sheaths quickly, giving a clean, elevated appearance. The foliage is borne largely in the plant's upper reaches, thus making it suitable for planting in association with shorter subjects.*Semiarundinaria* will tolerate sun and shade and should be planted on a damp site with a high humus content. Most species will withstand lows of –10°C/14°F and *S.fastuosa* will survive  –22°C/–8°F, soon regenerating even if cut to the ground.

*S.fastuosa* (Marliac ex Mitford) Mak. ex Nak. (*Arundinaria fastuosa* (Marliac ex Mitford) Houz.; *Arundinaria narihira* Mak.; *Phyllostachys fastuosa* (Marliac ex Mitford) Nichols.). NARIHADAKE.
Culms splendidly erect, 3–12m × 2–7cm, very hollow, lined purple-brown, glabrous, with white powder below the nodes; sheaths thick, glabrous except sometimes beneath, usually wine-coloured and shining insides. Leaves 9–21 × 1.5–2.7cm, glabrous; sheaths glabrous. S Japan. var. *viridis* Mak. Culms green, leaves longer than in type. Z7.

*S.tranquillans* Koidz. See × *Hibanobambusa tranquillans*.

*S.yamadorii* Muroi.
More slender than *S.fastuosa* with longer branches, some leaves noticeably larger, to 24 × 5cm, and culm sheaths hairy at base. Japan. Z8.

*S.yashadake* (Mak.) Mak.
Differs from *S.yamadorii* in being rather more stiffly erect, with more open growth, short branches and paler, somewhat smaller leaves. Japan. Z8.

**Sesleria** Scop. (For Leonardo Sesler (*d* 1785), Venetian physician and owner of a botanic garden.) MOOR GRASS. Gramineae. Some 23 species of grasses, mostly caespitose perennials. Leaves flat, plicate or involute; ligule short, truncate or obtuse. Inflorescence spike-like, cylindrical to globose, often tinted blue, grey or white; bracts at base of inflorescence well-developed, spikelets laterally compressed, with 2–5 florets; glumes unequal, membranous, 1–3-veined; lemma 5-veined, membranous, with 2–5 usually aristulate teeth at apex; palea as long or longer than lemma. Europe, centered on Balkans.

CULTIVATION    Small, rather unassuming grasses, largely from mountainous or moorland habitats, often on chalk. They are grown for their attractive foliage and short flower spikes. The two Blue Moor Grasses, *S.albicans* and *S.caerulea*, are commonly confused, or held to be synonymous with each other, or the latter treated as a variety of the former. *S.albicans* has blue glaucous leaves in varying degrees (sometimes plain green) and small spikes of an arresting mauve or steely blue. *S.caerulea* forms a low disorderly tuft of blunt-tipped leaves. These are silvery blue above and fresh green beneath. Its inflorescences, often reported to be mauve, tend in many cases to be green ripening parchment. *S.nitida*, with pointed pale blue-green leaves and rather grey flower spikes, has also been offered under the name *S.glauca*. Whatever their true identity, these are among the best low-growing blue grasses, tolerating a wide range of soils and exposures, preferably in full sunlight. They work especially well planted on large rockeries, among paving or at the edges of borders and contrast beautifully with plants such as *Ophiopogon planiscapus* 'Nigrescens', *Plantago major* 'Rubrifolia', *Arthropodium candidum* 'Maculatum' and *Athyrium nipponicum* 'Pictum'. Propagation by division is easy.

**S.albicans.** Kit. ex Schultes (*S.caerulea* ssp. *varia* (Jacq.) Hayek; *S. caerulea* ssp. *calcarea* (Celak.) Hegi). BLUE MOOR GRASS.
Culms 10–45cm, slender. Leaves (1.5–) 2.5–3(–5)mm wide, grey-green or green, flat or plicate, the uppermost not more than 1cm; sheaths glabrous. Inflorescence 10–30 × 5–9mm, rather lax, mauve to steely blue; glumes 4–7mm, ovate-lanceolate, acuminate, usually unawned, glabrous, rarely ciliate; lemma 4–6mm, ovate- lanceolate, sparsely ciliate on margins and veins, up to 5-dentate, the middle tooth with an awn. W & C Europe. ssp. *albicans*. Leaves 2.5–3(–5)mm wide with 17–19 veins. Glumes 4–6mm; lemma 4–5mm, middle awn not more than 0.5mm. Balkans. ssp. *angustifolia* (Hackel & G. Beck) Deyl. (*S.caerulea* var. *angustifolia* Hackel & G. Beck). Leaves 1.5–2mm wide, with usually 15 veins. Glumes 5–7mm; lemma to 6mm, middle awn 1–2mm. Balkans.

**S.argentea** (Savi) Savi (*S.cylindrica* DC.).
Culms to 50cm, leafy at least to the middle. Leaves 3–5mm wide, glabrous, scabrid on margins, the uppermost 3–10cm; sheaths glabrous. Inflorescence 35–55(–126) × 6–8(–10)mm, rather white or grey; glumes 5–7mm, narrowly lanceolate, ciliate on the keel; awn 1–2.5mm; lemma 4.5–5.5mm, glabrous; middle awn 1–1.5mm. Italy & SE France; N & E Spain.

**S.autumnalis** (Scop.) F. W. Schultz (*S. elongata* Host).
AUTUMN MOOR GRASS.
Culms to 40cm, slender. Leaves up to 4mm wide, usually flat, glabrous, glaucescent, strongly scabrid, the uppermost 4–8cm or more above the middle of the stem; sheaths glabrous. Inflorescence 34–100 × 5–7mm, lax, greyish or whitish; glumes 5–6mm, narrowly lanceolate, sparsely ciliate at base; middle awn 0.5–1.5mm, the laterals 0.5mm. N & E Italy to C Albania. Z5.

**S.caerulea** (L.) Ard. (*S.uliginosa* Opiz; *S.caerulea* ssp. *uliginosa* (Opiz) Hayek). BLUE MOOR GRASS.
Like *S.albicans* but culms stouter, tufted, rather low-lying; leaves lying rather flat in a loose, spreading tussock, bright blue and glaucous above,   usually green beneath, 2–8 × 0.2–0.5cm, flat, tip blunt. Inflorescence 10–14 × 7–9mm, dense, shortly cylindric, glumes ciliate on margin and veins; lemma with awn to 1mm. C Sweden and NW Russia southwards to Bulgaria. Z4.

*S.caerulea* var. *angustifolia* Hackel & G. Beck. See *S.albicans* ssp. *angustifolia*.
*S. caerulea* ssp. *calcarea* (Celak.) Hegi. See *S.albicans*.

*S.caerulea* ssp. *varia* (Jacq.) Hayek. See *S.albicans*.
*S.cylindrica* DC. See *S.argentea*.
*S. elongata* Host. See *S.autumnalis*.
*S. filifolia* Hoppe. See*S.rigida*.
*S.haynaldiana* Schur. See*S.rigida*.

**S.heufleriana** Schur. BLUE-GREEN MOOR GRASS.
Culms to 70cm. Leaves 2–3(–7)mm wide, strongly pruinose above when young, the uppermost 1.5–3(–5)cm; sheaths glabrous. Inflorescence usually 10–30 × 7–11mm; glumes usually 3.5–4.5mm, narrowly lanceolate, sparsely ciliate; awn 0.5–2.5mm; lemma pubescent; awns (1–) 2 (–4)mm. Eastern C Europe. ssp. *heufleriana*. Laxly caespitose. Leaves flat or convolute, rarely folded. Glumes to 4.5mm; lemma and palea to 4.5mm. Throughout the range of the species. ssp. *hungarica* (Ujhelyi) Deyl. (*S.hungarica* Ujhelyi). Densely caespitose. Leaves flat or folded. Glumes to 6mm. Lemma 5mm. Czechoslovakia & Hungary. Z4.

*S.hungarica* Ujhelyi. See *S.heufleriana* ssp. *hungarica*.

**S.nitida.** Ten. GREY MOOR GRASS.
Culms to 70cm. Leaves 3.5–6mm wide, flat, rarely convolute, glabrous, pale blue-green, glaucescent, with scabrid margins, the uppermost 3.5–7.5cm, above the middle of the stem; sheaths glabrous or sparsely pubescent. Inflorescence 20–30 × 9–14mm, dense, grey-black; glumes 5–5.5mm, lanceolate, ciliate on the veins, with 3–5 awns, the middle 1–2mm, the laterals 0.5mm; lemma 5–6mm, ovate-lanceolate, usually glabrous; middle awn 1–2mm, the laterals 0.5mm. C & S Italy and Sicily. Z4.

**S.rigida.** Heuffel ex Reichenb. (*S. filifolia* Hoppe; *S.haynaldiana* Schur.).
Culms *c*15cm. Leaves 1–2(–3)mm wide, plicate, glaucous blue-grey above, the uppermost 0.5–1(–2)cm. Inflorescence 15–20 × 7–8mm, greyish or whitish; glumes 5–6mm, acute or shortly awned; lemma (3–) 4–5mm, ovate-lanceolate, usually glabrous between the veins, sparsely ciliate; middle awn 0.5–1mm. Romania and N & C parts of Balkan peninsula. ssp. *rigida*:. Mature leaves with 7–13 veins, hairy above. Culms shorter than leaves; sheaths usually hairy. Throughout the range of the species, except extreme south. ssp. *achtarovii*. (Deyl) Deyl. Mature leaves with at least 13 veins, glabrous or subglabrous above. Culms longer than leaves; sheaths glabrous. S Bulgaria.

*S.uliginosa* Opiz. See *S.caerulea*.

**Setaria** Palib. (From Latin *seta*, a bristle, referring to the bristles of the inflorescence.) Gramineae. Some 100 species of annual or perennial grasses of varying habit. Leaves elliptic to ovate, rarely sagittate. Inflorescence paniculate, setaceous; spikelets 2-flowered, short-stipitate, gibbous, subtended by 1 to many rough bristles; lower flower male or sterile, upper flower hermaphrodite; glumes shorter than spikelet, papery; lower lemma sulcate, to half length of spikelet; upper lemma, palea rugose, becoming coriaceous. Tropics, subtropics, warm temperate zones.

CULTIVATION Natives of grassland and woodland, *Setaria* species are amongst the most ornamental of grasses, grown primarily for their large spikes which often arch gracefully under their own weight at the top of erect stems; they must be cut and air dried when green if they are not to become too brittle on drying. A number of the more tender species also have attractively pleated foliage. Annuals are easily grown in sun in any well drained soil, from seed sown *in situ* in spring. Tender species such as *S.plicata*, need a well drained, fertile and moisture-retentive loam with minimum glasshouse temperatures of 7–10°C/45–50°F, and plentiful water when in growth. Propagate by seed; perennials also by division.

**S.glauca** (L.) Palib. (*S.lutescens* Hubb.; *Panicum glaucum* L.). YELLOW BRISTLE GRASS; GLAUCOUS BRISTLE GRASS.
Annual to 75cm. Culms solitary or loosely clumped, upright. Leaves linear, flat, to 30 × 1cm, glaucous, glabrous; lower sheaths smooth, keeled; ligule a pubescent fringe. Inflorescence cylindric, erect, very bristly, to 13× 0.8cm, branches bearing 5–10 yellow or red-tinged bristles; spikelets elliptic, obtuse, to 3mm; upper glume to 2mm. Summer–autumn. Old World warm temperate. Z6.

**S.italica** (L.) Palib. (*Panicum italicum* L.; *Panicum germanicum* Mill.; *Chaetochloa italica* (L.) Scribn.). FOXTAIL; ITALIAN MILLET; JAPANESE MILLET.
Annual, to 1.5m. Culms stout, upright, branching from base. Leaves linear to narrow-lanceolate, flat, to 45 × 2cm, glabrous, scabrous; sheaths pubescent. Inflorescence erect to pendent, cylindric or lobed, compact, to 30 × 3cm; spikelets broad-elliptic, obtuse, to 3mm, persistent, bristled; bristles to 12mm; upper flower abscising at maturity, variously white, cream, yellow, purple-red, brown to black; upper glume to 4/5 length of spikelet. Autumn. Warm temperate Asia. Z6. An ancient grain crop.

*S.lutescens* Hubb. See *S.glauca*.

**S.palmifolia** (Koenig) Stapf (*Panicum palmifolium* Koenig; *Panicum plicatum* Willd. non Lam.; *Chaetochloa palmifolia* A. Hitchc. & Chase). PALM GRASS.
Coarse perennial, to 3m. Culms very stout, erect; nodes pubescent. Leaves narrow-elliptic, to 90 × 13cm, glabrous to pubescent, scabrous, apex long-acute, base attenuate. Inflorescence ovoid to cylindric, to 90 × 30cm; branches spreading to straight, to 30cm, bristles solitary, inconspicuous; spikelets ovate to elliptic, to 3mm; upper glume to 2mm; upper lemma smooth. Summer. Tropical Asia. One of a group of tall species with similar leaves; others suitable for cultivation include *S.chevalieri* Stapf, *S.megaphylla* (Steud.) Schinz, *S.paniculifera* (Steud.) Fourn., *S.poiretiana* (Schult.) Kunth and *S.pumila* Roem. & Schult. Z9.

**S.persica** hort. ex Rchb.
Close to *S.glauca*, but differing in its purple-brown, spike-like flower heads.

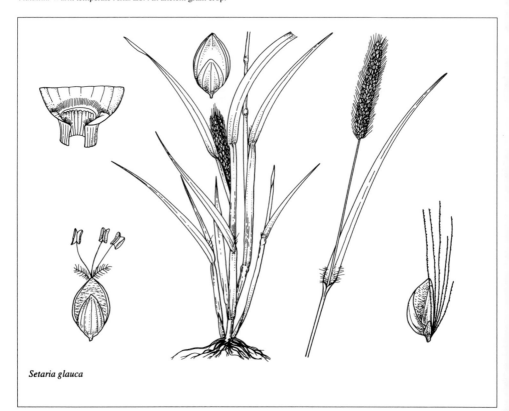

*Setaria glauca*

**S.plicata** (Lam.) Cooke (*Panicum plicatum* Lam. non Willd.). Resembles *S.plicatilis*, but spikelets and upper glumes larger, upper lemma transversely wrinkled. India. Z9.

**S.plicatilis** (Hochst.) Hackel (*Panicum plicatile* Hochst.; *Panicum plicatum* hort.).
Perennial, to 1.5m. Culms loosely clumped, slender to stout. Leaves linear to linear-lanceolate, to 35 × 2.5cm, closely plicate, rough, base attenuate; ligule a pubescent fringe; sheaths laterally flattened, scabrous. Inflorescence linear to cylindric, to 30 × 6cm, bristles to 16mm, exceeding branchlets, green to pink-tinged, silky; spikelets ovate to elliptic-oblong, to 3mm; upper glumes to 2mm; upper lemma smooth. Summer. E Africa (mts). Z9.

**S.verticillata** (L.) Palib.
Differs from *S.viridis* in panicle bristles restrorsely barbed.

**S.viridis** (L.) Palib. (*Panicum viride* L.). GREEN BRISTLE GRASS; ROUGH BRISTLE GRASS.
Annual, to 60cm. Culms upright, slender, loosely clumped. Leaves to 20 × 1cm, glabrous, slightly scabrous; apex long-acute; ligule a pubescent fringe. Inflorescence linear-cylindric, spike-like, pseudo-whorled, compact, erect, very bristly, to 10×1cm, tinged green to purple; bristles to 3 together, to 1cm; spikelets elliptic-oblong, to 3mm, obtuse; upper glume to 3mm; upper lemma finely rugose, equal to upper glume. Summer–autumn. Warm temperate Eurasia. Z6.

**Shibataea** Mak. ex Nak. (For Keita Shibata (1897–1949), Japanese botanist and biochemist.) Gramineae. 8 species of squat shrubby bamboos; rhizomes basically running but plants often appear clumped. Culms slender, almost solid, flattened on one side; sheaths thin, mostly deciduous, unmarked, glabrous, with small or no bristles; nodes prominent; branches 3–5, very short, almost equal in length. Leaves tessellate, 1–2 per branch, without sheaths but with long narrow membranous bracts at the lower nodes. Far East.

CULTIVATION  *S.kumasasa* is unique among dwarf bamboos for its squat, shrubby appearance and dense, elliptic leaves, which resemble a luxuriant Butcher's Broom (*Ruscus*). Use either as a specimen set off by gravel or moss in a Japanese garden or for medium-height groundcover, planted 60cm/24in. apart. It responds well to clipping, readily producing flushes of fresh green growth, and may even be trimmed into rounded shapes. In a damp, shaded site rich in humus, it will develop a slowly expanding clump of culms to 1m/39in. tall. New growth is produced in spring and will be impaired if the plants are allowed to dry out. *S.kumasasa* will tolerate temperatures as low as −23°C/−10°F.

**S.kumasasa** (Zoll. ex Steud.) Mak. ex Nak. (*Phyllostachys kumasasa* (Zoll. ex Steud.) Munro; *Phyllostachys ruscifolia* (Sieb. ex Munro) Satow). OKAME-ZASA.
Habit compact, bushy. Culms 0.5–1.8m × 2–5mm, rather thick, internodes short and obscurely flexuous; sheaths fairly persistent, glabrous, ciliate, blade minute. Leaves 5–11 × 1.3–2.5cm, elliptic-ovate, noticeably short and broad, long-stalked, scaberulous, dark green, glabrous above, glabrescent beneath, often withering at the apex, ligule prominent, narrow. China, Japan. Z6.

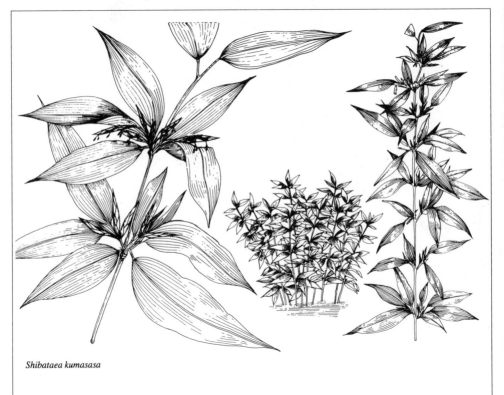

*Shibataea kumasasa*

**Sinobambusa** Mak. ex Nak. (From *sino*, Chinese, and *bambusa*, bamboo.) Gramineae. Some 20 species of medium-sized bamboos. Rhizomes running. Culms terete, with very long internodes; sheaths deciduous; nodes prominent, hairy when young. Leaves tessellate, scaberulous; sheaths with smooth bristles. China, Vietnam. Z9.

CULTIVATION   Elegant bamboos found in damp woodland sites in Southern China, they will not withstand temperatures below –8°C/18°F and in northern temperate zones perform best when grown in tubs in a cold greenhouse or conservatory. Pot in a free-draining soil rich in humus, site in bright indirect light or dappled shade in a draught-free position. Thin out exhausted culms. They may be placed outside during the summer; if kept under glass, regular syringing of foliage is beneficial. The rhizomes may become invasive and caution should be exercised if planting in borders or the open garden. Propagate by division.

***S.tootsik*** (Sieb. ex Mak.) Mak.
Culms 5–12m× 2–6cm, glabrous; sheaths unmarked, hairy, especially at the base, ciliate, with prominent auricles and bristles, ligule tall, blades narrow. Leaves 8–20 × 1.3–3cm, sometimes pubescent beneath; sheaths hairless with auricles and bristles. China. f. ***albostriata*** Muroi. Leaves striped white or yellow.

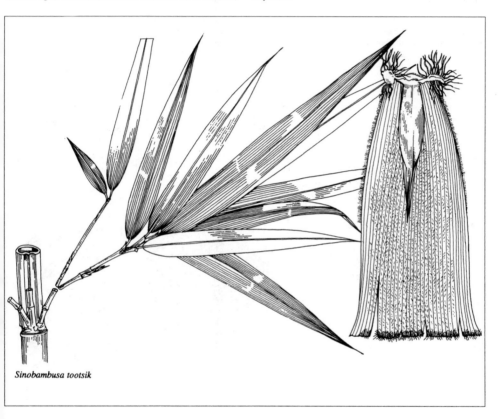

*Sinobambusa tootsik*

**Sorghastrum** Nash. (From *Sorghum*, and Latin suffix *astrum*, a poor imitation, referring to its close resemblance to *Sorghum*.) Gramineae. Some 16 species of annual or perennial grasses. Culms clumped, flimsy to robust. Leaves narrow, flat or rolled; ligules scarious. Inflorescence narrow, terminal, digitate or paniculate; spikelets solitary, 2-flowered; upper flower fertile, lower flower reduced to a sterile lemma; glumes coriaceous, lower glume firm; lemmas enclosed by glume, upper lemma linear to oblong, entire to bilobed, awn long, twisted. Tropical Africa, Americas. Z4.

CULTIVATION   US natives from prairies, open woods and dry slopes, grown for their foliage (grey-blue in *S.nutans* cultivars, becoming yellow or bright orange in autumn or winter) and their copper-brown flower spikes with long bristly awns and conspicuous bright yellow pollen sacs (these last well if cut and dried). They tolerate most soils in full sun and will withstand drought and spring mowing. Suited to specimen or group plantings and naturalizing in wild and meadow gardens. Propagate by seed or division in spring.

*S.avenaceum* (Michx.) Nash. See *S.nutans*.

**S.nutans** (L.) Nash (*S.avenaceum* (Michx.) Nash). INDIAN GRASS; WOOD GRASS.
Coarse perennial, to 2m. Culms clumped. Leaves linear, to 60 × 1cm, scabrous, attenuate; sheaths glabrous, auricles erect; ligules membranous. Panicle terminal, narrow, compact, to 35cm, tinged golden-yellow to brown, hairy; spikelets sessile, lanceolate, to 0.8cm, hirsute, awn to 1.5cm; lemmas hyaline; anthers yellow. Autumn. E & C US. Sometimes offered under the name *Chrysopogon nutans*. 'Osage': seed cultivar representing typical prairie form of the species. Leaves relatively wide, often glaucous blue. 'Sioux Blue': very erect; leaves powder blue, selected from 'Osage' at Longwood Gardens.

**S.secundum** (Elliott) Nash.
Culms to 2m. Leaves less than 5mm diam., flat or somewhat involute. Panicle to 40cm, 1-sided, narrow; spikelets about 7mm, pilose, tinged brown. US (S Carolina to Florida & Texas).

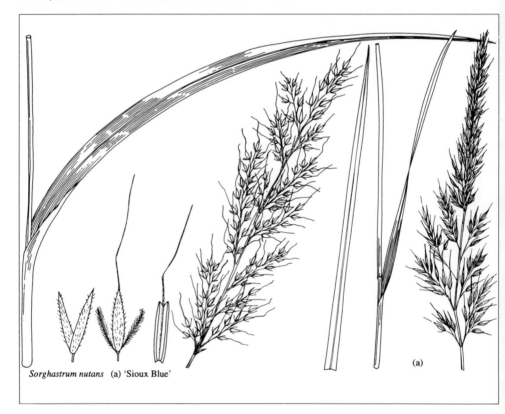

*Sorghastrum nutans*   (a) 'Sioux Blue'

**Sorghum** Moench. (From Italian *sorgo*, the name for this plant.) Gramineae. MILLET. Some 20 species of annual or perennial, sometimes rhizomatous grasses. Culms clumped, usually tall and robust. Inflorescence densely paniculate, more or less cylindric; branches racemose, internodes pubescent; spikelets in pairs, one sessile, fertile, one stipitate, male or sterile; fertile spikelet awned, glumes equal, coriaceous, lower glume flattened or rounded below; stipitate spikelets awnless, lower glume coriaceous, convex, becoming 2-keeled; upper lemma awnless, or to 2-lobed and awned. Old World Tropics, subtropics, 1 sp. endemic to Mexico. Z8.

CULTIVATION    *S.bicolor* is an important cereal, used for human and animal food, also a source of paper, brushes and syrup; it is cultivated for fodder and grain in most warm temperate zones, dry tropics. Native of tropical and subtropical savannah and forest margin, *Sorghum* is occasionally grown in botanic collections of economic plants and as an ornamental for its lush foliage and decorative, drooping panicles, sometimes in subtropical bedding schemes. Propagate by seed sown in late winter/early spring under glass and set out after danger of frost is passed. If overwintered, *Sorghum* will require warm glasshouse protection in cool temperate zones. Grow in full sun in large pots with a fertile and moisture-retentive soil. Water plentifully when in growth and keep almost dry when leafless in winter.

*S.bicolor* (L.) Moench (*S.vulgare* Pers.; *Holcus sorbus* L.; *Holcus bicolor* L.; *Andropogon sorghum* (L.) Brot.). SORGHUM; GREAT MILLET; KAFIR CORN.
Annual, to 6m. Leaves flat, to 90 × 10cm, midrib white; ligules conspicuous. Panicles lax to compact, very dense, to 60 × 25cm; fertile spikelets to 1cm; glumes coriaceous to scarious, green to white or pink to purple, brightly coloured when mature, drying buff; lemma awned or awns absent. Grains large, brown to black. Summer. China, S Africa.

*S.halapense* (L.) Pers. (*Holcus halapensis* (L.) Brot.; *Andropogon halapensis* (L.) Brot.). JOHNSON GRASS; ALEPPO GRASS; MEANS GRASS.
Perennial, to 1.5cm. Leaves linear, to 60 × 2.5cm, flat, midrib tinged white. Panicles compact, lanceolate to oblong, to 25 × 8cm; fertile spikelets elliptic, to 5mm, cream to yellow, awn present or absent; lower glume coriaceous, awn to 1.5cm. Summer–autumn. Mediterranean.

*S.vulgare* Pers. See *S.bicolor*.

**Spartina** Schreb. (From Greek *spartion,* a grass used for cordage.) MARSH GRASS; CORD GRASS. Gramineae. Some 15 species of rhizomatous, perennial, sometimes, halophytic, perennial grasses. Culms rigid to erect. Leaves flat to convolute, tough; ligules ciliate. Inflorescence racemose; racemes digitate, secund; spikelets sessile, adpressed to spreading, pectinate, distichous; spikelets cuneate, laterally flattened, 1-flowered, abscising; glumes unequal, keeled, upper glume longer than flower, lower glume shorter than flower, membranous; acute to short-awned; palea shorter than lemma, 2-ribbed; lemma keeled, rigid, acute. W & S Europe, NW & S Africa, America, S Atlantic Is. Z5.

CULTIVATION   An adaptable species found in freshwater swamps, salt marshes and wet prairies, *S.pectinata* is grown for its graceful habit, slender inflorescences (good for cutting and drying) and for the beautiful yellow autumn colour of its narrow, arching leaves. It prefers damp or even wet soils but is generally undemanding as to their composition and pH. Cord Grasses thrive in full sun and will quickly colonize the banks of ponds and streams, flowering and fruiting over the summer and autumn. Both species described here are spreaders and may become invasive: their typical forms are best suited to naturalizing in semi-wild gardens. The yellow-variegated *S.pectinata* 'Aureomarginata' is less vigorous and will provide pleasing foliage contrasts in more confined situations. Propagate by division.

*S.alterniflora* Loisel. SALTWATER CORDGRASS; SMOOTH CORD-GRASS; HERBA SALEA.
Culms robust, simple, erect, 0.6–2.5m. Leaves 25–45 × 1.3–1.5cm, light green; blades flat at base when young, tip involute, glabrous or minutely scabrous at the margins, narrowing finely to an involute tip. Inflorescence of racemose spikes, forming a narrow, oblong, straight panicle to 35cm; spikes 5–15 × 2–2.5cm, few to several, erect, remote, adpressed; spikelets 1-flowered, somewhat remote or crowded along the rachis, sparsely pilose or glabrous; glumes glabrous or sparsely pilose on the keel, the lower about two-thirds as long as the upper, narrow, the upper glume obtuse, a little longer than the lemma; lemmas 0.8–1cm, glabrous or sparsely pilose, firm, keeled, the lateral nerves obscure. N America.

*S.cynosuroides*    'Aureomarginata'.    See    *S.pectinata* 'Aureomarginata'.
*S.michauxiana* A. Hitchc. See *S.pectinata*.

*S.pectinata* Link (*S.michauxiana* A. Hitchc.). PRAIRIE CORD GRASS; FRESHWATER CORD GRASS; SLOUGH GRASS.
To 2m. Stems tufted, robust; rhizome scaly, hard, clumped. Leaves elongate, flat, to 120 × 1.5cm, convex beneath, narrow-acute, scabrous, tapering. Inflorescence narrow, compact, erect; spikes to 30, to 10cm; glumes with toothed keels, upper glume with awn to 0.8cm. Autumn–winter. N America. 'Aureomarginata' ('Variegata'): leaves arching, olive green edged golden yellow; flower spikes narrow, stamens purple and hanging. Z5.

*Spartina pectinata*

**Spodiopogon** R. Br. (From Greek *spodios*, ashen, and *pogon*, beard, referring to the grey-hairy inflorescences.) Gramineae. Some 9 species of perennial, rarely annual grasses. Culms slender to stout. Leaves linear to lanceolate, flat, occasionally appearing petiolate; ligules membranous. Inflorescence paniculate, open to compact; branches flexible, racemose; spikelets paired, one sessile, one stipitate, abscising entire when mature, near glabrous to long-pubescent, 2-flowered, upper flower hermaphrodite, lower flowers male; callus glabrous to short pubescent; glumes pubescent, lower glume papery, convex, ribs raised; lemmas hyaline, upper lemmas bifid, awned. Temperate, subtropical Asia. Z7.

CULTIVATION    *S.sibiricus* is grown for its attractive hairy panicles produced in late summer and autumn, and for its comparatively broad, often purple-stained leaves – these are held horizontally with the result that individual plants grown well in full sun will make remarkably rounded clumps. It tolerates light shade and thrives on a wide range of soils given adequate moisture. Cut back to the ground in late autumn. Propagate by seed, or division in spring.

*S.sibiricus* Trin. (*Andropogon sibiricus* (Trin.) Steud.).
Perennial, to 1.5m, usually shorter. Culms upright; rhizome creeping, scaly. Leaf blades held more or less perpendicular to culms, giving the plant a strong horizontal orientation, linear-lanceolate, to 38 × 1.8cm, green, often tinged purple-red in autumn, usually glabrous, long-acute. Inflorescence narrow-lanceolate, open to compact, to 20cm, pubescent; spikelets narrow-ovate, to 0.5cm; glumes densely white-pubescent, awns to 1.3cm. Summer. Siberia, Manchuria, Korea, China, Japan.

**Sporobolus** R. Br. (Gk. *spora*, seed and *ballein*, to throw, a reference to the caryopsis which is free from the lemma and palea). DROPSEED; RUSHGRASS. Gramineae. Some 100 species of annual and perennial grasses. Inflorescence an open or contracted panicle; spikelets 1-flowered, small, stalked; glumes 1-nerved shorter than or equalling thin-textured, 1-nerved, awnless lemma; caryopsis free from lemma and palea. Cosmopolitan.

CULTIVATION    Fine-textured, clump-forming grasses grown for their foliage and flowers, the Dropseeds will grow on most well drained soils in full sunlight and are suited to use either as specimens in mixed borders (especially if back-lighted or given a dark background) or in drifts, naturalized in meadow gardens. Among the most elegant and refined of grasses, *S.heterolepis* produces delicate open panicles in mid to late summer, held high above the foliage on very slender stalks. Unusually for a grass, the inflorescence is strongly scented, some have said of coriander, others of burnt popcorn. The foliage, a fine light green in summer, turns orange in late autumn. All Dropseeds are fairly deep-rooted and drought-tolerant. Once mature and established (possibly several years), they are long-lived and trouble-free. Propagate by seed or division.

*S.cryptandrus* (Torr.) A. Gray. SAND DROPSEED.
Culms to 1m in small clumps, erect to spreading. Leaves to 0.6cm across, flat or rolling when dry; sheaths with an apical tuft of long pale hairs. Inflorescence to 25cm, lax, terminal, branches spreading; spikelets to 0.3cm, pale to leaden grey, crowded toward tips of principal branches. N America. Z5.

*S.heterolepis* (A.Gray) A.Gray. PRAIRIE DROPSEED; NORTHERN DROPSEED.
Densely tufted, narrowly upright perennial. Culms 30–76cm, slender. Leaves mid-green, fine, mostly basal, about half as long as culm; blades 20–60 × 0.2cm, involute, upper surface, midrib and margins rough or sparsely pilose; sheaths glabrous with some hairs at apex. Inflorescence 8–30cm, fragrant, long-exserted, grey-green, narrowly pyramidal, branches distant, alternate or fascicled, 2.5–10cm, erect or ascending, finally spreading, bearing dark green spikelets toward their tips; spikelets short-pedicellate, 4–6mm long, 1-flowered; glumes unequal, acuminate or aristate, lower glume awl-shaped, 2–3mm, upper glume lanceolate, 4–5mm, slightly longer than lemma, often awn-pointed; lemmas 1-nerved, glabrous, obtuse or acute. Summer and autumn. N America. Z4.

*S.pulchellus* R. Br.
Culms to 32cm, clumped. Leaves mostly in basal third of stems, blades somewhat rigid, flat or keeled, ciliate. Inflorescence to 8cm, pyramidal, branches slender, spreading in whorls; spikelets to 0.3cm, stalked, lustrous, glumes obtuse, obscurely keeled. Australia. Z9.

**Stenotaphrum** Trin. (From Greek *stenos*, narrow, and *taphros*, trench, referring to the depressions in the raceme axis.) Gramineae. Some 7 species of annual or perennial grasses. Stems creeping or ascending, rooting at nodes. Leaves linear to lanceolate, flat or folded, obtuse, blades held perpendicular to stem and the broad, green, compressed and overlapping sheaths; ligules ciliate. Inflorescence of short terminal and axillary racemes, embedded in swollen inflorescence axis; spikelets subsessile, 2-flowered, lower flower male, upper flower hermaphrodite, glumes membranous, short, or upper glume equal to spikelet; lower lemma leathery, upper lemma chartaceous. New & Old World Tropics. Z9.

CULTIVATION    *Stenotaphrum* is widely used for lawns in the dry and humid tropics and subtropics, the two colour forms opposed with great effect in the sweeping curves and rhythmic geometric compositions in the gardens of Roberto Burle Marx. In cooler regions *Stenotaphrum* is used as a cascading basket plant or as a fast-spreading groundcover in the warm glasshouse or in subtropical bedding, especially its attractive variegated cultivar.. Grow in rich, moisture-retentive soil in sun. Propagate by plugs rooted in moist sand with bottom heat.

*S.americanum* Schrank. See *S.secundatum*.

*S.secundatum* (Walter) Kuntze (*S.americanum* Schrank). ST AUGUSTINE GRASS; BUFFALO GRASS.
Perennial, to 30cm. Stems branching, creeping and rooting. Leaves linear-oblong, flat to folded, to 15 × 1.5cm, glabrous, obtuse; sheaths flattened, glabrous. Inflorescence ascending, stout, straight or curved, to 10cm; spikelets 3 per raceme, to 0.5cm, light green. Summer– autumn. Tropical America, W Africa, Pacific Is. 'Variegatum' (var. *variegatum* Hitchc.): leaves longitudinally striped pale green and ivory.

**Stipa** L. (From Latin *stipa*, tow or oakum, from the feathery inflorescences.) NEEDLE GRASS; SPEAR GRASS; FEATHER GRASS. Gramineae. Some 300 species of bristled, perennial, rarely annual grasses. Culms clumped. Leaves narrow, rough, convolute or plicate, occasionally flat, wide, veins prominent above. Inflorescence narrow-paniculate; spikelets stipitate, flattened, flowers solitary, narrow-ellipsoid; callus bearded; glumes subequal, membranous; lemma membranous to leathery, shorter than glumes, apex usually entire, awned, awn hygroscopic, twisted near base, forming a column, and straight, slender above, forming a glabrous to pubescent seta, usually twice geniculate; margins overlapping, convolute, entire to bifid.Temperate, warm temperate regions.

CULTIVATION  Native to steppes, and valued in gardens for their strong form; the robust basal clump of foliage give rise to long slender stems bearing often large and conspicuous inflorescences which dry and dye particularly well, and frequently persist into the winter months. These are large grasses, making specimen features in herbaceous borders and grass gardens. Four are particularly noteworthy: *S.arundinacea*, an Australasian native, acquires orange and yellow-brown leaf tints at the end of the season, lasting well into winter; *S.calamagrostis* carries windswept plumes high above its foliage and late into the year; *S.gigantea* produces spectacular open heads of long-awned, shimmering spikelets; the awns of *S.pennata* are longer still and feathery. Most are hardy to at least −15°C/5°F, and are easily grown in any moderately fertile garden soil in sun. The European species sometimes suffer in intensely hot, humid summers. Propagate by seed or division.

**S.arundinacea** Hook. f. NEW ZEALAND WIND GRASS; PHEASANT'S TAIL GRASS.
Perennial, to 1.5m. Culms clumped, upright or arching; rhizome short, scaly, creeping. Leaves to 30×0.6cm, coriaceous, involute, dark green, becoming orange- or pale brown-striped from late summer through winter, margin ciliate; ligules truncate. Panicle pendent; spikelets stipitate, tinged purple; glumes equal, lanceolate, acuminate, fertile glume sessile; awn slender, to 8mm; stamens 1. E Australia, New Zealand. 'Autumn Tints': leaves flushed red in late summer. 'Gold Hue': leaves flushed golden yellow in late summer. Z8.

**S.barbata** Desf.
Perennial, to 70cm. Leaves convolute, sickle-shaped, glabrous above, sparsely pubescent below; sheaths sparsely pubescent or glabrous; upper leaves enclosing inflorescence base; ligule ciliate. Panicle narrow; glumes linear-lanceolate; lemma to 14mm, glabrous, awn to 19cm, base pubescent S Europe. Z8.

**S.calamagrostis** (L.) Wahl. (*Achnatherum calamagrostis* (L.) P.Beauv.). SILVER SPIKE GRASS.
Perennial, to 120cm. Culms clumped, robust. Leaves to 30 × 0.5cm, attenuate; ligule inconspicuous. Panicle to 80cm, loose, appearing secund and windswept, feathery; spikelets to 9mm, tinged purple; glumes lanceolate, acute; lemma to 4mm, coriaceous, pubescent below, hairs to 4mm, awned, awn to 1cm, straight or curved. Summer. C & S Europe. Z7. 'Lemperg': selected form.

**S.capillata** L.
To 80cm. Culms upright, clumped. Leaves thread-like, convolute, blue-green, glaucous, scabrous; sheaths smooth; ligule to 20mm, lower ligules to 2mm. Panicle loose; branches unequal; glumes to 35mm, apex attenuate; lemma to 12mm, awned, awn to 20mm rough, glabrous. Summer–autumn. S & C Europe, Asia. Z7.

**S.elegantissima** Labill.
Perennial to 1m, tussock-forming. Leaves to 3mm diam., acuminate, yellow-green, inrolled above base; outer sheaths pale; ligule to 5mm, membranous, truncate or torn. Panicle initially ovoid, spreading with age, to 20cm; spikelets to 1.3cm, narrowly ellipsoid, tinged purple, with a single floret; glumes unequal, linear, acute to acuminate, lower to 1.3cm, exceeding upper, 3-nerved, sparsely villous; lemma to 8mm, linear, scabrous; awn to 4.5cm, terminal, pilose. Temperate Australia.

**S.gigantea** Link (*Macrochloa arenaria* (Brot.) Kunth). GIANT FEATHER GRASS.
Perennial, to 250cm. Culms clumped. Leaves to 70 × 0.3cm, involute when dry, smooth to slightly rough; sheaths glabrous, auricle ciliate; ligule very short. Panicle to 50cm, loose, open and arching, held high above foliage, very spectacular; spikelets shimmering straw-yellow; glumes subequal, 3 ribbed, to 3.5cm; lemma to 18mm, pubescent below, awned, apex bidentate, awn to 12cm, rough. C & S Spain, Portugal, NW Africa. Z8.

**S.lasiagrostis** Nichols. See *S.calamagrostis*.

**S.pennata** L. EUROPEAN FEATHER FRASS.
Perennial, to 60cm. Culms occasionally sparsely pubescent. Leaves erect to weeping in a tight bunch, to 60 × 0.6cm, smooth, glabrous above; sheaths pubescent above, margins involute; ligule to 5mm. Panicle very loose, rather sparse; spikelets yellow; glumes to 5cm; lemma to 2cm, awn to 28cm, plumose. Summer. S & C Europe to Himalaya. Z7.

**S.pulcherrima** K. Koch.
Perennial, tufted. Culms to 1m, smooth, glabrous. Leaf sheaths glabrous, smooth; leaves to 4mm diam., flat, involute, smooth or scabrous; ligule to 3mm. Glumes to 8cm, long-acuminate; lemma to 2.5cm; awn to 50cm. C & S Europe.

**S.pulchra** Hitchc. PURPLE NEEDLEGRASS.
Culms to 1m. Leaves long, narrow, flat or involute; ligule about 1mm. Panicle to 20cm, nodding, loose, with slender, spreading branches; glumes narrow, long-acuminate, 3-nerved, the first about 2cm, the second to 0.4cm; lemma to 1.3cm, fusiform; awn to 9cm. W US.

**S.spartea** Trin.
Culms about 1m. Leaves to 30 × 5cm, flat, involute; ligule to 5mm, rather firm. Panicle to 20cm, narrow, nodding; branches slender, bearing 1 or 2 spikelets; glumes to 4cm; lemma to 2.5cm, subcylindric; awn to 20cm, stout. US.

**S.splendens** Trin. CHEE GRASS.
Perennial, to 2.5m. Culms stout, robust, clumped. Leaves scabrous; ligule elongate, to 6mm. Panicle to 50cm, tinged purple to white; spikelets to 4mm; glumes unequal, to 6mm; lemma as long as glumes, awn bent. Summer. C Asia, Siberia. Z7.

**S.tenacissima** L. (*Macrochloa tenacissima* (L.) Kunth. ESPARTO GRASS.
Perennial, to 2m. Culms very stout. Leaves convolute, to 1mm diam., glabrous, smooth above, minutely pubescent beneath; auricles erect; ligule ciliate. Inflorescence narrow, compact, to 35cm; glumes to 3cm; lemmas to 1cm, hairy; apex 2-lobed, awned, awn to 4cm, plumose below bend. W Mediterranean. Z8.

**S.tenuissima** Trin.
Culms to 70cm, slender, borne in large tufts; ligule 2mm. Leaves to 30cm or more, filiform, wiry, closely involute. Panicle to 30cm, narrow, soft, nodding; glumes about 1cm; lemma 2–3mm, oblong-elliptic, glabrous; awn about 5cm. US (Texas & New Mexico), Mexico, Argentina.

**S.tirsa** Steven.
Perennial. Culms very stout. Leaves to 1m, convolute; ligule inconspicuous; leaf sheaths glabrous, short, stiffly pubescent above, densely pubescent below, apex bristled. Panicle compact; glumes tapering; lemma to 2cm, awn to 4cm, bristle-tip hairy. Summer. C, S, E Europe. Z7.

**Stipa**  (a) *S. gigantea*  (b) *S. calamagrostis*

**Stipagrostis** Nees (From Latin *stipa*, tow, and Greek *agrostis* a type of grass, referring to the sometimes feathery inflorescence.) Gramineae. Some 50 species of perennial grasses. Culms erect, sometimes clumped. Leaves narrow, elongated, flat or convolute, sometimes pungent; ligules a ciliate fringe. Inflorescence narrow-paniculate; spikelets solitary, stipitate, laterally flattened; callus pubescent; glumes to 11-ribbed; lemmas convolute, leathery, 3-ribbed, with 3 plumose, abscising awns. Middle East, C Asia, Pakistan.

CULTIVATION *S.pennata* occurs in a range of arid and semi-arid habitats, in sandy steppes and semi-desert, sometimes on shifting dunes. It may have potential as an ornamental for dry gardens. Grow in full sun in well drained, low-fertility soils. Propagate by seed or division.

*S.pennata* (Trin.) De Winter (*Arista pennata* Trin.).
Perennial to 50cm. Culms branched basally, glabrous, smooth. Leaves linear, convolute, tó 0.2cm wide, scabrous, short-pubescent above, acuminate; ligule short-pubescent. Panicle very loose, to 20cm; spikelets plumose, to 1.5cm; glumes unequal, to 1.2cm; lemma to 0.6cm, glabrous; awns to 1.5cm. C, S & W Asia, SE Russia. Z8.

**Thamnocalamus** Munro. (From Greek *thamnos*, bush, and *kalamos*, reed.) Gramineae. Some 6 species of clumping bamboos. Culms terete, thick-walled, glabrous; sheaths deciduous with auricles and dark scabrous bristles; branches many. Leaves tessellate, glabrous, sheaths glabrous with auricles and bristles. Himalaya, S Africa.

CULTIVATION   Vigorous clumping bamboos native to damp sites in woodlands and high savannah. Attaining 3m/10ft in most gardens, they are valuable either as large specimen plants at the waterside or for screen-planting in wet areas. Most species will tolerate temperatures as low as –20°C/–4°F and are unfussy as to soil and light, if provided with a good water supply. The rather more tender *T.tessellatus* may become invasive in very mild areas. In full sunlight it will produce pink-streaked shoots which become tall, stiffly erect, pale green culms flushed purple and sheathed papery-white; in their second year, the culms bear dense whorls of short branches. Propagate by division.

*T.aristatus* (Gamble) Camus (*T.spathiflorus* ssp. *aristatus* (Gamble) D. McClintock; *Arundinaria aristata* Gamble). Culms 3–10m × 1–6cm, yellow-green, speckled brown with age; sheaths sometimes with dark markings, sparsely hairy, bristly at the base, ligule short, blade short; nodes with white waxy bloom below; branches and branchlets usually red-lined. Leaves 6.5–13.5 × 0.6–1.8cm, scaberulous, with a hairy callus at the base of the stalk. NE Himalaya. Z7.

*T.crassinodus* (*Fargesia crassinoda* Yi). Culms 3–4m × 1–2cm, blue-green with a thick white or blue-grey bloom at first, later smooth, purple- or yellow-green, nodal ridges swollen on one side, sheath scars prominent; culm sheaths large, thickly papery, parchment-like, soon standing away from nodes then falling; branches (1)3–6 per node, slightly flexuous. Leaves 4.5–9 × 0.5–1cm, smooth, grey-green beneath, margins coarse, tessellation distinct. Tibet. Includes 'Kew Beauty' with very slender branchlets and small narrow leaves. Z7.

*T.falcatus* (Nees) Camus. See *Drepanostachyum falcatum*.

*T.falconeri* Hook. ex Munro. See *Himalayacalamus falconeri*.

*T.spathaceus* (Franch.) Söderstr. See *Fargesia murielae*.

*T.spathiflorus* (Trin.) Munro (*Arundinaria spathiflora* Trin.). Differs from *T.aristatus* in its somewhat flexuous grey culms, white-bloomed at first and later flushed pink in strong sunlight; sheaths with fewer or no hairs, ciliate, blades often longer; leaves greyer with no callus. NW Himalaya. Z7.

*T.spathiflorus* ssp. *aristatus* (Gamble) D. McClintock. See *T.aristatus*.

*T.tessellatus* (Nees) Söderstr. & R.P. Ellis (*Arundinaria tessellata* (Nees) Munro). BERGBAMBOES. Culms 2.5–7m × 1–6cm, the lowest nodes close together, unbranched below and leafless in the first year; sheaths fairly persistent, usually exceeding internodes, pale, papery, ligule large, fringed, blades long and narrow; nodes with a pink-purple ring below; branches numerous, short, tufted, erect, tinted purple. Leaves 5–21 × 0.5–1.9cm, finely tapered, scaberulous on one margin, dark green, ligule tall, fringed. S Africa. Z8.

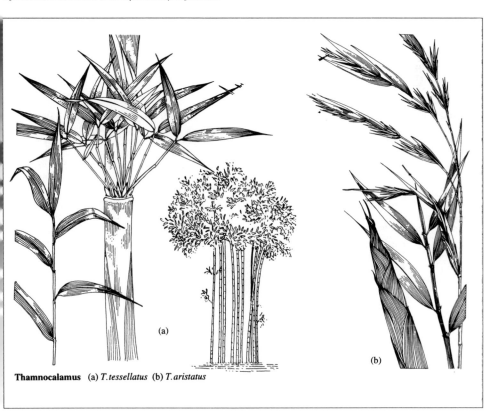

**Thamnocalamus**   (a) *T.tessellatus*   (b) *T.aristatus*

**Themeda** Forssk. (From the Arabic name for this plant, *Thoemed.*) Gramineae. About 19 species of annual or perennial grasses. Leaves linear; ligule much reduced. Inflorescence of solitary racemes in sheathing spatheoles; racemes composed of 2 homogamous pairs of spikelets and 2–4 sessile spikelets, homogamous spikelets persistent, sterile; sessile spikelets terete; lower glume rarely grooved, upper lemma stipitiform, awnless. Old World Tropics & Subtropics; E Asia.

CULTIVATION  *T.triandra* is a large, tender, clump-forming grass grown for its handsome flower heads invested with rust-coloured bracts. Its general treatment is as for *Cymbopogon*. From East Asia, the hardy *T.japonica* Tanaka (*T.triandra* var. *japonica*) is little known in cultivation, but undoubtedly merits a place in the garden. It tolerates most well drained soils – even dry conditions – and performs best in full sunlight, attaining heights of 1–1.5m. The culms are erect to arching and radiate in a dramatic, irregular arc, clothed with slender leaves which turn bright golden-orange in autumn and retain their colour throughout the winter. Clustered in late summer along the upper portions of the culms, the flowers are relatively insignificant. Propagate by seed, or division in spring.

*T.australis* (R. Br.) Stapf. See *T.triandra*.
*T.japonica* Tanaka. See under CULTIVATION above.

***T.triandra*** Forssk. (*T.australis* (R. Br.) Stapf).
Tufted perennial to 2m. Leaves compressed, to 30 × 0.8cm. Inflorescence to 30cm; racemes 2–8, spatheoles to 3.5cm, rus-set; homogamous spikelets equal, lower glume narrowly elliptic, to 1.5cm, glabrous to hirsute; sessile spikelet solitary, to 11mm, lower glume brown; awn to 7cm. Old World Tropics & Subtropics. (*T.triandra* var. *japonica*. See *T.japonica* above.) Z9.

**Thysanolaena** Nees. (From Greek *thysanos*, tassel or fringe, and *laina*, cloak, referring to the pubescent fringe on the fertile lemmas.) Gramineae. 1 species, a perennial grass, to 3.6m. Culms clumped, erect, thickened near base. Leaves narrow-lanceolate to oblong-lanceolate, to 60×6cm, flat, glabrous; ligule scarious. Inflorescence an ovoid panicle to 60×30cm, green, tinged yellow or brown; spikelets lanceolate, laterally compressed, to 0.3cm, 2-flowered, abscising with rachis; glumes membranous, shorter than spikelets, blunt; flowers stipitate, dimorphic, brittle below pedicel; upper flower hermaphrodite, lemma 3-ribbed, ciliate, acuminate, lower flower sterile, palea absent; lemma membranous, to 3-ribbed, glabrous, acuminate. Summer–autumn. Tropical Asia. Z9.

CULTIVATION  Native to tropical scrub, and frequently grown in tropical regions as a screening plant, *Thysanolaena* is occasionally grown in tubs or borders in large glasshouses, needing warm humid conditions, a rich soil and plentiful water with a position in full sun. Propagate by seed or division.

*T.acarifera* Arn. & Nees. See *T.maxima*.
*T.agrostis* Nees. See *T.maxima*.
*T.latifolia* (Roxb.) Honda. See *T.maxima*.

***T.maxima*** (Roxb.) Kuntze (*T.agrostis* Nees; *T.acarifera* Arn. & Nees; *T.latifolia* (Roxb.) Honda). TIGER GRASS.

**Trichloris** Fourn. (From Latin *tri*, three-fold, and *Chloris*, a genus of grasses, referring to the 3-awned lemmas.) Gramineae. 2 species of perennial grasses. Culms tufted, upright or ascending. Leaves linear; ligules ciliate. Inflorescence digitate, racemose; spikelets short-stipitate, to 5-flowered, dorsally flattened, breaking up at maturity above glumes; one fertile flower and occasionally a smaller male flower, spikelet axis vestigially awned at apex; glumes scale-shaped to lanceolate, papery, upper glume equalling flower, persistent; apex acute to short-awned; lemma rounded, 3-ribbed, midrib prominent, often scabrous, apex branching to 3 awns, central awn longest. Summer–autumn. S America. Z9.

CULTIVATION  Native to dry plains and hillsides, *Trichloris* is occasionally grown for the erect, densely flowered and feathery inflorescence. Frost-tender, it may be treated as an annual in cool temperate zones, sown in early spring under glass, and planted out in a warm, sunny border.

***T.crinita*** (Lagasca) L. Parodi (*T.mendocina* (Philippi) Kurtz; *Chloris crinita* Lagasca).
To 90cm. Leaves to 45 × 0.4cm, scabrous, glaucous; leaf base pubescent; sheaths flattened, keeled. Inflorescence erect, compact, bristly; spikes to 13cm; spikelets to 2-flowered; lemma linear, to 0.3cm; awns to 1.3cm.

*T.mendocina* (Philippi) Kurtz. See *T.crinita*.

**Tricholaena** Schräd. (From the Greek *thrix*, hair, and *chlaina*, cloak.) Gramineae. About 4 species of perennial, rarely annual grasses. Leaves glaucous, frequently inrolled; ligule a ciliate rim. Inflorescence an open panicle; spikelets silky, oblong, compressed, awnless; lower glume somewhat reduced, upper glume equalling spikelet, membranous, emarginate to acute, lower floret male. Africa and Mediterranean to India. Z9.

CULTIVATION    As for *Trichloris*.

*T.repens* (Willd.) A. Hitchc. See *Rhynchelytrum repens*.
*T.rosea* Nees. See *Rhynchelytrum repens*.

*T.teneriffae* (L. f.) Link.
Tussock-forming perennial; culms hard, slender, to 60cm. Leaves glaucous, flat or inrolled, 2–15 × 0.5cm, apex pun-gent. Panicles usually narrowly oblong, to 15cm; spikelets to 4mm, pale green or purple-tinted, with long white hairs; lower glume scale-like, upper glume ovate, mucronate, lower lemma resembling upper glume. S Europe, Africa, India.

**Tridens** Roem & Schult. (Latin, 3-toothed.) Gramineae. 18 species of tufted perennials. Inflorescence an open, contracted or spike-like panicle; spikelets somewhat compressed, plump, lemmas rounded, nerves usually pubescent below, apex usually emarginate to bidentate. Americas, 1 species in Angola.

CULTIVATION    *T.flavus*, the Purple Top, is a North American native much loved for the 'purple top' it places on fallow fields and meadows in late summer and early autumn. Its delicately airy, purple panicles add interest when the plant is used as a single specimen or in groupings, and work well in cut arrangements, fresh or dried. The foliage takes on purple-bronze tints in autumn. This grass is ideally suited to naturalizing in meadow gardens, where it will remain attractive from late summer through to the following spring. It prefers full sun, but is undemanding as to soil. Easily propagated from seed (it may self-sow), also by division in spring or autumn.

*T.flavus* Hitch. (*Triodia flava*; *Poa flava*). PURPLE TOP.
Clump-forming, erect to 90cm. Leaves to 30 × 1.8cm, rather coarse, mid-green, tinted purple-bronze in autumn. Panicles open, standing tall above foliage, stained deep metallic purple-red. Summer–autumn. E US. Z5.

**Typha** L. (Name used by Theophrastus for these or possibly other aquatic plants.) BULRUSH REEDMACE; CAT TAIL. Typhaceae. 10 species of aquatic and marginal perennial herbs, stout, rhizomatous, monoecious; stems erect, simple, corm-like at base rooting into bed or bank. Leaves distichous, usually basal, thick or spongy particularly at base, elongate, linear to ligulate, somewhat basally twisted, sheaths enveloping stem. Inflorescence a terminal spike occupying uppermost quarter of scape, dense, narrow-oblong to cylindric, appearing thickly brown-velvety, terminating in a long, slender point, male flowers above female with simple or forked hairs or scales, sometimes naked; perianth absent; stamens 1–3 occasionally to 8; female flowers on branched stalks with floss like hairs, occasionally scales; perianth absent; ovary superior, stipitate, unilocular; ovule solitary. Fruit dry, usually dehiscent with slender stipe formed by gynophore and style, 1-seeded. Cosmopolitan.

CULTIVATION Grown for their bold, poker-like seed-heads, useful in dried and fresh flower arrangements, *Typha* species may be cultivated in rich, wet soils or up to 30cm of water in sun or part-shade. They are suitable for larger gardens and lakesides (except for *T.minima*, a charming dwarf ideal for small ponds), where the invasive habit is easily accommodated; alternatively, plant into containers. They are useful bank stabilizers. For floral decorations, pick the spikes early in the season, whilst male flowers are still in bloom, and air dry upright or flat; seal with lacquer. Specimens picked too late in the season may explode without warning in the warmth of the home, shedding disproportionately large quantities of down.

*T.angustata* Bory & Chaub. See *T.domingensis*.

***T.angustifolia*** L. LESSER BULRUSH; NARROW-LEAVED REEDMACE; SOFT FLAG.
To 2m. Leaves 3–6mm wide, dark green; sheaths closed at base, auriculate. Inflorescence dark rust brown; scape shorter than leaves; male and female flowers separated on scape by 3–8cm; scales of male flowers simple or forked; scales of female flowers dark brown opaque, spathulate; stigmas broad, flat-topped. America, Europe, N Africa, N & C Asia. Z3.

***T.domingensis*** (Pers.) Steud. (*T.angustata* Bory & Chaub.).
To 3m, robust. Sheath open at base, tapering into yellow-green lamina. Scape equal to or slightly shorter than leaves; male and female flowers separated on stalk by 1–6cm; scales of male flowers linear, often apically laciniate; scales of female flowers obovate, apiculate, pale brown, translucent; stigmas linear. America, Europe, Asia.Z5.

***T.latifolia*** L. CAT'S TAIL; BULRUSH; NAILROD; GREAT REEDMACE; MARSH BEETLE.
To 2m, robust. Leaves 8–20mm wide; sheaths usually open at throat, tapering into lamina, glaucous. Scape shorter than leaves; inflorescence chestnut-brown, male and female parts contiguous or separated by less than 2.5cm; male part 4–16cm, pubescent; female part 8–15 × 2–3cm, dark brown mottled white later, with flowers lacking scales, pubescent from base of gynophore; stigmas lanceolate, fleshy, brown, extending beyond hairs. N America, Europe, Asia, N Africa. Z3.

***T.laxmannii*** Lepech. (*T.stenophylla* Fisch. & C.A. Mey.).
To 1.5m, slender. Leaves to 1.5mm wide, occasionally wider; sheath usually open at throat, auriculate. Scape shorter than or equal to leaves; inflorescence brown, male and female parts separated by 1–6cm; male 10–25cm, sometimes more; flowers pubescent; scaly; female portion 5–10cm, pale brown; flowers pubescent, lacking scales; stigmatic surface flat, elliptic, extending beyond hairs. Seed to 1mm. Eurasia. Z4.

***T.minima*** Hoppe. DWARF CAT TAIL; DWARF-REED MACE.
40–80cm, slender. Leaves to 2mm wide, grey-green to sea green. Scape shorter than leaves; male part contiguous with female or separated, lacking scales and hairs; female part 2–5cm, cylindric to oblong, dark brown, sometimes bracteate; flowers scaly, pubescent; stigmas linear extending beyond swollen tipped hairs. Eurasia. Z6.

***T.shuttleworthii*** Koch & Sonder.
To 1.5m, somewhat robust. Leaves 5–10mm wide sometimes more; sheath open at throat, tapering into lamina. Scape shorter than leaves; male part of inflorescence contiguous with female; flowers pubescent; female part 5–15cm, brown becoming silver-grey; flowers with hairs inserted in zone at base of gynophore; stigmas lanceolate, fleshy. S Europe. Z5.

*T.stenophylla* Fisch. & C.A. Mey. See *T.laxmannii*.

**Uncinia** Pers. (From Latin *uncinatus*, hooked at the end, referring to the hook from the inflorescence axis, projecting beyond the utricle, which facilitates dispersal.) HOOK SEDGE. Cyperaceae. Some 35 species of tufted herbs, with basal sheaths. Stems slender, more or less 3-angled. Leaves grass-like, linear, shallowly grooved above, margins scaberulous. Inflorescence simple, the terminal portion staminate; female flowers enclosed in a utricle (prophyll), which is closed throughout its length; styles 3. Fruit a trigonous nut, rhachilla produced beyond the mouth of the utricle and terminating in a glume which is sharply reflexed to form a hook. Malesia, Pacific, C & S America, S Indian & Atlantic Is.

CULTIVATION   As for the New Zealand *Carex* species, whose burnished copper and bronze tones they surpass – i.e. moist, neutral to acid soils and some protection from hard frosts, winter wet and harsh winds.

*U.egmontiana* Hamlin.
Whole plant dull red-green or brilliant copper. Culms to 40cm, densely tufted, smooth, angled to terete. Leaves 4–5, to 30cm × 1.5mm, flat, somewhat scabrous above and on margins. Spikes slender, to 12cm; pistillate flowers 10–17; glumes membranous, lanceolate, acute, red, persistent; utricles subtrigonous, 6 × 1mm, striate, contracted below into a stipe to 1.5mm, narrowed above into a beak to 1.5mm. New Zealand.

*U.rubra* Boott.
Loosely tufted or shortly rhizomatous, the entire plant dark red to rust or copper with bronze-yellow tints. Stems to 35cm × 1mm, rigid, scabrous on angled. Leaves 2–5 per stem, to 35cm × 2mm, abruptly acuminate, rigid, flat or involute, margins somewhat scabrous. Spikes to 6cm × 5mm, pistillate flowers 6–13; glumes to 6mm, obtuse or subacute, coria-ceous, red, lighter on margin; utricles to 6mm, fusiform or convex, striate, somewhat yellow, contracted below into a 1mm stipe, narrowed above into a beak to 2mm. New Zealand.

*U.uncinata* (L.f.) Kukenthal.
Loosely to densely tufted, dark bronze-green to rusty brown throughout. Stems to 4.5cm, smooth, tufted, angled. Leaves 5–10 per stem, to 45 × 0.4cm, flat, strongly scrabrous above and on margins. Spikes to 20cm × 3mm, usually bracteate; pistillate flowers 60–120+; glumes subacute, coriaceous, somewhat yellow to very dark brown; utricles convex, to 5 × 1mm, smooth, shiny, somewhat yellow, contracted below into a stipe to 1.5mm, narrowed above into a 1mm beak. New Zealand.

**Uniola** L. (Latin Name (adopted) of an unknown plant.) SPANGLE GRASS; SPIKE GRASS. Gramineae. 4 species of rhizomatous, perennial grasses. Culms stoloniferous, clumped. Leaves linear to narrow-lanceolate, flat, rolled when dry, scabrous; ligules ciliate. Inflorescence racemose, paniculate, loose or compressed; racemes crowded, overlapping, borne on a central axis; spikelets sessile, flattened, few- to many-flowered, abscising when mature, occasionally tinged purple; lower flowers sterile; glumes rigid, keeled, persistent; palea with winged keels; lemmas to 9-ribbed, leathery, glabrous. S & E US to Ecuador. Z7.

CULTIVATION   *U.paniculata* occurs on salt flats and sand dunes, and is sometimes used as a sand binder. Its invasive tendencies are more than offset by its very decorative flowers, similar in effect to those of *Briza* and, of course, *Chasmanthium* (formerly a member of this genus); these can be used in fresh and dried arrangements. It prefers a moist, sandy soil and is tolerant of some shade. Propagate by seed or division.

*U.latifolia* Michx. See *Chasmanthium latifolium*.

*U.paniculata* L. NORTH AMERICAN SEA OATS; SEASIDE OATS; SEA OATS.
To 240cm. Rhizome creeping. Culms loosely clumped, robust, upright. Leaves to 7.5 × 0.8cm, rigid. Inflorescence dense, erect to pendent, to 5 × 10cm; spikelets ovate to oblong, short-stipitate, laterally flattened, to 24-flowered, to 3 × 1.3cm, golden brown; lemmas lanceolate, to 1cm, overlapping. Summer. E US to W Indies.

**Vetiveria** Bory. (From Tamil *vettivēru*: *veti*, coarse, *ver*, grass.) Gramineae. 10 species of perennial grasses. Culms clumped, flimsy to robust; ligule a line of hairs. Inflorescence narrow-paniculate; primary branches whorled, each branch bearing a raceme; spikelets to 20 per raceme, laterally flattened, 2-flowered, acute, paired, one sessile, hermaphrodite, spiny, the other stipitate, male; glumes tough, enclosing lemma, lower glume papery to coriaceous, upper glume short-awned; upper lemma 2-dentate, awned, membranous. Tropical Asia. Z9.

CULTIVATION Occurring naturally on flood plains and on streams and riverbanks, *V.zizanoides* is occasionally grown in botanical collections of useful plants. Its roots are used in perfume manufacture (producing oil of vetiver) and for weaving mats, baskets etc., which become fragrant when moistened. Cultivate as for *Cymbopogon*.

*V.zizanoides* (L.) Nash (*Andropogon muricatus* Retz.; *Andropogon zizanoides* (L.) Urban). VETIVER; KHUS KHUS GRASS; KHAS KHAS).
To 180cm or over. Rhizomes strongly perfumed. Culms robust, erect, simple, densely clumped. Leaves linear, to 90 × 1cm, rigid, margins scabrid; sheaths glabrous. Inflorescence lanceolate-oblong, to 45cm, condensed; racemes to 8cm, slender; spikelets to 0.5cm, awnless; lower glume tough, minutely spiny beneath.

**Yushania** Keng f. (For the Yushan mountains in Taiwan.) Gramineae. 2 species of bamboos, separated from *Fargesia* principally by their rhizomes – these are clumping, but can also elongate to form fresh clumps. The leaves are larger and the flowers differ. E Asia. Z8.

CULTIVATION As for *Fargesia*.

*Y.anceps* (Mitford) Yi (*Arundinaria anceps* Mitford; *Arundinaria jaunsarensis* Gamble; *Chimonobambusa jaunsarensis* (Gamble) Bahadur & Naithani; *Sinarundinaria anceps* (Mitford) Chao & Renvoize).
Culms 2–5m × 0.5–1.2cm, not developing branches in the first year, branchlets eventually numerous and pendulous; sheaths not long persistent, hairless, sometimes with auricles and bristles; ligules fringed. Leaves 6–14 × 0.5–1.8cm, not or hardly scaberulous, mid-green, stalks occasionally purple-hued; sheaths with bristles and often auricles. C & NW Himalaya. 'Pitt White' (*Arundinaria niitikayamensis* Lawson, non Hayata). grows to twice the height of type and comes true from seed; a graceful grower but not very hardy.

*Y.aztecorum* McClure & E.W. Sm. See *Otatea acuminata*.
*Y.humilis* (Mitford) W.C. Lin. See *Pleioblastus humilis*.

*Y.maling* (Gamble) Stapleton (*Arundinaria maling* Gamble; *Arundinaria racemosa* auct. non Munro; *Fargesia maling* (Gamble) Simon ex D. McClintock; *Sinarundinaria maling* (Gamble) Chao & Renvoize).
Culms 3–10m × 2–5cm, markedly rough at least until old; sheaths persistent, bristly hairy and ciliate, without auricles or bristles, blades long and narrow. Leaves 8.5–18 × 0.8–1.6cm, scaberulous, sometimes hairy beneath; ligules short. NE Himalaya.

*Yushania anceps*

**Zea** L. (Name adopted from Greek *zea*, a kind of cereal.) Gramineae. 4 species of annual, rarely perennial grasses. Female inflorescences axillary, raceme enveloped by spathes (often known as husks), occasionally tipped by a male raceme; internodes bearing a solitary sessile spikelet (ear), distichous; spikelets almost enclosed by internodes; callus truncate, flat; glumes equal, rounded or emarginate, lower glume crustaceous, smooth, lower flower sterile; style very long, silky; male inflorescence tassels, terminal racemes digitate to paniculate; spikelets paired, narrow, 2-flowered, one sessile, one stipitate, flowers male; glumes membranous; lemmas translucent. C America.

CULTIVATION  Maize has been the principal cereal in Mexico, Central America and many South American countries since pre-Colombian times. It is now the most important cereal grown in tropical and warm areas, for human and animal food, cooking oil, also made into beer and spirits. Its milled flour is made into dough; crushed grains are known as hominy grits. Forms with hard endosperm explode when heated to produce popcorn (var. *praecox*). There is archaeological evidence for the cultivation of primitive forms over 5000 years ago. Unless as an escape from cultivation, Sweetcorn is not known in the wild but is widely cultivated throughout tropical and temperate regions of the world. The sugar content of the seed or kernels decreases as maturity approaches but the origin of sweetcorn as a distinct variety is probably a more recent development.

Sweetcorn is a warm-season crop and will not withstand frost. Depending on the cultivar and growing conditions it requires from 70 to 110 frost-free days from planting to harvest. The optimum soil temperature for seed germination is between 21–27°C/70–80°F with a minimum of 10°C/50°F. Optimum mean air temperatures for growth lie within the range 21–30°C/70–85°F although the crop can be grown at mean temperatures as low as 16°C/60°F and as high as 35°C/95°F. Flowering is influenced by day length and is promoted by short days, although cultivars have been selected for flowering over a wide range of latitudes covering both tropical and temperate situations.

A sunny but well-sheltered site is required. Sweetcorn can be grown on a wide range of soil types but fertile medium-textured loams with a high organic content are preferable. The soil pH should be in the range 5.5–6.8. It takes up a large area but can be undercropped with lower-growing vegetables such as marrows, courgettes, french beans and most salad crops.

*Z.mays* can be sown outdoors during late spring when the risk of frost has passed and soil temperatures exceed 10°C/50°F. However, in most areas it is preferable to raise transplants in peat blocks or pots under gentle heat. In cool-temperate regions these should be hardened off before transplanting in the open; where possible continued protection from cloches until plants become too tall will ensure an earlier harvest. Clear plastic film can also be used for raising soil temperatures following direct sowings but must either be removed or slit at a later stage to permit development of the seedlings. Plants should be established in blocks, approximately 35cm/4in. apart; this arrangement favours pollination whereby pollen from the tassels is transferred by wind and gravity to the silk of the female flowers lower in the leaf canopy. Plants of different cultivars should not be planted together as cross-pollination can affect flavour and sugar content. Plants should be earthed up slightly when about 30cm/12in. high to increase their stability. In more exposed situations it may even be advisable to support plants with individual canes. Watering should be carried out during dry weather, particularly during the flowering cob-filling stages. Each plant usually produces one or two pods which should be harvested by snapping from the stem at the time silks begin to die off and turn brown. At this time the kernels, which can be examined prior to harvesting by peeling open the protective leaves, should contain a milky juice. The full sweetness and flavour is lost once the kernel starts to develop a more solid texture.

The main pest is the frit fly but slugs can also cause severe damage to young seedlings. The many agricultural pests are not usually a problem in the garden, except where plots are close to maize fields. Smut is the most likely to be seen, and in the US, Stewart's wilt may occur after mild winters.

*Z.mays* is also grown as an ornamental for foliage (notably 'Variegata' and 'Albovariegata') and for the cobs, which are used in dried arrangements. The multi-coloured cobs of Indian Corn are beautiful and time-honoured props of autumnal festivities such as Thanksgiving, alongside gourds and pumpkins.Treat as a frost-tender annual in cool temperate zones and sow in late winter under glass, or, in regions with long, hot summers, *in situ* in early spring.

*Z.curagua* Molina. See *Z.mays*.

*Z.gracillima* hort. See *Z.mays* var. *gracillima*.

**Z.mays** L. MAIZE; CORN; MEALIE; SWEET CORN.
Annual, to 4m+. Culms robust, erect, to 6cm diam., rooting at basal nodes. Leaves 2-ranked, to 90 × 11cm lanceolate, acuminate, arching, somewhat undulate; ligule to 5mm, truncate; sheaths overlapping. Male inflorescence to 20×20cm, erect; spikelets to 1.5cm. Female inflorescence borne in leaf axils, to 20cm, enclosed in tough spathes; spikelets borne in rows to 30 on a thickened axis (cob), styles to 40cm; grain to 1cm, flattened, larger than glumes when mature. Summer–autumn. Cultigen, first cult. in Mexico. All modern cultivars of sweetcorn are hybrids. Recent developments include increases in sugar levels to produce what have been referred to as 'supersweets' although these suffer from reduced vigour and require warmer growing conditions. Early- maturing cultivars include 'Aztec', 'Butter Imp' (small

cobs), 'Earlibelle', 'Earliking', 'Jubilee', 'Kelvedon Glory', 'Stardust', 'Platinum Lady', 'White Sunglow', and supersweets 'Candle', 'Early Extra Sweet', 'Sweet Desire' and 'Sweet Nugget'. Mid-season cultivars include 'Flavour King', 'Merton' (very large cobs), 'Reward', 'Silver Queen' (white cobs), and 'Sundance', with supersweets 'Illini Extra Sweet', 'Reward', 'Sundance', 'Sweet Dreams' and 'Sweet 77' (very large cob). Ornamental cultivars include the following: 'Albovariegata': leaves striped white. 'Cutie Blues': grains small, dark blue. 'Fiesta': grains long, coloured white, yellow, red, blue and purple in patches. 'Harlequin': leaves striped green, red, grains deep red. 'Indian Corn': grains multicoloured. 'Strawberry Corn': grains small, burgundy; husks yellow. 'Quadricolor': leaves striped green, white, yellow, pink. 'Variegata': leaves longitudinally striped pale yellow, tinted pink. var. **gracillima** Körn. Dwarf form; leaves narrow. var. **japonica** Alph. Wood. To 120cm. Leaves striped white. Z7.

**Zizania** L. (From Greek *zizanion*, name for a weed of wheatfields, possibly referring to *Lolium temulentum*.) WILD RICE; WATER OATS. Gramineae. 3 species of annual or perennial aquatic grasses. Leaves linear, flat; ligule membranous. Inflorescence terminal, paniculate, branches unisexual; spikelets oblong, 1-flowered, female spikelets above male spikelets with palea 2-ribbed, lemmas leathery, 3-ribbed, awned, male spikelets with palea 3-ribbed, lemma membranous, 5-ribbed, acuminate to short-awned; stamens 6. N America, E Asia. *Z.aquatica* is a local food plant – its grains harvested for flour; it also provides shelter for waterbirds on lake margins; the culms of *Z.latifolia*, swollen by the fungal smut *Ustilago esculenta*, are used as a vegetable in China.

CULTIVATION   Occurring in marshland and as a marginal aquatic, *Z.aquatica*, a tall handsome annual with soft luxuriant foliage and magnificent heads of flower, is suitable for similar situations in cultivation; sow seed in early spring in pots immersed in water, and plant out when danger of frost is passed, alternatively sow directly on wet or shallowly submerged mud outdoors in mid to late spring. *Z.latifolia* needs warm glasshouse protection in cool temperate zones.

**Z.aquatica** L. (*Z.palustris* L.). ANNUAL WILD RICE; WATER RICE; CANADIAN WILD RICE; WILD RICE.
Annual, to 3.6m. Culms robust, upright, sometimes decumbent at base. Leaves to 1.2m × 3cm, scabrous, deep green; sheaths glabrous, rolled. Panicles pyramidal, to 75cm, light brown, loosely and symmetrically branched; branches to 20cm; female spikelets to 0.5cm, awn to 8cm, male spikelets larger, to 1cm, awns absent. N America. Z6.

*Z.caducifolia* Hand.-Mazz. See *Z.latifolia*.

**Z.latifolia** (Griseb.) Turcz. (*Z.caducifolia* Hand.-Mazz.). MANCHURIAN WILD RICE; WATER RICE.
Stoloniferous perennial, to 3.6m. Culms clumped. Leaves linear-lanceolate, to 1.5m × 3.5cm, glabrous, scabrous; sheaths smooth, spongy; ligules to 2.5cm. Panicle narrow-pyramidal, to 60cm; female spikelets to 2cm, awn to 3cm, male spikelets to 1.3cm, tinged purple, awn to 1cm. Summer–autumn. E Asia. Z9.

*Z.palustris* L. See *Z.aquatica*.

# Names no longer in use

**Achnatherum** P.Beauv.
*A.brachytrichum* (Steud.) P.Beauv. See *Calamagrostis brachytricha*.
*A.calamagrostis* (L.) P. Beauv. See *Stipa calamagrostis*.

**Arista**
*A.pennata* Trin. See *Stipagrostis pennata*.

**Arthrostylidium** Rupr.
*A.longifolium* (Fourn.) Camus. See *Otatea acuminata*.

**Baldingera** P.Gaertn., B.Mey. & Schreb.
*B.arundinacea* (L.) Dumort. See *Phalaris arundinacea*.

**Chaetochloa** Scribn.
*C.italica* (L.) Scribn. See *Setaria italica*.
*C.palmifolia* A. Hitchc. & Chase. See *Setaria palmifolia*.

**Chloris** Sw.
*C.crinita* Lagasca. See *Trichloris crinita*.

**Chrysopogon** Trin.
*C.nutans* Benth. See *Sorghastrum nutans*.

**Clinelymus** (Griseb.) Nevski
*C.sibiricus* (L.) Nevski. See *Elymus sibiricus*.

**Cymophyllus** Mack. ex Britt. & A. Br.
*C.fraseri* (Andr.) Mack. See *Carex fraseri*.
*C.fraserianus* (Ker-Gawl.) Kartesz & Gandhi. See *Carex fraseri*.

**Dendrocalamopsis** (L.C. Chia & H.L. Fung) Keng f.
*D.oldhamii* (Munro) Keng f. See *Bambusa oldhamii*.

**Dichromena** Michx.
*D.ciliata* Pers. See *Rhynchospora nervosa* ssp. *ciliata*.
*D.nervosa* Vahl. See *Rhynchospora nervosa*.

**Digraphis** Trin.
*D.arundinacea* (L.) Trin. See *Phalaris arundinacea*.

**Dichanthelium** (Hitch. & Chase) Gould
*D.clandestinum* See *Panicum clandestinum*.

**Erianthus** Michx.
*E.alopecuroides* (L.) Ell. See *Saccharum alopecuroides*.
*E.brevibarbis* Michx. See *Saccharum brevibarbe*.
*E.contortus* Ell. See *Saccharum contortum*.
*E.giganteus* (Walt.) P. Beauv. See *Saccharum giganteum*.
*E.ravennae* (L.) P. Beauv. See *Saccharum ravennae*.

**Gymnothrix** P.Beauv.
*G.caudata* Schräd. See *Pennisetum macrourum*.
*G.latifolia* (Spreng.) Schult. See *Pennisetum latifolium*.

**Holoschoenus** Link.
*H.australis* (Murr) Rchb. See *Scirpoides holoschoenus*.
*H.vulgaris* Link. See *Scirpoides holoschoenus*.

**Leleba** Nak.
*L.multiplex* (Lour.) Nak. See *Bambusa multiplex*.
*L.oldhamii* (Munro) Nak. See *Bambusa oldhamii*.
*L.vulgaris* (Schrad. ex Wendl.) Nak. See *Bambusa vulgaris*.

**Lophochloa** Rchb.
*L.cristata* (L.) Hylander. See *Rostraria cristata*.
*L.phleoides* (Vill.) Rchb. See *Rostraria cristata*.

**Macrochloa** Kunth.
*M.arenaria* (Brot.) Kunth. See *Stipa gigantea*.
*M.enacissima* (L.) Kunth. See *Stipa tenacis-sima*.

**Mariscus** Vahl
*M.cyperoides* (L.) Urban. See *Cyperus cyperoides*.
*M.sieberianus* Nees. See *Cyperus cyperoides*.

**Nippocalamus** Nak.
*N.argenteostriatus* (Reg.) Nak. See *Pleioblastus argenteostriatus*.
*N.argenteostriatus* var. *distichus* (Mitford) Nak. See *Pleioblastus pygmaeus* var. *distichus*.
*N.chino* (Franch. & Savat.) Nak. See *Pleioblastus chino*.
*N.chino* var. *akebono* (Mak.) Mak. See *Pleioblastus akebono*.
*N.fortunei* (Van Houtte) Nak. See *Pleioblastus variegatus*.
*N.humilis* (Mitford) Nak. See *Pleioblastus humilis*.
*N.pygmaeus* (Miq.) Nak. See *Pleioblastus pygmaeus*.
*N.simonii* (Carr.) Nak. See *Pleioblastus simonii*.

**Piptatherum** P.Beauv.
*P.multiflorum* (Cav.) P.Beauv. See *Oryzopsis miliacea*.

**Sinarundinaria** Nak.
*S.anceps* (Mitford) Chao & Renvoize. See *Yushania anceps*.
*S.falcata* (Nees) Chao & Renvoize. See *Drepanostachyum falcatum*.
*S.falconeri* (Hook. ex Munro) Chao & Renvoize. See *Himalayacalamus falconeri*.
*S.hookeriana* (Munro) Chao & Renvoize. See *Himalayacalamus hookerianus*.
*S.jaunsarensis.* See *Yushania anceps*.

*S.maling* (Gamble) Chao & Renvoize. See *Yushania maling.*

*S.murielae* (Gamble) Nak. See *Fargesia murielae.*

*S.nitida* (Mitford) Nak. See *Fargesia nitida.*

**Sinocalamus** McClure

*S.giganteus* (Munro) Keng f. See *Dendrocalamus giganteus.*

*S.oldhamii* (Munro) McClure. See *Bambusa oldhamii.*

**Tetragonocalamus** Nak.

*T.angulatus* (Munro) Nak. See *Chimonobambusa quadrangularis.*

**Triodia** R. Br.

*T.flava* See *Tridens flava.*

**Typhoides** Moench.

*T.arundinacea* (L.) Moench. See *Phalaris arundinacea.*

**Yadakea** Mak.

*Y.japonica* (Sieb. & Zucc. ex Steud.) Mak. See *Pseudosasa japonica.*

# Index of Popular Names

GREAT QUAKING GRASS   *Briza maxima*
GREAT REEDMACE   *Typha latifolia*
GREAT WATER GRASS   *Glyceria maxima*
GREATER WOODRUSH   *Luzula sylvatica*
GREEN BRISTLE GRASS   *Setaria viridis*
GREEN MOOR GRASS   *Sesleria heufleriana*
GREY FESCUE   *Festuca glauca*
GREY MOOR GRASS   *Sesleria nitida*

HAIR GRASS   *Aira* (*A. elegantissima*); *Deschampsia*;
  *Eleocharis acicularis*; *Koeleria*
HAIRY BROME   *Bromus ramosus*
HAIRY CUP GRASS   *Eriochloa villosa*
HAIRY FINGER GRASS   *Digitaria sanguinalis*
HAIRY MELIC GRASS   *Melica ciliata*
HAIRY MUHLY GRASS   *Muhlenbergia pubescens*
HAIRY WOOD-RUSH   *Luzula pilosa*
HAKONE GRASS   *Hakonechloa macra*
HARD RUSH   *Juncus inflexus*
HARDING GRASS   *Phalaris aquatica*
HARDY ORIENTAL FOUNTAIN GRASS   *Pennisetum
  orientale*
HARDY PAMPAS GRASS   *Saccharum ravennae*
HARE'S TAIL   *Eriophorum vaginatum*
HARE'S TAIL   *Lagurus*
HASSOCKS   *Deschampsia cespitosa*
HEDGE BAMBOO   *Bambusa multiplex*
HEDGEHOG FESCUE   *Festuca punctoria*
HERBE SALEA   *Spartina alterniflora*
HOG MILLET   *Panicum miliaceum*
HOOK SEDGE   *Uncinia*
HOTEI-CHIKU   *Phyllostachys aurea*
HUNANGAMOLIO GRASS   *Chionochloa conspicua*

IDIOT GRASS   *Oplismenus hirtellus*
INDIAN GRASS   *Sorghastrum nutans*
INDIAN MILLET   *Oryzopsis hymenoides*; *Panicum
  miliaceum*
INTERMEDIATE WHEATGRASS   *Elymus hispidus*
INYOU-CHIKUZOKU   × *Hibanobambusa tranquillans*
ITALIAN MILLET   *Setaria italica*

JAPANESE BLOOD GRASS   *Imperata cylindrica*
JAPANESE LOVE GRASS   *Eragrostis japonica*
JAPANESE MILLET   *Setaria italica*
JAPANESE SEDGE GRASS   *Carex oshimensis*
JAPANESE SILVER GRASS   *Miscanthus sinensis*
JET SEDGE   *Carex atrata*
JOB'S TEARS   *Coix lacryma-jobi*
JOHNSON GRASS   *Sorghum halapense*
JOINTED RUSH   *Juncus articulatus*
JUNE GRASS   *Koeleria pyramidata*; *Poa pratensis*

KAFIR CORN   *Sorghum bicolor*
KE-OROSHIMA-CHIKU   *Pleioblastus pygmaeus*
KENTUCKY BLUE GRASS   *Poa pratensis*
KHAS KHAS   *Vetiveria zizanoides*
KHUS KHUS GRASS   *Vetiveria zizanoides*
KIKKOUCHIKU   *Phyllostachys edulis* f. *heterocycla*
KLEBERG GRASS   *Dichanthium annulatum*
KOREAN FEATHER REEDGRASS   *Calamagrostis
  brachytricha*
KOU-CHIKU   *Phyllostachys sulphurea*
KU-CHIKU   *Phyllostachys bambusoides*
KUMA ZASA   *Sasa veitchii*

KURAKKAN   *Eleusine coracana*
KURO-CHIKU   *Phyllostachys nigra*
KWEEK   *Cynodon dactylon*
KYO-CHIKU   *Dendrocalamus giganteus*

LACE GRASS   *Eragrostis capillaris*
LADIES' HAIR GRASS   *Briza media*
LAMB'S TAIL GRASS   *Alopecurus pratensis*
LARGE BLUE FESCUE   *Festuca amethystina*
LARGE BLUE HAIR GRASS   *Koeleria glauca*
LATE BLOOMING FOUNTAIN GRASS   *Pennisetum*
  'Moudry'
LAZY MAN'S GRASS   *Eremochloa ophiuroides*
LEATHERLEAF SEDGE   *Carex buchananii*
LEMON GRASS   *Cymbopogon citratus*
LESSER BULRUSH   *Typha angustifolia*
LESSER QUAKING GRASS   *Briza minor*
LITTLE BLUESTEM   *Schizachyrium scoparium*
LOOSE SILKY BENT   *Apera spica-venti*
LOVE GRASS   *Eragrostis*
LOVELY BAMBOO   *Pseudosasa amabilis*
LYME GRASS   *Elymus*; *Leymus arenarius*

MACE SEDGE   *Carex grayi*
MADAKE   *Phyllostachys bambusoides*
MAIDEN GRASS   *Miscanthus sinensis* 'Gracillimus'
MAIDENHAIR GRASS   *Briza media*
MAIZE   *Zea mays*
MANA GRASS   *Cymbopogon nardus*
MANCHURIAN WILD RICE   *Zizania latifolia*
MANILA GRASS   *Zoysia matrella*
MANNA GRASS   *Glyceria*
MANY-FLOWERED WOOD-RUSH   *Luzula multiflora*
MARRAM GRASS   *Ammophila*
MARSH BEETLE   *Typha latifolia*
MARSH GRASS   *Spartina*
MASCARENE GRASS   *Zoysia tenuifolia*
MAURITANIA VINE REED   *Ampelodesmos*
MEADOW CAT'S TAIL   *Phleum pratense*
MEADOW FESCUE   *Festuca elatior*
MEADOW FOXTAIL   *Alopecurus pratensis*
MEADOWGRASS   *Poa*
MEALIE   *Zea mays*
MEANS GRASS   *Sorghum halapense*
MEDAKE   *Pleioblastus simonii*
MEDITERRANEAN BARLEY   *Hordeum hystrix*
MEL GRASS   *Ammophila*
MELIC   *Melica*
MELICK   *Melica*
MELLIC   *Melica*
METAKE   *Pseudosasa japonica*
MEXICAN LOVE GRASS   *Eragrostis mexicana*
MEXICAN WEEPING BAMBOO   *Otatea acuminata*
MILLET   *Panicum miliaceum*; *Sorghum*
MINIATURE FOUNTAIN GRASS *Pennisetum
  alopecuroides* 'Little Bunny'
MINIATURE PAPYRUS   *Cyperus prolifer*
MIYAKO-ZASA   *Sasa nipponica*
MOOR GRASS   *Sesleria*
MOSO BAMBOO   *Phyllostachys edulis*
MOSQUITO GRASS   *Bouteloua gracilis*
MOUNTAIN MELIC   *Melica nutans*
MOUNTAIN SEDGE   *Carex montana*
MOUSOU-CHIKU   *Phyllostachys edulis*
MUHLY   *Muhlenbergia* (*M.rigens*)

SORGHUM    *Sorghum bicolor*
SPANGLE GRASS    *Chasmanthium latifolium*;
    *Uniola    paniculata*
SPEAR GRASS    *Poa; Stipa*
SPIKE GRASS    *Uniola*
SPIKE RUSH    *Eleocharis*
SPLITBEARD BLUESTEM    *Andropogon ternarius*
SPRING SEDGE.    *Carex caryophyllea*
SQUARE BAMBOO    *Chimonobambusa quadrangularis*
SQUIRRELTAIL BARLEY    *Hordeum jubatum*
ST AUGUSTINE GRASS    *Stenotaphrum secundatum*
ST LUCIE'S GRASS    *Cynodon dactylon*
STAR GRASS    *Chloris truncata; Cynodon dactylon*
STIFF BROME    *Bromus madritensis*
STRONG RED CREEPING FESCUE    *Festuca rubra*
STRONG-SCENTED LOVE GRASS    *Eragrostis cilianensis*
SUGAR CANE    *Saccharum officinarum*
SUGARCANE PLUME GRASS    *Saccharum giganteum*
SWAMP FOXTAIL GRASS    *Pennisetum alopecuroides*
SWAMP MEADOWGRASS    *Poa palustris*
SWAMP SEDGE.    *Carex acutiformis*
SWEET CALAMUS    *Acorus calamus*
SWEET CORN    *Zea mays*
SWEET FLAG    *Acorus calamus*
SWEET GRASS    *Glyceria*
SWEET HAY    *Glyceria maxima*
SWEET VERNAL GRASS    *Anthoxanthum odoratum*
SWITCH CANE    *Arundinaria gigantea*
SWITCH GRASS    *Panicum virgatum*

TALL FESCUE    *Festuca elatior*
TALL PURPLE MOOR GRASS    *Molinia caerulea*
TALL SWITCH GRASS    *Panicum virgatum* 'Strictum'
TEFF    *Eragrostis tef*
TENDER FOUNTAIN GRASS    *Pennisetum setaceum*
TENDER PURPLE FOUNTAIN GRASS    *Pennisetum
    setaceum* 'Rubrum'
TENDER WHITE-FLOWERED FOUNTAIN GRASS
    *Pennisetum villosum*
THREE-WAY SEDGE    *Dulichium arundinaceum*
TIGER GRASS    *Thysanolaena maxima*
TIMOTHY    *Phleum pratense*
TOAD RUSH    *Juncus bufonius*
TOBOSA GRASS    *Hilaria mutica*
TOE TOE    *Cortaderia richardii*
TONKIN CANE    *Pseudosasa amabilis*
TOOTHACHE GRASS    *Ctenium aromaticum*
TOOWOMBA CANARY GRASS    *Phalaris aquatica*
TOR GRASS    *Brachypodium pinnatum*
TORTOISESHELL BAMBOO    *Phyllostachys edulis* f.
    *heterocycla*
TOTTER    *Briza media*
TREMBLING GRASS    *Briza media*
TUFTED FESCUE    *Festuca amethystina*
TUFTED HAIR GRASS    *Deschampsia cespitosa*
TUFTED SEDGE.    *Carex elata*
TUMBLE GRASS    *Eragrostis spectabilis*
TUSSOCK GRASS    *Deschampsia cespitosa*
TWITCH    *Agropyron repens*

UMBRELLA BAMBOO    *Fargesia murielae*
UMBRELLA PLANT    *Cyperus alternifolius*
UPRIGHT BROME    *Bromus erectus*
URUGUAY GRASS    *Chloris berroi*
URUGUAY PENNISETUM    *Pennisetum latifolium*
UVA GRASS    *Gynerium sagittatum*

VELD GRASS    *Ehrharta erecta*
VELVET BENT    *Agrostis canina*
VELVET BENT GRASS    *Agrostis canina*
VELVET GRASS    *Holcus lanatus*
VERNAL GRASS    *Anthoxanthum*
VETIVER    *Vetiveria zizanoides*
VIRGINIA WILD-RYE    *Elymus virginicus*

WALL BROME    *Bromus madritensis*
WALLABY GRASS.    *Danthonia setacea*
WATER GRASS    *Paspalum dilatatum*
WATER OATS    *Zizania*
WATER RICE    *Zizania aquatica; Z.latifolia*
WAVY HAIR GRASS    *Deschampsia flexuosa*
WEEPING LOVE GRASS    *Eragrostis curvula*
WEEPING SEDGE    *Carex pendula*
WESTERN MEADOW SEDGE    *Carex pensa*
WHEATGRASS    *Agropyron*
WHITE FLOWERING FOUNTAIN GRASS    *Pennisetum
    alopecuroides* 'Caudatum'
WILD CANE    *Gynerium sagittatum*
WILD RICE    *Zizania (Z.aquatica)*
WILD RYE    *Elymus*
WIND GRASS    *Apera interrupta*
WINDMILL GRASS    *Chloris*
WIRE GRASS    *Eleusine indica*
WITCH GRASS    *Panicum capillare*
WOOD BROME    *Bromus ramosus*
WOOD CLUB RUSH    *Scirpus sylvaticus*
WOOD FALSE BROME    *Brachypodium sylvaticum*
WOOD FESCUE    *Festuca altissima*
WOOD GRASS    *Sorghastrum nutans*
WOOD MEADOWGRASS    *Poa nemoralis*
WOOD MELIC    *Melica uniflora*
WOOD MILLET    *Milium effusum*
WOOD SEDGE    *Carex sylvatica*
WOOD SMALL-REED    *Calamagrostis epigejos*
WOOD-RUSH    *Luzula*
WOOL GRASS    *Scirpus cyperinus*
WOOLLY FOXTAIL GRASS    *Alopecurus lanatus*

YARD GRASS    *Eleusine indica*
YELLOW BLUE STEM    *Bothriochloa ischaemum*
YELLOW BRISTLE GRASS    *Setaria glauca*
YELLOW-GROOVE BAMBOO    *Phyllostachys
    aureosulcata*
YORKSHIRE FOG    *Holcus lanatus*

ZEBRA GRASS    *Miscanthus sinensis* 'Zebrinus'

# Bibliography

## General

### FLORAS

Allan, H. H. et al. 1961–80. *Flora of New Zealand* (Wellington: Government Printer).

Britton, N. L. and Brown, A. (ed. Gleason, H.A.) 1952. *The New Britton & Brown Illustrated Flora of the North Eastern United States* (New York Botanical Garden).

*European Garden Flora.* 1984– (Cambridge: Cambridge University Press).

*Flora Europaea.* 1964–80 (Cambridge: Cambridge University Press).

Flora of North America Editorial Committee. 1992. *Flora of North America* (New York: Oxford University Press).

Ohwi, J. 1965. *Flora of Japan* (Washington, DC: Smithsonian Institution).

Stace, C.A. 1991. *The New Flora of the British Isles* (Cambridge: Cambridge University Press).

### SYSTEMATICS

Bentham, G. 1883. Gramineae. In G. Bentham and J.D. Hooker, *Genera Plantarum* (London: Reeve), vol. 3, part 2.

Chapman, G.P. and Peat, W.E. *An Introduction to the Grasses* (Wallingford: CAB International).

Clayton, W.D. and Renvoize, S.A. 1986. *Genera Graminum: Grasses of the World. Kew Bulletin Additional Series 13* (London: HMSO).

Gould, F.W. 1968. *Grass Systematics* (New York: McGraw-Hill).

Prat H. 1960. Vers une classification naturelle des Graminées. *Bulletin Societé Botanique de France* 107: 32–79.

Renvoize, S.A. and Clayton, W.D. 1992. Classification and evolution of the grasses. In Chapman, G.P. (ed.), *Grass Evolution and Domestication* (Cambridge: Cambridge University Press).

Söderstrom, T. R. et al. (eds) 1987. *Grass: Systematics and Evolution* (Washington, D.C.: Smithsonian Institution Press).

Watson, L. and Dallwitz, M.J. 1992. *The Grass Genera of the World* (Wallingford: CAB International).

### BIOLOGY

Brown, W.V. 1958. Leaf anatomy in grass systematics. *Botanical Gazette* 119: 170–78.

—— 1960. The morphology of the grass embryo. *Phytomorphology* 10: 215–23.

Chapman, G.P. (ed.) 1990. *Reproductive Versality in the Grasses* (Cambridge: Cambridge University Press).

—— 1992. *Grass Evolution and Domestication* (Cambridge: Cambridge University Press).

Metcalfe C.R. 1960. *Anatomy of the Monocotyledons,* vol. 1, *Gramineae* (Oxford: Clarendon Press).

Sharman, B. C. 1945. Leaf and bud initiation in the Gramineae. *Botanical Gazette* 106: 269–89.

Turpin, P.J.F. 1819, Mémoire sur l'inflorescence des Graminées et des Cyperées, comparée avec celle des autres végétaux sexiferes; suivi de quelques observations sur les disques. *Mémoires Musée Hist. Nat. Paris* 4: 67.

## Grasses and grass-like plants

### GENERAL AND HORTICULTURAL

Corley, W.L. 1975. *Ornamental Grasses for Georgia.* Georgia Agricultural Experimental Station Research Report 217 (Athens, GA).

Darke, F.P. 1990. *Ornamental Grasses at Longwood Gardens* (Kennett Square, PA: Longwood Gardens Inc.).

Darke, F.P. 1994. *For Your Garden: Ornamental Grasses* (Boston: Little, Brown).

Foerster, K. 1978. *Einzug der Graser und Farne in die Garten* (Berlin: Verlag J. Neumann-Neudman).

Frederick, W.H. 1972. Use of ornamental grass in your garden. *Horticultural Society of New York Bulletin* 3(2): 2, 6.

Frederick, W.H. and Simon, R.A. 1965. Grass. *Horticulture* 63(8): 14–15, 26.

Greenlee, J. 1992. *The Encyclopedia of Ornamental Grasses* (Philadelphia: Rodale Press).

Grounds, R. 1989. *Ornamental Grasses.* (London: Christopher Helm).

Knowles, E. 1967. Grasses. *Gardener's Chronicle and New Horticulturist* 162(2): 14.

Loewer, H.P. 1977. *Growing and Decorating with Grasses* (New York: Walker and Co.).

—— (ed.) 1988. *Ornamental Grasses.* A Brooklyn Botanic Garden handbook, revised 1990. *BBG Record* 44 (3).

Meyer, M.H. 1975. *Ornamental Grasses: decorative plants for home and garden* (New York: Charles Scribner's Sons).

Meyer, M.H. and Mower, R.G. 1986. Ornamental grasses for home and garden. *Information Bulletin* 64 (Ithaca, NY: Cornell Cooperative Extension Publications).

Oakes, A.J. 1990. *Ornamental Grasses and Grasslike Plants* (New York: Reinhold).

Oehme, W. and Van Sweden, J. with Frey, S. 1990. *Bold Romantic Gardens.* (Reston, VA: Acropolis Books Ltd.).

Ottesen, C. 1989. *Ornamental Grasses: The Amber Wave* (New York: McGraw-Hill Publishing Co.).

Paterson, A. 1964. Grasses for the flower garden. *Gardener's Chronicle and New Horticulturist* 155(14): 301–2.

—— 1969. Decorative grasses. *Annual of Amateur Gardening*: 32–3.

Pesch, B.B. (ed.) 1988. Ornamental grasses. *Brooklyn Botanic Garden Record* 44, No.3.

Reinhardt, T. A. and M. and Moskowitz, M. 1989. *Ornamental Grass Gardening: Design Ideas, Functions and Effects* (London: Macdonald; Los Angeles: H.P. Books).

Steinegger, D.H., Sherman, R.C. and Janssen, D.E. 1979. An evaluation of native and exotic grass species for ornamental use in Nebraska. *Nebraska Agricultural Experimental Station* SB 546.

*Taylor's Guide to Groundcovers, Vines and Grasses.* 1987. (Boston: Houghton Mifflin Co.).

Taylor, N.J. 1992. *Ornamental Grasses, Bamboos, Rushes and Sedges* (London: Ward Lock).

### SYSTEMATICS AND REGIONAL SURVEYS

Blomquist, H.L. 1948. *The Grasses of North Carolina.* (Durham, NC: Duke University Press).

Bor, N.L. 1960. *The grasses of Burma, Ceylon, India and Pakistan* (New York: Pergamon Press)

Brown, L. 1979. *Grasses: An Identification Guide* (Boston: Houghton, Mifflin).

*Collins Guide to the Grasses, Sedges, Rushes and Ferns of Britain and Northern Europe.* 1984. (London: Collins).

Crampton, B. 1974. *Grasses in California* (University of California Press).

Deam, C.C. 1929. *Grasses of Indiana* (Indianapolis: Indiana Department of Conservation).

Fassett, N.C. 1951. *Grasses of Wisconsin* (Madison: University of Wisconsin Press).

Gupta, P.K. and Baum B.R. 1989. Stable classification and nomenclature in the Triticeae: desirability, limitations and prospects. *Euphytica* 41: 191–7.

Harrington, H.D. 1977. *How to identify grasses and grasslike plants* (Chicago: The Swallow Press).

Hitchcock, A.S. 1950. *Manual of the Grasses of the United States* (Washington, D.C.: US Government Printing Office).

Hubbard, C.E. 1992. *Grasses.* New edition, revised by J.C.E. Hubbard (London: Penguin).

Kükenthal, G. 1955–6. Cyperaceae–Scirpoideae–Cypereae. In Engler, *Pflanzenreich* IV, 20: 101 (Stuttgart: Engelman).

Mosher, E. 1918. The grasses of Illinois. *University of Illinois Agricultural Experimental Station Bulletin* 205.

Norton, J.B.S. 1930. Maryland grasses. *Maryland Agricultural Experimental Station Bulletin* 323.

Pohl, R.W. 1954. *How to know grasses* (Dubuque, IA: Wm. C. Brown and Co.).

Silveus, W.A. 1933. *Texas grasses* (San Antonio: The Clegg Company).

### INDIVIDUAL GENERA

#### Anthoxanthum

Valdes, B. 1973. Revision de las Especies anuales del Genero *Anthoxanthum. Lagascalia* 3: 99–141.

#### Bouteloua

Gould, F.W. 1979. The genus *Bouteloua. Annals of the Missouri Botanical Garden* 66: 348–416.

#### Bromus

Scholz, H., 1970. Zur Systematik der Gattung *Bromus* Subgenus *Bromus. Wildenowia* 6: 139–59.

#### Calamagrostis

Wasiljew, W.N. 1960. Das System der Gattung *Calamagrostis* Roth. *Feddes Repertorium* 63: 229–51.

#### Carex

Kükenthal, G. 1909. *Carex.* In Engler, *Das Pflanzenreich.*

#### Cortaderia

Hornback, B. 1994. *In Praise of Pampas Grass* (Pacific Horticulture).

Stapf, O., 1905. The Pampas Grasses, *Cortaderia,* Stapf. *Flora and Sylva* 3: 171–6.

#### Cymbopogon

Soekarno, S. 1977. The genus *Cymbopogon. Reinwardtia* 9: 225–375.

#### Cyperus

Kükenthal, G. 1935. *Cyperus.* In Engler, *Das Pflanzenreich,* 101.

#### Digitaria

Henrard, J.T., 1950. A monograph of the genus *Digitaria.*

**Echinochloa**

Gould, F.W., Ali, M.A. & Fairbrothers, D.E. 1972. A revision of *Echinochloa* in the United States. *American Midland Naturalist* 87: 36–59.

**Festuca**

Hackel, E. 1882. *Monographia Festucarum Europaeum.*

Tzvelev, N.N. 1971. K sistematike i filogenii ovsyananits (*Festuca* L.) flory SSSR: I. *Botanicheskii Zhurnal* 56: 1252–62.

**Hyparrhenia**

Clayton, W.D. 1969. A revision of the genus *Hyparrhenia. Kew Bulletin* Additional Series II (London: HMSO).

**Koeleria**

Domin, K. 1907. Monographic der Gattung *Koeleria. Bibliotheca Botanica* 65.

**Melica**

Hempel, W. 1971. Vorarbeiten zu einer Revision der Gattung *Melica* L. *Feddes Repertorium* 81: 657–86, 84: 533–68 (1973).

**Phalaris**

Anderson, D.E., 1961. Taxonomy and distribution of the genus *Phalaris. Iowa State Journal of Science* 36: 1–96.

**Rhynchospora**

Simpson, D. 1993. *Rhynchospora nervosa. Kew Magazine* 10: 225.

**Scirpus**

Beetle, A.A. 1947. *Scirpus,* L. *North American Flora* 18(8): 481–504.

**Spartina**

Mabberley, D.G. 1956. Taxonomy and distribution of the genus *Spartina. Iowa State Journal of Science* 30: 471–574.

Saint-Yves, A.,1932. Monographia Spartin-arum. *Candollea* 5: 19–100.

## Bamboos

GENERAL AND HORTICULTURAL

Camus, E.G. 1913. *Les Bambusées* (Paris: Paul Lechavier).

Chao, C.S. 1989. *A Guide to Bamboos Grown in Britain* (Royal Botanic Gardens, Kew).

Cobin, M. 1947. Notes on the propagation of the sympodial or clump type of bamboos. *Proceedings Florida State Horticultural Society* 60: 181–4.

Crampton, D. 1994. *Discovering Bamboos. The Garden* 119: 6.

Creech, J.L. 1957. Hardiness of the running bamboos. *National Horticultural Magazine* 36(4): 335–9.

Crouzet, Y. 1981. *Les Bambous.*

Farrelly, D. 1984. *The Book of Bamboo* (San Francisco: Sierra Club Books).

Freeman-Mitford, A. 1896. *The Bamboo Garden* (London: Macmillan).

Haubrich, R. 1981. Handbook of bamboos cultivated in the United States. *Journal of the American Bamboo Society.*

Haun, J.R. and Young, R.A. 1961. *Bamboo in the United States.* U.S. Ministry of Agriculture Handbook, No. 193.

Hodge, W.H. and Bisset, D.A. 1957. Running bamboos for hedges. *National Horticultural Magazine* 36(4): 335–9.

Houzeau de Lehaie, J., 1906–08. Les Bambous, Nos. 1–10.

Lawson, A.H. 1968. *Bamboos: A Gardener's Guide to their Cultivation in Temperate Climates* (London: Faber & Faber).

Li, Hui Lin, 1942. Bamboo and Chinese civilisation. *Journal of the New York Botanical Garden* 43: 213–23.

Lin, W.C. 1962, 1964. Studies on the propagation by level cuttings of various bamboos, parts 1 and 2. *Bulletin Taiwan Forestry Research Institute* Nos 80 and 105.

McClure, F.A. 1938. Notes on the bamboo culture with special reference to South China. *Hong Kong Nat.* 9, 4–18.

—— 1955. Propagation of bamboo by whole-culm cuttings. *Proceedings American Society for Horticultural Science* 65, 283–8.

—— 1956a. Bamboo in the economy of oriental peoples. *Economic Botany* 10(4): 335–61.

—— 1956b. Bamboo culture in the South Pacific. *South Pacific Quarterly Bulletin* 6: 38–40.

—— 1966. *The Bamboos, a fresh perspective.* (Cambridge, MA: Harvard University Press)

Martin F. and Demoly, J.-P 1979. Bambous, 1 Clé des principaux bambous cultivés en climat tempéré. *Bulletin de l'Ass. des Parcs botaniques de France* 1: 7–17.

Okamura H. 1986. *The Horticultural Bamboo Species in Japan* (Kobe: published by the author)

Recht, C. and Wetterwald, M.F. 1992. *Bamboos* (London: B.T.Batsford Ltd).

Rivière, A and Rivière, C. 1879. Les Bambous. *Bulletin Societé Acclim.* 5: 221–53, 290–322, 392–421, 460–78, 501–26, 597–645, 666–721, 758–828. Reprinted 1878 as a separate, continuously paged publication.

Satow, E. 1899. The cultivation of bamboos in Japan. *Transactions of the Asiatic Society of Japan* 27: 1–127.

Shepherd, F.W. 1961. The Propagation of *Arundinaria japonica*. Ministry of Agriculture Experimental Horticulture, No.5 (London: H.M.S.O).

Thomas, G.S. 1957. Bamboos. *Journal of the Royal Horticultural Society* 82: 247–55.

Wang Dajun and Shen Shao-Jin. 1987. *Bamboos of China* (London: Christopher Helm).

White, D.G. 1947a. Longevity of bamboo seed under different storage conditions. *Trop. Agr.* 24: 51–3.

—— 1947b. Propagation of bamboo by branch cuttings. *Proceedings American Society for Horticultural Science* 50: 392–4.

Young, R. A. and Haun, J.R. 1961. *Bamboo in the United States* (Washington, DC: US Government Printing Office).

SYSTEMATICS

Demoly, J.-P. 1991a. Confusions à éviter. *Bulletin de l'Association des Parcs botanique de France* 14: 30.

—— 1991b. Notes et nouveautés nomenclaturales. *Ibid.* 14:31.

Franchet, M.A. 1889. Note sur deux nouveaux genres de Bambusées. *Journal Botanique* 3: 277–84.

Gamble, J.S. 1888. Notes on the small bamboos of the genus *Arundinaria*. *Indian Forester* 14: 306–14.

—— 1896. The Bambuseae of British India. *Annals Royal Botanic Garden Calcutta* 7.

Holttum, R.E. 1946. The classification of Malayan bamboos. *Journal of the Arnold Arboretum* 27: 340–46.

—— 1956. Classification of bamboos. *Phytomorphology* 6: 73–90.

McClintock, D. 1979. Bamboos: some facts and thoughts on their naming and flowering. *Plantsman* 1(i): 31–50.

—— 1992. The shifting sands of bamboo genera. *The Plantsman* 14:169–77.

Munro, W. 1868. A monograph on the Bambusaceae. Report from *Transactions of the Linnean Society of London* 26(1).

Suzuki, S. 1978. *Index to Japanese Bambusaceae.*

INDIVIDUAL GENERA

Bambusa

Holttum, R.E., 1956. On the identification of the common hedge bamboo of SE Asia. *Kew Bulletin* 11: 207–11.

McClure, F.A. 1946. The genus *Bambusa* and some of its first-known species. *Blumea*, Supp. lll, 90–112.

Chusquea

Parodi, L.R., 1945. Sinopsis de las graminas chilenas del genero *Chusquea*. *Revisita Universitaria, Universidad Catolica de Chile* 30(1): 61–71.

Dendrocalamus

Holttum, R.E. 1958. The bamboo of the Malay Peninsula. *Gardens' Bulletin Singapore* 16: 89–104.

Drepanostachyum

Gamble, J.S. 1921. Flowering of *Arundinaria falcata* in the Temperate House. *Kew Bulletin Miscellaneous Information* 302–6.

See also Stapf (1904) under *Himalayacalamus*.

Himalayacalamus

Stapf, O. 1904. Himalayan bamboos. *Arundinaria falconeri* and *A. falcata*. *Gardener's Chronicle* 35: 304–7.

Stapleton, C.M.A. 1993. *Himalayacalamus hookerianus* (a new combination) in flower in Edinburgh. *The Bamboo Society (European Bamboo Society Great Britain) Newsletter* 17: 20–21.

——1994. The Blue-Stemmed Bamboo, *Himalayacalamus hookerianus*. *The New Plantsman* I.i.

W.W. 1898. *Arundinaria hookeriana*. *Gardener's Chronicle* Ser.3, 23: 2.

Otatea

McClure, F.A. 1973. Genera of bamboos native to the New World. *Smithsonian Contributions to Botany* 9: 116–19.

Calderon, C.E. & Soderstrom, T.R. 1980. The genera of Bambusoideae (Poaceae) of the American continent. *Smithsonian Contributions to Botany* 44: 21.

Haubrich, R. 1980. The American Bamboos, Part 1. *Journal of the American Bamboo Society* 1: 29–30.

Dunmire, J.R. 1981. Weeping Mexican Bamboos. *Pacific Horticulture* 42: 18.

Phyllostachys

Haubrich, R. 1980. Handbook of Bamboos cultivated in the United States, Part 1: the genus *Phyllostachys*. *Journal of the American Bamboo Society* 1: 48–96.

McClure, F.A. 1945. The vegetative characters of the bamboo genus *Phyllostachys* and descriptions of eight new species introduced from China. *Journal of the Washington Academy of Science* 35: 276–93.

—— 1957. Bamboos of the genus *Phyllostachys* under cultivation in the United States. US Department of Agriculture Handbook No.114.

Wang Cheng-Ping et al. 1980. A taxonomical study of *Phyllostachys*. *Acta Phytotaxonomica Sinica* 18: 15–19, 168–93.

Sasa

Haubrich, R. 1982. The Sasas. *Journal of the American Bamboo Society* 2: 24–38.

Sasaella

McClintock, D., 1983. On the nomenclature and the flowering in Europe of the bamboo *Sasaella ramosa (Arundinaria vagans)*. *Kew Bulletin* 38: 191–5.

**BIOLOGY**

Bean, W.J. 1903. The flowering of bamboos. *Gardeners Chronicle* Ser. 3, 34: 169.

Bibral, B. 1899. The flowering of seedlings of *Dendrocalamus strictus*. *Indian Forester* 25: 305–6.

Blatter, E. 1929. Indian bamboos brought up to date. *Indian Forester* 55: 541–62, 586–612.

—— 1929, 1930, 1931. The flowering of bamboos I–III. *Journal of the Bombay Natural History Society* 33:899–921; 34: 135–141, 447–67.

Chaturvedi, B. 1947. Aerial rhizomes in bamboo culms. *Indian Forester* 73: 543; pl. 33, fig. 1.

Gupta, M. 1952. Gregarious flowering of *Dendrocalamus strictus*. *Indian Forester* 78: 547–50.

Hanke, D.E. 1990. Seeding the bamboo revolution. *Nature* 334: 291–2.

Hughes, R.H. 1951. Observations of cane (*Arundinaria*) flowers, seed and seedlings in the North Carolina coastal plain. *Bulletin Torrey Botanical Club* 78: 113–21.

Janzen, D.H. 1976. Why bamboos wait so long to flower. *Annual Reviews of Ecology and Systematics* 7: 347–91.

Kawamura, S. 1927. On the periodical flowering of the bamboo. *Japan Journal of Botany* 3: 335–49.

Lin, Wei-chih. 1974. Studies on the morphology of bamboo flowers. *Bulletin of the Taiwan Forestry Institute* 68.

McClintock, D. 1967. The flowering of bamboos. *Journal of the Royal Horticultural Society* 92: 520–25.

Metcalfe C.R. 1956. Some thoughts on the structure of bamboo leaves. *Botanical Magazine, Tokyo* 69: 391–400.

Nadgauda, R.S., Parasharami, V.A. and Mascarenhas, A.F. 1990. Precocious flowering and seeding behaviour in tissue-cultured bamboos. *Nature* 334: 335–6.

Porterfield, W. M. 1926. The morphology of the bamboo flower with special reference to *Phyllostachys nidularia*, Munro. *China Journal of Science and Arts* 5: 256–60.

—— 1930a. Morphology of the growing point in bamboo. *Bulletin Yenching University Deptartment of Biology* 1: 7–15.

—— 1930b. The mechanism of growth in bamboo. *China Journal* 13: 86–91, 146–53.

Ueda, K. 1960. Studies in the physiology of bamboo. *Resources Bureau Reference Data* No. 34. Tokyo.